Mechanisms and Machine Science

Volume 53

Series editor

Marco Ceccarelli
LARM: Laboratory of Robotics and Mechatronics
DICeM: University of Cassino and South Latium
Via Di Biasio 43, 03043 Cassino (Fr), Italy
e-mail: ceccarelli@unicas.it

More information about this series at http://www.springer.com/series/8779

Clément Gosselin · Philippe Cardou
Tobias Bruckmann · Andreas Pott
Editors

Cable-Driven Parallel Robots

Proceedings of the Third International
Conference on Cable-Driven Parallel Robots

Springer

Editors
Clément Gosselin
Département de génie mécanique
Université Laval
Quebec City, QC
Canada

Tobias Bruckmann
Chair for Mechatronics
Universität Duisburg-Essen
Duisburg, Nordrhein-Westfalen
Germany

Philippe Cardou
Département de génie mécanique
Université Laval
Quebec City, QC
Canada

Andreas Pott
Fraunhofer-Institut für Produktionstechnik
   und Automatisierung IPA
Stuttgart, Baden-Württemberg
Germany

ISSN 2211-0984 ISSN 2211-0992 (electronic)
Mechanisms and Machine Science
ISBN 978-3-319-61430-4 ISBN 978-3-319-61431-1 (eBook)
DOI 10.1007/978-3-319-61431-1

Library of Congress Control Number: 2017945285

© Springer International Publishing AG 2018
This work is subject to copyright. All rights are reserved by the Publisher, whether the whole or part of the material is concerned, specifically the rights of translation, reprinting, reuse of illustrations, recitation, broadcasting, reproduction on microfilms or in any other physical way, and transmission or information storage and retrieval, electronic adaptation, computer software, or by similar or dissimilar methodology now known or hereafter developed.
The use of general descriptive names, registered names, trademarks, service marks, etc. in this publication does not imply, even in the absence of a specific statement, that such names are exempt from the relevant protective laws and regulations and therefore free for general use.
The publisher, the authors and the editors are safe to assume that the advice and information in this book are believed to be true and accurate at the date of publication. Neither the publisher nor the authors or the editors give a warranty, express or implied, with respect to the material contained herein or for any errors or omissions that may have been made. The publisher remains neutral with regard to jurisdictional claims in published maps and institutional affiliations.

Printed on acid-free paper

This Springer imprint is published by Springer Nature
The registered company is Springer International Publishing AG
The registered company address is: Gewerbestrasse 11, 6330 Cham, Switzerland

Preface

This book is a compendium of the articles presented at the Third International Conference on Cable-Driven Parallel Robots, also known by its diminutive CableCon2017, held at Université Laval, Quebec City, Canada. The first two conferences of this series were both held in Germany, respectively, in Stuttgart, in 2012, and in Duisburg, in 2014. It is therefore the first time that the conference leaves the European continent, which we hope will be an occasion to foster new links with researchers from the Americas.

Some readers may be left wondering as to the nature of the cable-driven parallel robots mentioned in the conference title. In general, these parallel robots are made of a rigid mobile platform attached to a fixed frame by several cables acting in parallel, their lengths being controlled by servo-actuated winches. These robots and their variants are the topic of CableCon2017. In the past decade, cable-driven parallel robots have attracted a renewed interest from the research community and from industry. This may be seen from the number of researchers who took part in the first editions of CableCon, but also from scientific literature and from the various industrial projects that were undertaken during these years. This interest stems from several advantages that are widely recognised to favour cable-driven parallel robots over others: large workspace, low cost, good dynamic properties, reconfigurability, portability, and compatibility with vision systems.

Yet, as much as these advantages are enticing, several issues have hindered the development of effective cable-driven parallel robots. Some of these issues have been the subject of significant progress, e.g. workspace determination, cable tension resolution, and winch design. Others still pose important challenges to researchers, despite remarkable efforts to solve them, e.g. forward displacement analysis, vibration control, accuracy, interferences. Moreover, cable-driven parallel robots remain unknown or have only been partially tested in several applications where they promise great leaps in efficiency.

In this context, we believe that CableCon2017 can provide a stimulating forum for the exchange of ideas, of potential applications, and of key challenges that remain to be addressed, just as were the first two editions of the conference. We deem the articles included in this book to be of excellent quality, which allows us to foresee fruitful presentations and discussions. The articles are distributed into four themes: modelling; displacement and workspace analysis; trajectory planning and control; design and applications. Under these themes, one should find all the main engineering challenges that need to be resolved to allow cable-driven parallel robots to reach their full potential. We hope that this conference can be useful in taking one more step towards this goal.

Finally, we would like to express our gratitude to all the authors for their valuable contributions and to all the reviewers and scientific committee members for their expertise and selfless efforts in maintaining the standards of the conference.

May 2017

Tobias Bruckmann
Philippe Cardou
Clément Gosselin
Andreas Pott

Organization

Program Chairs

Clément Gosselin   Université Laval, Canada
Philippe Cardou   Université Laval, Canada

Program Committee

Sunil Agrawal   Columbia University, USA
Marc Arsenault   Laurentian University, Canada
Tobias Bruckmann   Universität Duisburg-Essen, Germany
Philippe Cardou   Université Laval, Canada
Stéphane Caro   CNRS - LS2N, France
Clément Gosselin   Université Laval, Canada
Marc Gouttefarde   CNRS - LIRMM, France
Jean-Pierre Merlet   Inria, France
Leila Notash   Queen's University, Canada
Andreas Pott   Fraunhofer IPA, Germany
Dieter Schramm   Universität Duisburg-Essen, Germany

Contents

Modelling

Modelling of Flexible Cable-Driven Parallel Robots Using a Rayleigh-Ritz Approach 3
Harsh Atul Godbole, Ryan James Caverly, and James Richard Forbes

Assumed-Mode-Based Dynamic Model for Cable Robots with Non-straight Cables 15
Jorge Ivan Ayala Cuevas, Édouard Laroche, and Olivier Piccin

Manipulator Deflection for Optimum Tension of Cable-Driven Robots with Parameter Variations 26
Leila Notash

Sensitivity Analysis of the Elasto-Geometrical Model of Cable-Driven Parallel Robots 37
Sana Baklouti, Stéphane Caro, and Eric Courteille

CASPR-ROS: A Generalised Cable Robot Software in ROS for Hardware 50
Jonathan Eden, Chen Song, Ying Tan, Denny Oetomo, and Darwin Lau

A Polymer Cable Creep Modeling for a Cable-Driven Parallel Robot in a Heavy Payload Application 62
Jinlong Piao, XueJun Jin, Eunpyo Choi, Jong-Oh Park, Chang-Sei Kim, and Jinwoo Jung

Bending Fatigue Strength and Lifetime of Fiber Ropes 73
Martin Wehr, Andreas Pott, and Karl-Heinz Wehking

Bending Cycles and Cable Properties of Polymer Fiber Cables for Fully Constrained Cable-Driven Parallel Robots . 85
Valentin Schmidt and Andreas Pott

Displacement and Workspace Analysis

A New Approach to the Direct Geometrico-Static Problem of Cable Suspended Robots Using Kinematic Mapping 97
Manfred Husty, Josef Schadlbauer, and Paul Zsombor-Murray

Determination of the Cable Span and Cable Deflection of Cable-Driven Parallel Robots 106
Andreas Pott

Geometric Determination of the Cable-Cylinder Interference Regions in the Workspace of a Cable-Driven Parallel Robot 117
Antoine Martin, Stéphane Caro, and Philippe Cardou

Twist Feasibility Analysis of Cable-Driven Parallel Robots 128
Saman Lessanibahri, Marc Gouttefarde, Stéphane Caro, and Philippe Cardou

Initial Length and Pose Calibration for Cable-Driven Parallel Robots with Relative Length Feedback 140
Darwin Lau

Static Analysis and Dimensional Optimization of a Cable-Driven Parallel Robot 152
Matthew Newman, Arthur Zygielbaum, and Benjamin Terry

Improving the Forward Kinematics of Cable-Driven Parallel Robots Through Cable Angle Sensors 167
Xavier Garant, Alexandre Campeau-Lecours, Philippe Cardou, and Clément Gosselin

Direct Kinematics of CDPR with Extra Cable Orientation Sensors: The 2 and 3 Cables Case with Perfect Measurement and Ideal or Elastic Cables 180
Jean-Pierre Merlet

Trajectory Planning and Control

Randomized Kinodynamic Planning for Cable-Suspended Parallel Robots 195
Ricard Bordalba, Josep M. Porta, and Lluís Ros

Rest-to-Rest Trajectory Planning for Planar Underactuated Cable-Driven Parallel Robots 207
Edoardo Idá, Alessandro Berti, Tobias Bruckmann, and Marco Carricato

Dynamically-Feasible Elliptical Trajectories for Fully Constrained 3-DOF Cable-Suspended Parallel Robots 219
Giovanni Mottola, Clément Gosselin, and Marco Carricato

**Dynamic Transition Trajectory Planning of Three-DOF
Cable-Suspended Parallel Robots**.............................. 231
Xiaoling Jiang and Clément Gosselin

**Transverse Vibration Control in Planar Cable-Driven
Robotic Manipulators**... 243
Mitchell Rushton and Amir Khajepour

**Application of a Differentiator-Based Adaptive Super-Twisting
Controller for a Redundant Cable-Driven Parallel Robot**............ 254
Christian Schenk, Carlo Masone, Andreas Pott, and Heinrich H. Bülthoff

**Tension Distribution Algorithm for Planar Mobile Cable-Driven
Parallel Robots**... 268
Tahir Rasheed, Philip Long, David Marquez-Gamez, and Stéphane Caro

**Improvement of Cable Tension Observability Through
a New Cable Driving Unit Design**.............................. 280
Mathieu Rognant and Eric Courteille

A Fast Algorithm for Wrench Exertion Capability Computation...... 292
Giovanni Boschetti, Chiara Passarini, Alberto Trevisani,
and Damiano Zanotto

Design and Applications

**Design and Analysis of a Novel Cable-Driven Haptic Master
Device for Planar Grasping**................................... 307
Kashmira S. Jadhao, Patrice Lambert, Tobias Bruckmann,
and Just L. Herder

**On the Design of a Three-DOF Cable-Suspended Parallel Robot
Based on a Parallelogram Arrangement of the Cables**............... 319
Dinh-Son Vu, Eric Barnett, Anne-Marie Zaccarin,
and Clément Gosselin

On Improving Stiffness of Cable Robots........................ 331
Carl A. Nelson

**Optimal Design of a High-Speed Pick-and-Place Cable-Driven
Parallel Robot**.. 340
Zhaokun Zhang, Zhufeng Shao, Liping Wang, and Albert J. Shih

**On the Improvements of a Cable-Driven Parallel Robot
for Achieving Additive Manufacturing for Construction**............. 353
Jean-Baptiste Izard, Alexandre Dubor, Pierre-Elie Hervé, Edouard Cabay,
David Culla, Mariola Rodriguez, and Mikel Barrado

Concept Studies of Automated Construction Using Cable-Driven Parallel Robots 364
Tobias Bruckmann, Christopher Reichert, Michael Meik, Patrik Lemmen, Arnim Spengler, Hannah Mattern, and Markus König

Inverse Kinematics for a Novel Rehabilitation Robot for Lower Limbs 376
Abdelhak Badi, Maarouf Saad, Guy Gauthier, and Philippe Archambault

On the Design of a Novel Cable-Driven Parallel Robot Capable of Large Rotation About One Axis 390
Alexis Fortin-Côté, Céline Faure, Laurent Bouyer, Bradford J. McFadyen, Catherine Mercier, Michaël Bonenfant, Denis Laurendeau, Philippe Cardou, and Clément Gosselin

Preliminary Running and Performance Test of the Huge Cable Robot of FAST Telescope 402
Hui Li, Jinghai Sun, Gaofeng Pan, and Qingge Yang

Author Index 415

Modelling

Modelling of Flexible Cable-Driven Parallel Robots Using a Rayleigh-Ritz Approach

Harsh Atul Godbole[1(✉)], Ryan James Caverly[2], and James Richard Forbes[1]

[1] Department of Mechanical Engineering, McGill University,
817 Sherbrooke Street West, Montreal, QC H3A 0C3, Canada
harsh.godbole@mail.mcgill.ca
[2] Department of Aerospace Engineering, University of Michigan,
1320 Beal Avenue, Ann Arbor, MI 48109, USA
james.richard.forbes@mcgill.ca

Abstract. This paper investigates the use of the Rayleigh-Ritz method to model single degree-of-freedom flexible cable-driven parallel robots (CDPRs) using a set of time-dependent basis functions to discretize cables of varying length. An energy-based model simplification is proposed to further facilitate reduction in the computational load when performing numerical simulations involving the proposed model. Open-loop system responses are used to compare the effect of the energy-based model simplification. Frequency responses are used to compare the influence of the number of basis functions used and to provide a comparison to a lumped-mass model.

1 Introduction

Accurate models representing the dynamics of flexible cable-driven parallel robots (CDPRs) are useful for developing accurate simulations, designing high-performance closed-loop controllers, and analyzing the closed-loop stability properties of these systems. All models are approximations, and the amount of detail put into a given CDPR model is highly dependent on the intended application. Rigid cable models are sufficient when cables are short and cable elasticity and mass may be neglected. For longer cables, the dynamics of the cable itself and cable sag play a significant role in the dynamics of the system, and should therefore be modelled. High-tension flexible CDPRs with long cables feature significant longitudinal cable vibrations, which can significantly affect the ability to accurately track a desired CDPR payload pose. For example, the CableRobot Simulator at the Max Planck Institute for Biological Cybernetics operates with high-tension cables that transmit forces to a payload with a human passenger [11]. High-frequency longitudinal cable vibrations must not be detected by the passenger and, therefore, must be be suppressed in some way [11]. This paper considers high-tensioned CDPRs that are similar in nature to the CableRobot Simulator, in the sense that the high-frequency longitudinal vibrations cannot be neglected.

The work of [8] models flexible CDPRs using a lumped-mass model in a three dimensional setting. Alternatively, [2,3] uses a similar lumped-mass model incorporating actuator dynamics and cable wrap around the winch to ensure conservation of energy and establish a passive input-output map. Significant limitations of these models include the fact that they can only operate in a reduced workspace, where the lumped masses may not come into contact with the winches, and they require many lumped masses to accurately reproduce the true natural frequencies of the cable. The work of [7] shows that accurate representation of cable vibrations is possible using finite elemental analysis. However, repeated mesh generations and calculations for different orientations of the payload cause a massive computational overhead, and are not practical for dynamic simulation purposes. Static stiffness of long cables exhibiting properties such as cable sag and its dependencies on cable tension is studied in [10]. Also, the effect of cable sag on the stiffness of CDPRs and its minimization is investigated by [1]. However, this paper focuses on high-tension cables that exhibit negligible cable sag and dynamically varying stiffness due to their varying lengths. The Rayleigh-Ritz model presented in this paper is inspired by the work of [13,15]; however, this paper differs from [13,15] by rigorously analyzing of selection of Ritz basis functions and their nonlinear time dependencies, as well as comparing differences between Rayleigh-Ritz and lumped-mass cable models.

The novel contributions of this paper are (1) the derivation of a nonlinear dynamic model of a flexible CDPR using a Rayleigh-Ritz discretization of the flexible cables, (2) an approximate model that simplifies the nonlinear dynamic model without comprising accuracy, and (3) a comparison of Rayleigh-Ritz and lumped-mass models of a CDPR. Although six degree-of-freedom flexible CDPRs, such as the CableRobot Simulator, are of great interest in practice, a single degree-of-freedom flexible CDPR is investigated in this paper. This simplifies the derivations and analyses, which will then be used as building blocks for more complicated and physically relevant CDPRs in future work. The remainder of the paper proceeds as follows. In Sect. 2, a Rayleigh-Ritz model of a flexible CDPR is derived and an approximate model is proposed. Open-loop numerical simulations of both models are presented in Sect. 3 that demonstrate the accuracy and numerical efficiency of the approximate model. A comparison of the linear frequency responses of Rayleigh-Ritz and lumped-mass models is performed in Sect. 4. Section 5 provides concluding remarks.

2 Flexible CDPR Modelling Using Rayleigh-Ritz Discretization

This section presents a rigorous and systematic approach to modelling a single degree-of-freedom flexible CDPR using a Rayleigh-Ritz discretization.

2.1 Time-Dependent and Spatial Variables

Consider a cable wrapped around a winch as shown in Fig. 1(a). The inertial reference frame $\mathscr{F}_i$ is defined by basis vectors $\underline{i}_1$, $\underline{i}_2$, $\underline{i}_3$. The reference frame

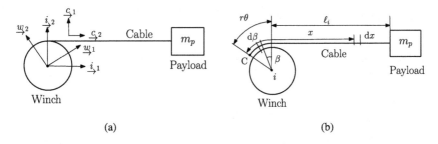

Fig. 1. A single degree-of-freedom unconstrained CDPR (referred to as a half-system).

$\mathscr{F}_c$ describes the orientation of the cable relative to $\mathscr{F}_i$, and is defined by basis vectors $\underrightarrow{c}_1$, $\underrightarrow{c}_2$, $\underrightarrow{c}_3$, where $\underrightarrow{c}_2$ is aligned with the cable; and the reference frame $\mathscr{F}_w$ describes the orientation of the winch, and is defined by the basis vectors $\underrightarrow{w}_1$, $\underrightarrow{w}_2$, $\underrightarrow{w}_3$. The length of the portion of cable between the point of contact on the winch and the payload is given by ℓ_i, as shown in Fig. 1(b). The cable is defined by a nominal initial length L, and the point C, which corresponds to the initial point of contact between the cable and the winch. The winch rotation is described by the angle θ. The length of cable wrapped around the winch is given by $r\theta$ and the length of cable between the payload and the point of contact on the winch is given by $\ell_i = L - r\theta$.

A time-independent spatial coordinate x is used to represent the position of each constituent mass element of the cable along its length relative to the point C on the cable. The angle β is a spatial variable used to define the angular position of a mass element on the winch, as shown in Fig. 1(b). The coordinate x_p is the position of the payload relative to the centre of the winch in the $\underrightarrow{i}_1$ direction.

The elongation of the cable at any point between the winch and the payload is given by $w(x,t) = \boldsymbol{\Psi}(x,\ell_i(t))\mathbf{q}_e(t)$. The time-dependent elastic coordinates of the cable are $\mathbf{q}_e(t) \in \mathbb{R}^n$. The matrix $\boldsymbol{\Psi}(x,\ell_i(t)) \in \mathbb{R}^{1 \times n}$ contains the Ritz basis functions, also referred to simply as basis functions, which will be written as $\boldsymbol{\Psi}(x,\ell_i)$ for the remainder of this paper. The selection of these basis functions is arbitrary, provided they satisfy the boundary conditions of the problem at hand, and are conventionally taken to be time-independent coordinates. However, in the interest of selecting computationally-efficient basis functions based on the exact eigenfunctions used to describe the axial vibration of cables, the basis functions are selected as [15]

$$\boldsymbol{\Psi}_{oe}(x,\ell_i) = \left[\sin\left(\frac{\pi(x-L+\ell_i)}{2\ell_i}\right) \ \sin\left(\frac{2\pi(x-L+\ell_i)}{2\ell_i}\right) \ \sin\left(\frac{3\pi(x-L+\ell_i)}{2\ell_i}\right) \ \cdots \right]. \quad (1)$$

The basis functions in Eq. (1) are time-dependent due to the presence of ℓ_i, which is justified in Sect. 3. The unconstrained generalized coordinates for a single winch, cable and payload system are $\mathbf{q}^\mathsf{T} = \begin{bmatrix} \theta & \mathbf{q}_e^\mathsf{T} & x_p \end{bmatrix}$.

2.2 Dynamic Model of Half-System

A model of the half-system is derived in this section using Lagrange's Equation. In the inertial frame, the position of an elemental mass, dm, of the portion of the cable attached to the winch is given by $\mathbf{r}_i^{dm\,i} = \begin{bmatrix} -r\sin\beta & r\cos\beta & 0 \end{bmatrix}^\mathsf{T}$. The velocity of this element is expressed in $\mathscr{F}_i$ as $\mathbf{v}_i^{dm\,i/i} = \begin{bmatrix} -r(\cos\beta)\dot{\theta} & -r(\sin\beta)\dot{\theta} & 0 \end{bmatrix}^\mathsf{T}$. The kinetic energy of the length of the cable wrapped around the winch, C_w, is given by

$$\frac{1}{2}\int_{C_w}(\underset{\rightarrow}{v}^{dm\,i/i})\cdot(\underset{\rightarrow}{v}^{dm\,i/i})dm = \frac{1}{2}\int_0^{r\theta}(\mathbf{v}_i^{dm\,i/i})^\mathsf{T}(\mathbf{v}_i^{dm\,i/i})\rho A dx = \frac{1}{2}\dot{\mathbf{q}}^\mathsf{T}\mathbf{M}_{cw}\dot{\mathbf{q}},$$

where $\mathbf{M}_{cw} = \mathrm{diag}\{r^3\rho A\theta, \mathbf{0}\} \in \mathbb{R}^{(n+2)\times(n+2)}$ and $\mathbf{0}$ is a matrix filled with zeros of compatible dimensions. Note that this mass matrix is not positive definite as expected because $\mathbf{M}_{cw}$ only represents a part of the entire mass matrix of the half-system. In a similar manner, the kinetic energy of the winch and payload can be shown to be $T_w = \frac{1}{2}\dot{\mathbf{q}}^\mathsf{T}\mathbf{M}_w\dot{\mathbf{q}}$, and $T_p = \frac{1}{2}\dot{\mathbf{q}}^\mathsf{T}\mathbf{M}_p\dot{\mathbf{q}}$, respectively where $\mathbf{M}_w = \mathrm{diag}\{J_w, \mathbf{0}\} \in \mathbb{R}^{(n+2)\times(n+2)}$, $\mathbf{M}_p = \mathrm{diag}\{\mathbf{0}, m_p\} \in \mathbb{R}^{(n+2)\times(n+2)}$, J_w is the inertia of the winch, and m_p is the mass of the payload. As seen in Fig. 1(b), the position of any elemental mass dm in the portion of the cable between the payload and the winch relative to the winch centre, expressed in $\mathscr{F}_i$ is $\mathbf{r}_i^{dm\,i} = \begin{bmatrix} x - r\theta + \mathbf{\Psi}\mathbf{q}_e & r & 0 \end{bmatrix}^\mathsf{T}$. The corresponding velocity of every mass element relative to the centre of the winch is given by $\mathbf{v}_i^{dm\,i/i} = \begin{bmatrix} -r\dot{\theta} + \mathbf{\Psi}\dot{\mathbf{q}}_e + \dot{\mathbf{\Psi}}\mathbf{q}_e & 0 & 0 \end{bmatrix}^\mathsf{T}$. From Eq. (1), the time derivative of $\mathbf{\Psi}$ is $\dot{\mathbf{\Psi}} = \left(-r\dot{\theta}\right)\frac{\partial \mathbf{\Psi}}{\partial \ell_i}$, where $\frac{\partial \mathbf{\Psi}}{\partial \ell_i} = \begin{bmatrix} -\frac{\pi x}{2\ell_i^2}\cos\left(\frac{\pi x}{2\ell_i}\right) & -\frac{2\pi x}{2\ell_i^2}\cos\left(\frac{2\pi x}{2\ell_i}\right) & -\frac{3\pi x}{2\ell_i^2}\cos\left(\frac{3\pi x}{2\ell_i}\right) \end{bmatrix}$. The kinetic energy of the unwrapped cable is

$$\frac{1}{2}\int_{C_{\ell_i}}(\underset{\rightarrow}{v}^{dm\,i/i})\cdot(\underset{\rightarrow}{v}^{dm\,i/i})dm = \int_{C_{\ell_i}}(-r\dot{\theta}_1\frac{\partial\mathbf{\Psi}}{\partial\ell_i}\mathbf{q}_e + \mathbf{\Psi}\dot{\mathbf{q}}_e - r\dot{\theta}_1)^2 dm.$$

The kinetic energy of the cable can be rewritten as a quadratic function of the generalized coordinate rates, $T_c = \frac{1}{2}\dot{\mathbf{q}}^\mathsf{T}\mathbf{M}_{ce}\dot{\mathbf{q}}$, where

$$\mathbf{M}_{ce} = \int_{L-\ell_i}^{L}\rho A \begin{bmatrix} r^2\left(1 + 2\frac{\partial\mathbf{\Psi}}{\partial\ell_i}\mathbf{q}_e + \mathbf{q}_e^\mathsf{T}\left(\frac{\partial\mathbf{\Psi}}{\partial\ell_i}\right)^\mathsf{T}\frac{\partial\mathbf{\Psi}}{\partial\ell_i}\mathbf{q}_e\right) & -r\left(1 + \mathbf{q}_e^\mathsf{T}\left(\frac{\partial\mathbf{\Psi}}{\partial\ell_i}\right)^\mathsf{T}\right)\mathbf{\Psi} & 0 \\ * & \mathbf{\Psi}^\mathsf{T}\mathbf{\Psi} & 0 \\ * & * & 0 \end{bmatrix} dx. \tag{2}$$

The basis functions selected in Eq. (1) satisfy the boundary conditions of a cable fixed at one and and free at the other end. The integrals can be simplified using the coordinate transformation $z = x - L + \ell_i$. Furthermore, the evaluation of the integrals in Eq. (2) are carried out at a given instant of time, which results in $dz = dx$. This coordinate transformation transforms the nth term of the $\mathbf{\Psi}_{oe}$ row matrix to $\int_{L-\ell_i}^{L}\sin\left(\frac{n\pi(x-L+\ell_i)}{2\ell_i}\right)dx = \int_0^{\ell_i}\sin\left(\frac{n\pi z}{2\ell_i}\right)dz$. Applying this coordinate transformation to Eq. (2) and defining constant matrices $\boldsymbol{\Lambda}_i$ and $\boldsymbol{\Lambda}_{ii}$, $i = 1, 2$ yields

$$\mathbf{M}_{ce} = \rho A \begin{bmatrix} r^2 \left(\ell_i + 2\boldsymbol{\Lambda}_1 \mathbf{q}_e + \frac{1}{\ell_i} \mathbf{q}_e^\mathsf{T} \boldsymbol{\Lambda}_{11} \mathbf{q}_e \right) & -r \left(\ell_i \boldsymbol{\Lambda}_2 + \mathbf{q}_e^\mathsf{T} \boldsymbol{\Lambda}_{12} \right) & 0 \\ * & \ell_i \boldsymbol{\Lambda}_{22} & 0 \\ * & * & 0 \end{bmatrix}.$$

where $\boldsymbol{\Lambda}_1 = \int_0^{\ell_i} \frac{\partial \boldsymbol{\Psi}}{\partial \ell_i} dz$, $\boldsymbol{\Lambda}_{11} = \ell_i \int_0^{\ell_i} \left(\frac{\partial \boldsymbol{\Psi}}{\partial \ell_i} \right)^\mathsf{T} \frac{\partial \boldsymbol{\Psi}}{\partial \ell_i} dz$, $\boldsymbol{\Lambda}_{12} = \int_0^{\ell_i} \left(\frac{\partial \boldsymbol{\Psi}}{\partial \ell_i} \right)^\mathsf{T} \boldsymbol{\Psi} dz$, $\boldsymbol{\Lambda}_2 = \frac{1}{\ell_i} \int_0^{\ell_i} \boldsymbol{\Psi} dz$, $\boldsymbol{\Lambda}_{22} = \frac{1}{\ell_i} \int_0^{\ell_i} \boldsymbol{\Psi}^\mathsf{T} \boldsymbol{\Psi} dz$. The integrals above lead to simplified mass matrix expressions due to the somewhat unconventional choice of basis functions in Eq. (1). The computational load during simulation is reduced significantly as compared to the often used polynomial basis functions, due to improved conditioning of the mass matrix, especially in case of long cables. The potential energy of the system is stored in the cable in the form of strain energy. The energy stored in the elongated cable between the winch and the payload is given by

$$\frac{1}{2} \int_{L-\ell_i}^{L} EA \left(\frac{\partial (\boldsymbol{\Psi} \mathbf{q}_e)}{\partial x} \right)^2 dx = \frac{1}{2} \mathbf{q}_e^\mathsf{T} \int_0^{\ell_i} EA \left(\frac{\partial \boldsymbol{\Psi}}{\partial x} \right)^\mathsf{T} \left(\frac{\partial \boldsymbol{\Psi}}{\partial x} \right) dz \, \mathbf{q}_e = \frac{1}{2} \mathbf{q}^\mathsf{T} \mathbf{K} \mathbf{q},$$

where $\mathbf{K} = \mathrm{diag}\{0, EA\boldsymbol{\Lambda}_{33}/\ell_i, 0\}$ and $\boldsymbol{\Lambda}_{33} = \ell_i \int_0^{\ell_i} \left(\frac{\partial \boldsymbol{\Psi}}{\partial x} \right)^\mathsf{T} \left(\frac{\partial \boldsymbol{\Psi}}{\partial x} \right) dz$. To model the natural damping of the cable, the Rayleigh dissipation function of the system is chosen to be $R = \frac{1}{2} \dot{\mathbf{q}}^\mathsf{T} \mathbf{D} \dot{\mathbf{q}}$, where $\mathbf{D} = \mathrm{diag}\{0, c_1, c_2, \ldots c_n, 0\}$ and c_i, $i = 1, \ldots, n$ are the damping coefficients of each of the flexible modes of the cable. The equations of motion for the half-system derived using Lagrange's Method are

$$\mathbf{M}_e \ddot{\mathbf{q}} + \mathbf{D} \dot{\mathbf{q}} + \mathbf{K} \mathbf{q} = \hat{\mathbf{b}} \tau + \mathbf{f}_{\mathrm{non}}, \tag{3}$$

where $\mathbf{M}_e = \mathbf{M}_w + \mathbf{M}_p + \mathbf{M}_{cw} + \mathbf{M}_{ce}$, $\hat{\mathbf{b}}^\mathsf{T} = [1\ 0]$, $\mathbf{f}_{\mathrm{non}} = -\dot{\mathbf{M}}_e \dot{\mathbf{q}} + \frac{1}{2} \left(\frac{\tilde{\partial}}{\partial \mathbf{q}} \dot{\mathbf{q}}^\mathsf{T} \mathbf{M}_e \dot{\mathbf{q}} \right)^\mathsf{T} - \frac{1}{2} \left(\frac{\tilde{\partial}}{\partial \mathbf{q}} \mathbf{q}^\mathsf{T} \mathbf{K} \mathbf{q} \right)^\mathsf{T}$, $\dot{\mathbf{M}}_e = \dot{\mathbf{M}}_{cw} + \dot{\mathbf{M}}_{ce}$, $\dot{\mathbf{M}}_{cw} = \mathrm{diag}\{r^3 \rho A \dot{\theta}, \mathbf{0}\}$, and

$$\dot{\mathbf{M}}_{ce} = \rho A \begin{bmatrix} r^2 \left(2\boldsymbol{\Lambda}_1 \dot{\mathbf{q}}_e + \frac{r\dot{\theta}}{\ell_i^2} \mathbf{q}_e^\mathsf{T} \boldsymbol{\Lambda}_{11} \mathbf{q}_e + \frac{2}{\ell_i} \mathbf{q}_e^\mathsf{T} \boldsymbol{\Lambda}_{11} \dot{\mathbf{q}}_e - r\dot{\theta} \right) & \left(r^2 \dot{\theta} \boldsymbol{\Lambda}_2 - r \dot{\mathbf{q}}_e^\mathsf{T} \boldsymbol{\Lambda}_{12} \right) & 0 \\ * & -r\dot{\theta} \boldsymbol{\Lambda}_{22} & 0 \\ * & * & 0 \end{bmatrix}.$$

The partial derivative terms in $\mathbf{f}_{\mathrm{non}}$ are calculated while keeping the quadratic terms of the generalized coordinate rates constant. These terms simplify to,

$$\frac{1}{2} \left(\frac{\tilde{\partial}}{\partial \mathbf{q}} \dot{\mathbf{q}}^\mathsf{T} \mathbf{M} \dot{\mathbf{q}} \right)^\mathsf{T} = \frac{1}{2} \begin{bmatrix} \dot{\mathbf{q}}^\mathsf{T} \left(\frac{\partial \mathbf{M}_{cw}}{\partial \theta} + \frac{\partial \mathbf{M}_{ce}}{\partial \theta} \right)^\mathsf{T} \dot{\mathbf{q}} \\ \frac{\tilde{\partial} (\dot{\mathbf{q}}^\mathsf{T} \mathbf{M}_{ce} \dot{\mathbf{q}})}{\tilde{\partial} \mathbf{q}_e} \end{bmatrix}, \quad \frac{\partial \mathbf{M}_{cw}}{\partial \theta} = \mathrm{diag}\{r^3 \rho A, \mathbf{0}\},$$

$$\frac{\partial \mathbf{M}_{ce}}{\partial \theta} = \rho A \begin{bmatrix} r^2 \left(\frac{r}{\ell_i^2} \mathbf{q}_e^\mathsf{T} \mathbf{\Lambda}_{11} \mathbf{q}_e - r \right) & (r^2 \mathbf{\Lambda}_2) & \mathbf{0} \\ * & -r\mathbf{\Lambda}_{22} & \mathbf{0} \\ * & * & \mathbf{0} \end{bmatrix},$$

$$\frac{\tilde{\partial} \left(\dot{\mathbf{q}}^\mathsf{T} \mathbf{M}_{ce} \dot{\mathbf{q}} \right)}{\tilde{\partial} \mathbf{q}_e} = \rho A \left(\left(r\dot{\theta} \right)^2 \left(2\mathbf{\Lambda}_1 + \frac{2}{\ell_i} \mathbf{q}_e^\mathsf{T} \mathbf{\Lambda}_{11} \right) - 2r\dot{\theta} \left(\dot{\mathbf{q}}_e^\mathsf{T} \mathbf{\Lambda}_{12} \right) \right),$$

$$\frac{1}{2} \left(\frac{\tilde{\partial}}{\tilde{\partial} \mathbf{q}} \mathbf{q}^\mathsf{T} \mathbf{K} \mathbf{q} \right)^\mathsf{T} = \frac{1}{2} \begin{bmatrix} \mathbf{q}^\mathsf{T} \left(\frac{\partial \mathbf{K}}{\partial \theta} \right) \mathbf{q} \\ \mathbf{0} \end{bmatrix}, \quad \frac{\partial \mathbf{K}}{\partial \theta} = \mathrm{diag}\{0, EAr\mathbf{\Lambda}_{33}/\ell_i^2, 0\}.$$

Equation (3), along with its constituent matrices, are referred throughout this article as 'exact' equations of motion. Note that the payload is not yet constrained to the end of the cable and the appropriate constraints are added in Sect. 2.4.

2.3 Energy-Based Model Simplification

Although the cable model of the half-system developed in Sect. 2.2 with the Rayleigh-Ritz method maintains a high-degree of fidelity, it can be computationally inefficient in simulation. The approximate model presented in this section simplifies the model, while maintaining its high-fidelity nature.

The kinetic energy of the portion of the cable between the winch and the payload is given by $T_{ce} = \frac{1}{2} \int_{L-\ell_i}^{L} \left(-r\dot{\theta} + \mathbf{\Psi} \dot{\mathbf{q}}_e + \dot{\mathbf{\Psi}} \mathbf{q}_e \right)^2 \rho A \mathrm{d}x$. In numerical simulations, it is found that the term $\dot{\mathbf{\Psi}} \mathbf{q}_e$ is much smaller than $\mathbf{\Psi} \dot{\mathbf{q}}_e$ (on the order of 10^{-2} smaller). Hence, the proposed approximate model neglects this term in the kinetic energy expression and in effect, the matrices $\mathbf{\Lambda}_1$, $\mathbf{\Lambda}_{11}$ and $\mathbf{\Lambda}_{12}$ may be neglected. The approximation reduces the equations of motion in Eq. (3) to

$$\mathbf{M}_a \ddot{\mathbf{q}} + \mathbf{D} \dot{\mathbf{q}} + \mathbf{K} \mathbf{q} = \hat{\mathbf{b}} \tau + \mathbf{f}_{\mathrm{non}}, \tag{4}$$

where $\mathbf{M}_a = \mathbf{M}_w + \mathbf{M}_p + \mathbf{M}_{cw} + \mathbf{M}_{ca}$, $\dot{\mathbf{M}}_a = \dot{\mathbf{M}}_{cw} + \dot{\mathbf{M}}_{ca}$, $\dot{\mathbf{M}}_{cw} = \mathrm{diag}\{r^3 \rho A \dot{\theta}, \mathbf{0}\}$, $\mathbf{f}_{\mathrm{non}} = -\dot{\mathbf{M}}_a \dot{\mathbf{q}} + \frac{1}{2} \left(\frac{\tilde{\partial}}{\tilde{\partial} \mathbf{q}} \dot{\mathbf{q}}^\mathsf{T} \mathbf{M}_a \dot{\mathbf{q}} \right)^\mathsf{T} - \frac{1}{2} \left(\frac{\tilde{\partial}}{\tilde{\partial} \mathbf{q}} \mathbf{q}^\mathsf{T} \mathbf{K} \mathbf{q} \right)^\mathsf{T}$, and

$$\mathbf{M}_{ca} = \rho A \ell_i \begin{bmatrix} r^2 & -r\mathbf{\Lambda}_2 & \mathbf{0} \\ * & \mathbf{\Lambda}_{22} & \mathbf{0} \\ * & * & \mathbf{0} \end{bmatrix}, \quad \dot{\mathbf{M}}_{ca} = -\rho A r \dot{\theta} \begin{bmatrix} r^2 & -r\mathbf{\Lambda}_2 & \mathbf{0} \\ * & \mathbf{\Lambda}_{22} & \mathbf{0} \\ * & * & \mathbf{0} \end{bmatrix}.$$

The nonlinear terms reduce to

$$\frac{1}{2} \left(\frac{\tilde{\partial}}{\tilde{\partial} \mathbf{q}} \dot{\mathbf{q}}^\mathsf{T} \mathbf{M}_a \dot{\mathbf{q}} \right)^\mathsf{T} = \frac{1}{2} \begin{bmatrix} \dot{\mathbf{q}}^\mathsf{T} \left(\frac{\partial \mathbf{M}_{cw}}{\partial \theta} + \frac{\partial \mathbf{M}_{ca}}{\partial \theta} \right)^\mathsf{T} \dot{\mathbf{q}} \\ \mathbf{0} \end{bmatrix}, \quad \frac{\partial \mathbf{M}_{ca}}{\partial \theta} = -\rho A r \begin{bmatrix} r^2 & -r\mathbf{\Lambda}_2 & \mathbf{0} \\ * & \mathbf{\Lambda}_{22} & \mathbf{0} \\ * & * & \mathbf{0} \end{bmatrix},$$

$\frac{\partial \mathbf{M}_{cw}}{\partial \theta} = \mathrm{diag}\{\rho A r^3, \mathbf{0}\}$. All terms related to the stiffness matrix remain the same. The equations of motion in Eq. (4) are similar to those of a lumped-mass model [3], where the $\mathbf{1}_i$ and $\mathbf{1}_i^\mathsf{T} \mathbf{1}_i$ are replaced by $\mathbf{\Lambda}_2$ and $\mathbf{\Lambda}_{22}$ respectively, in the mass matrix.

2.4 Constraint Equations and the Null Space Method

In this section, two half-systems are constrained using the null space method [9]. Subscripts '1' and '2' refer to the constituent elements of Eqs. (3) or (4) that describe either the first or second half-system, as shown in Fig. 2. The unconstrained generalized coordinates of the system are $\mathbf{q}^T = \begin{bmatrix} \mathbf{q}_1^T & \mathbf{q}_2^T \end{bmatrix} = \begin{bmatrix} \theta_1 & \mathbf{q}_{e_1}^T & x_{p1} & \theta_2 & \mathbf{q}_{e_2}^T & x_{p2} \end{bmatrix}$. In order to constrain the system, each cable is attached to its respective payload and both payloads are constrained together. The velocity of the free end of the cable is $\dot{\rho}_1 = J_{\theta_1}\dot{\theta}_1 + \mathbf{J}_{e_1}\dot{\mathbf{q}}_{e_1}$, and $\dot{\rho}_2 = J_{\theta_2}\dot{\theta}_2 + \mathbf{J}_{e_2}\dot{\mathbf{q}}_{e_2}$, respectively for cables '1' and '2', where $J_{\theta_1} = r$ and $J_{\theta_2} = -r$ are rigid Jacobians, and $\mathbf{J}_{e_1} = \mathbf{\Psi}(\ell_i, \ell_i)$ and $\mathbf{J}_{e_2} = -\mathbf{\Psi}(\ell_i, \ell_i)$ are elastic Jacobians. Note that $\mathbf{J}_{e_1}$ and $\mathbf{J}_{e_2}$ are constant due to the chosen basis functions in $\mathbf{\Psi}_{oe}$. The constraints are summarized in the matrix $\mathbf{\Xi}$ as

$$\mathbf{\Xi}\dot{\mathbf{q}} = \mathbf{\Xi}\begin{bmatrix} \dot{\mathbf{q}}_1 \\ \dot{\mathbf{q}}_2 \end{bmatrix} = \mathbf{0}, \quad \mathbf{\Xi} = \begin{bmatrix} J_{\theta_1} & \mathbf{J}_{e_2} & -1 & 0 & 0 & 0 \\ 0 & 0 & 0 & J_{\theta_2} & \mathbf{J}_{e_2} & -1 \\ 0 & 0 & -1 & 0 & 0 & 1 \end{bmatrix}.$$

The first row of $\mathbf{\Xi}$ constrains the velocity of the first half-system payload to be $\dot{x}_{p1} = \dot{\rho}_1$. The second row of $\mathbf{\Xi}$ does the same for the second half-system. The third row constrains $\dot{\rho}_1 = \dot{\rho}_2$. The equations of motion for the constrained full system are

$$\mathbf{M}\ddot{\mathbf{q}} + \mathbf{D}\dot{\mathbf{q}} + \mathbf{K}\mathbf{q} = \hat{\mathbf{b}}\boldsymbol{\tau} + \mathbf{f}_{\text{non}} + \mathbf{\Xi}^T \boldsymbol{\lambda}, \tag{5}$$

where $\mathbf{M} = \text{diag}\{\mathbf{M}_1, \mathbf{M}_2\}$, $\mathbf{K} = \text{diag}\{\mathbf{K}_1, \mathbf{K}_2\}$, $\mathbf{D} = \text{diag}\{\mathbf{D}_1, \mathbf{D}_2\}$, $\hat{\mathbf{b}} = \text{diag}\{\hat{\mathbf{b}}_1, \hat{\mathbf{b}}_2\}$, $\boldsymbol{\tau}^T = \begin{bmatrix} \tau_1 & \tau_2 \end{bmatrix}$, and $\boldsymbol{\lambda}^T = \begin{bmatrix} \lambda_1 & \lambda_2 & \lambda_3 \end{bmatrix}$ are Lagrange multipliers. The independent constrained generalized coordinates of the full system are defined as $\mathbf{z}^T = \begin{bmatrix} \theta_1 & \mathbf{q}_{e_1}^T & \mathbf{q}_{e_2}^T \end{bmatrix}$. A reduction matrix $\mathbf{R}$ can be defined such that $\mathbf{q} = \mathbf{R}\mathbf{z}$, $\dot{\mathbf{q}} = \mathbf{R}\dot{\mathbf{z}}$ and $\mathbf{\Xi}\mathbf{R} = \mathbf{0}$ by

$$\mathbf{R}^T = \begin{bmatrix} 1 & 0 & J_{\theta_1} & J_{\theta_2}^{-1}J_{\theta_1} & 0 & J_{\theta_1} \\ 0 & 1 & \mathbf{J}_{e_1} & J_{\theta_2}^{-1}\mathbf{J}_{e_1} & 0 & \mathbf{J}_{e_1} \\ 0 & 0 & 0 & -J_{\theta_2}^{-1}\mathbf{J}_{e_2} & 1 & 0 \end{bmatrix}.$$

Premultiplying Eq. (5) by $\mathbf{R}^T$ and noting that $\mathbf{q} = \mathbf{R}\mathbf{z}$, $\dot{\mathbf{q}} = \mathbf{R}\dot{\mathbf{z}}$, and $\mathbf{\Xi}\mathbf{R} = \mathbf{0}$ yields

$$\mathbf{M}_{zz}\ddot{\mathbf{z}} + \mathbf{D}_{zz}\dot{\mathbf{z}} + \mathbf{K}_{zz}\mathbf{z} = \hat{\mathbf{b}}_{zz}\boldsymbol{\tau} + \mathbf{R}^T \mathbf{f}_{\text{non},zz}, \tag{6}$$

where the subscript 'zz' is used to represent the equivalent mass, stiffness and damping matrices along with expressions for nonlinear forces and actuator torques.

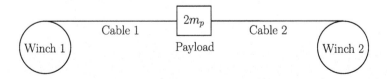

Fig. 2. The fully constrained single degree-of-freedom CDPR.

3 Nonlinear System Analysis

In this section, the constrained system is numerically simulated with mass matrix $\mathbf{M}_{ce}$ (Case 1) or $\mathbf{M}_{ca}$ (Case 2). The effect of the term $\dot{\mathbf{\Psi}}\mathbf{q}_e$ in the energy expression is investigated through open-loop simulations. The simulations use parameters of a payload mass of 1 kg ($m_p = 0.5$ kg), winch radius $r = 4$ cm, winch inertia $J_w = 1.39 \times 10^{-5}$ kg·m^2 and nominal cable length $L = 0.5$ m. The cable has a cross-sectional area of $A = 17.95 \times 10^{-6}$ m^2, a density of $\rho = (0.0385 \text{ kg/m})/(17.95 \times 10^{-6} \text{ m}^2) = 2200$ kg/m^3, and a modulus of elasticity of $E = 500$ MPa. The simulation is carried out for 5 seconds using an Runge-Kutta integrator of order 4, in C++ with a time step of 10 μs. The open source matrix library Armadillo [12] is used to optimize matrix calculations in C++. Both systems (Case 1 and Case 2) are given an initial payload velocity of 2 m/s. Figures 3(a) and (b) show system responses with no natural cable damping. Figures 3(c) and (d) show system responses when cable damping constants $c_i = 1 \times 10^{-5}$ N·s/m, $i = 1, \ldots, n$, are applied. Observing the open-loop response, it can be concluded that in the case of the undamped systems, the difference in the response of Case 1 and Case 2 is significant. However, with a nominal increase in the amount of natural damping the error in the open-loop

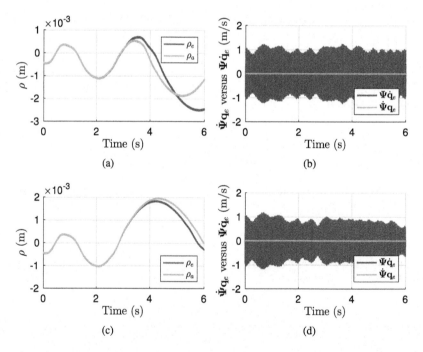

Fig. 3. Comparison of open-loop system undamped (a),(b) and damped (c),(d) responses with Case 1 and Case 2. The plots of (a) and (c) include payload position, ρ_e (Case 1) and ρ_a (Case 2), versus time. The plots of (b) and (d) compare $\dot{\mathbf{\Psi}}\mathbf{q}_e$ and $\mathbf{\Psi}\dot{\mathbf{q}}_e$ versus time.

system response reduces significantly. Simulations with the approximate model were completed in roughly 70–90% of the time it took to simulate the exact model, which highlights the numerical efficiency of the approximate model. Figures 3(b) and (c) also present a comparison of the magnitudes of the terms $\boldsymbol{\Psi}\dot{\mathbf{q}}_e$ and $\dot{\boldsymbol{\Psi}}\mathbf{q}_e$, when simulating Case 1. The term $\left|\dot{\boldsymbol{\Psi}}\mathbf{q}_e\right| \approx 10^{-2}\left|\boldsymbol{\Psi}\dot{\mathbf{q}}_e\right|$ in both damped and undamped cases, further validating the accuracy of the approximate model.

4 Linearized System Analysis

In this section, the constrained system is linearized at the equilibrium point $\mathbf{z} = \dot{\mathbf{z}} = \mathbf{0}$ and analyzed in the frequency domain. Similar to rigid robotic manipulators with closed loops, load-sharing parameters C_1 and C_2 are used [4–6]. The mapping $\boldsymbol{\tau} = \begin{bmatrix} C_1 J_{\theta_1} & C_2 J_{\theta_2} \end{bmatrix}^\mathsf{T} \tau_c = \mathbf{U}_\theta \tau_c$ converts Eq. (6) to

$$\mathbf{M}_{zz}\ddot{\mathbf{z}} + \mathbf{D}_{zz}\dot{\mathbf{z}} + \mathbf{K}_{zz}\mathbf{z} = \hat{\mathbf{b}}_{zz}\mathbf{U}_\theta \tau_c + \mathbf{R}^\mathsf{T}\mathbf{f}_{\text{non},zz}. \tag{7}$$

Load-sharing parameters allow the two input torques to be driven by a single input torque, which is useful in the following analysis. The input to this system is the generalized input torque τ_c, and the output is chosen to be the μ-tip velocity $\dot{\rho}_\mu = \mu\dot{\rho} + (1-\mu)\mathbf{U}_\theta^\mathsf{T}\dot{\boldsymbol{\theta}}$ [6,14], where $\dot{\boldsymbol{\theta}}^\mathsf{T} = \begin{bmatrix} \dot{\theta}_1 & \dot{\theta}_2 \end{bmatrix}$. To analyze the frequency response of linearized system, Bode plots of the linearized systems are investigated. The μ-tip velocity is chosen as the system's output, since it has been previously shown for lumped-mass models that the input-output map from τ_c to $\dot{\rho}_\mu$ is passive, which has advantages when designing a robust controller for the system. The phase of

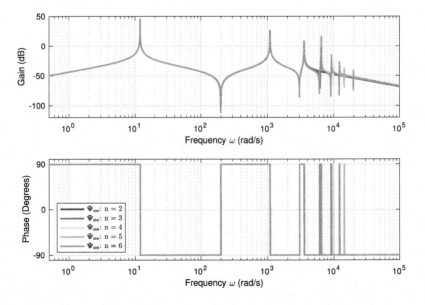

Fig. 4. Bode plot of $\tau_c \mapsto \dot{\rho}_\mu$ using $\mathbf{M}_{ce}$.

linear time-invariant passive systems is bounded by ±90°, which can be verified in the proposed Rayleigh-Ritz model. To perform the linearization, the nonlinear equations of Eq. (7) are rearranged in state-space form as

$$\dot{\mathbf{x}} = \mathbf{f}(\mathbf{x}, \tau_c) = \begin{bmatrix} \dot{\mathbf{z}} \\ \mathbf{M}_{zz}^{-1} \left(\hat{\mathbf{b}}_{zz} \mathbf{U}_\theta \tau_c + \mathbf{R}^\mathsf{T} \mathbf{f}_{\text{non},zz} - \mathbf{D}_{zz} \dot{\mathbf{z}} - \mathbf{K}_{zz} \mathbf{z} \right) \end{bmatrix},$$

$$\dot{\rho}_\mu = \mathbf{g}(\mathbf{x}) = \mu \dot{\rho} + (1-\mu) \mathbf{U}_\theta^\mathsf{T} \dot{\theta} = \begin{bmatrix} J_{\theta_1} \left(\mu + C_2 (1-\mu) \right) \mathbf{J}_{e_1} & -C_2 (1-\mu) \mathbf{J}_{e_2} \end{bmatrix} \dot{\mathbf{z}},$$

where $\mathbf{x}^\mathsf{T} = \begin{bmatrix} \mathbf{z}^\mathsf{T} & \dot{\mathbf{z}}^\mathsf{T} \end{bmatrix}$. The linearized equations are written as $\dot{\mathbf{x}} = \mathbf{A}\mathbf{x} + \mathbf{B}\tau_c$, and $\dot{\rho}_\mu = \mathbf{C}\mathbf{x}$, where the Jacobian matrices are $\mathbf{A} = \frac{\partial \mathbf{f}}{\partial \mathbf{x}}\big|_{\mathbf{x}=0}$, $\mathbf{B} = \frac{\partial \mathbf{f}}{\partial \tau_c}\big|_{\mathbf{x}=0}$, $\mathbf{C} = \frac{\partial \mathbf{g}}{\partial \mathbf{x}}\big|_{\mathbf{x}=0}$.

In this section, the convergence of the system's natural frequencies with increasing number of basis functions is analyzed for the proposed Rayleigh-Ritz model. In Fig. 4, the frequency response appears to converge towards the natural frequencies of the model on increasing the number of basis functions. Notice that even with a small number of basis functions (i.e., $n = 2$ or $n = 3$), the lower natural frequencies are accurately represented. It is shown in the frequency response plot of Fig. 5, that for the linearized lumped-mass method [3], the lowest natural mode of vibration is in agreement with the Rayleigh-Ritz method with a large number of lumped masses. However, the higher order natural frequencies are not in agreement even with a large number of lumped masses. They appear to converge towards the natural frequencies plotted by using the Rayleigh-Ritz method, although very slowly.

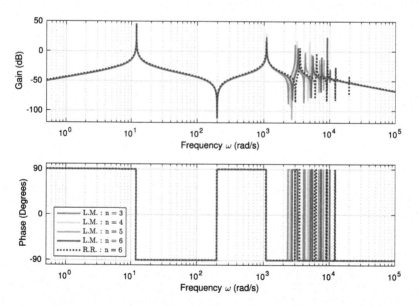

Fig. 5. Bode plot of $\tau_c \mapsto \dot{\rho}_\mu$ for the full system modelled using the lumped-mass (L.M.) method and Rayleigh-Ritz (R.R.) method with $\mathbf{M}_{ce}$.

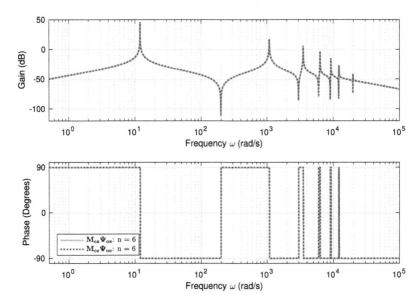

Fig. 6. Bode plot of $\tau_c \mapsto \dot{\rho}_\mu$ for the Rayleigh-Ritz method, with ($\mathbf{M}_{ca}$) and without ($\mathbf{M}_{ce}$) the energy-based model simplification.

It should also be noted that the higher natural frequencies of the lumped-mass model are significantly lower than those predicted by the Rayleigh-Ritz method. Figure 6 shows the effect of the model approximation on the natural frequencies of the linearized system. Six basis functions are used in each case. There is a strong agreement between the natural frequencies as predicted with and without the energy-based model simplification, which reinforces the accuracy of the energy-based model simplification.

5 Conclusion

The objective of this paper was to develop a cable model that accurately models the longitudinal cable vibrations of a flexible CDPR in a manner that is numerically efficient in simulation and useful for controller design. The Rayleigh-Ritz method developed uses time-dependent basis functions inspired from the exact solutions of axially vibrating cables and the work of [13, 15]. The basis functions were selected such that the dependency of the mass matrix on ℓ_i is minimized. Furthermore, an approximate model was proposed in order to improve computational efficiency in numerical simulation, which was shown to behave very similarly to the exact model in open-loop simulations when a small amount of natural damping was present in the cable. It was shown through linear frequency responses that the Rayleigh-Ritz model more accurately represents the natural frequencies of the system compared to the lumped-mass method, when using a small number of basis functions or lumped masses. This highlights a major advantage of using the Rayleigh-Ritz method to model flexible CDPMs, rather than lumped-mass methods. Future work includes extending this Rayleigh-Ritz cable model to three dimensions, in a similar manner to the lumped-mass model in [2].

References

1. Arsenault, M.: Stiffness analysis of a planar 2-DoF cable-suspended mechanism while considering cable mass. In: Cable-Driven Parallel Robots, pp. 405–421. Springer, Berlin (2013)
2. Caverly, R.J., Forbes, J.R.: Dynamic modeling and noncollocated control of a flexible planar cable-driven manipulator. IEEE Trans. Robot. **30**(6), 1386–1397 (2014)
3. Caverly, R.J., Forbes, J.R., Mohammadshahi, D.: Dynamic modeling and passivity-based control of a single degree of freedom cable-actuated system. IEEE Trans. Control Syst. Technol. **23**(3), 898–909 (2015)
4. Damaren, C.J.: Approximate inverse dynamics and passive feedback for flexible manipulators with large payloads. IEEE Trans. Robot. Autom. **12**(1), 131–138 (1996)
5. Damaren, C.J.: Modal properties and control system design for two-link flexible manipulators. Int. J. Robot. Res. **17**(6), 667–678 (1998)
6. Damaren, C.J.: An adaptive controller for two cooperating flexible manipulators. J. Robot. Syst. **20**(1), 15–21 (2003)
7. Du, J., Ding, W., Bao, H.: Cable vibration analysis for large workspace cable-driven parallel manipulators. In: Cable-Driven Parallel Robots, pp. 437–449. Springer, Berlin (2013)
8. Lambert, C., Nahon, M., Chalmers, D.: Implementation of an aerostat positioning system with cable control. IEEE-ASME Trans. Mechatron. **12**(1), 32–40 (2007)
9. Laulusa, A., Bauchau, O.A.: Review of classical approaches for constraint enforcement in multibody systems. J. Comput. Nonlin. Dyn. **3**(1), 011004 (2008)
10. Li, H.: On the static stiffness of incompletely restrained cable-driven robot. In: Cable-Driven Parallel Robots, pp. 55–69. Springer International Publishing (2015)
11. Miermeister, P., Lächele, M., Boss, R., Masone, C., Schenk, C., Tesch, J., Kerger, M., Teufel, H., Pott, A., Bülthoff, H.H.: The CableRobot simulator large scale motion platform based on cable robot technology. In: IEEE International Conference on Intelligent Robots, pp. 3024–3029 (2016)
12. Sanderson, C., Curtin, R.: Armadillo: a template-based C++ library for linear algebra. J. Open Source Softw. **1**(2), 26–32 (2016)
13. Walsh, A., Forbes, J.R.: Modeling and control of a wind energy harvesting kite with flexible cables. In: Proceedings of the American Control Conference, pp. 2383–2388 (2015)
14. Wang, D., Vidyasagar, M.: Passive control of a stiff flexible link. Int. J. Robot. Res. **11**(6), 572–578 (1992)
15. Zhang, Y., Agrawal, S.K., Hagedorn, P.: Longitudinal vibration modeling and control of a flexible transporter system with arbitrarily varying cable lengths. J. Vib. Control **11**(3), 431–456 (2005)

Assumed-Mode-Based Dynamic Model for Cable Robots with Non-straight Cables

Jorge Ivan Ayala Cuevas[1], Édouard Laroche[2](✉), and Olivier Piccin[3]

[1] INSA of Strasbourg, Strasbourg, France
jorge.ayalacuevas@insa-strasbourg.fr
[2] ICube Laboratory, Strasbourg University, Strasbourg, France
laroche@unistra.fr
[3] ICube Laboratory, INSA of Strasbourg, Strasbourg, France
olivier.piccin@insa-strasbourg.fr

Abstract. This paper presents an original method for deriving models of flexible cable robots including cable sagging based on assumed mode assumption. This method allows to derive low-order models that specially suit for control applications. The case of a winder and a planar cable without elongation but with sagging in the plane of movement is first considered. Then, the model of a planar robot with a punctual platform with three cables is presented. The model is written in the Lagrange framework for constrained systems. Simulation results for a three-cable robots are presented and discussed.

1 Introduction

Cable-driven parallel robots (CDPR) are a special class of parallel manipulators in which the end-effector is connected to the base through cables, the movement being provided by the winding and unwinding of cables. Compared to conventional serial or parallel manipulators, CDPR have interesting features: a large workspace capability, low inertia of moving components and reduced obstruction of the workspace. Their main drawback is common to all flexible manipulators in which the deflections and elongation of the links limit the precision when determining the position of the end-effector from the measurements of the joint positions.

A number of approaches considers straight inextensible cables [3,5]. Straight massless extensible cables are also often considered. In a simplistic case, the cable is modeled as the association of a rigid link with a spring which stiffness is inversely proportional to the cable length [8,13]. Models from continuum mechanics are also available in the literature that provide more accurate models of elastic cables [11]. When the mass of the cable is not negligible anymore, the sagging effect must be accounted for. In statics, this effect results in the catenary equation and is well documented [7,14]. Finite-element models are available for but they have the drawback of resulting in high order models [4]. More recently, Arsenault [2] and Yuan et al. [15] have considered elastic cables with sagging.

Following the dynamic stiffness matrix method [1], the stiffness matrix is first determined in statics and then introduced in the dynamic model.

The key idea of the original approach proposed herein is to consider cables as particular cases of flexible segments. When considering the control of systems composed of deformable segments modeled as Euler-Bernoulli beams, the assumed-mode approach is certainly the most standard and has been intensively used for serial robots [9, 10]. The segment deformations are first written as sums of contributions of a given base. Then, the geometry can be written as a function of a generalized position vector that includes deformation variables. The dynamic model, given by the Lagrange equation of motion in a standard way, accounts for the kinetic energy of the cable displacements. In this contribution, this approach is considered in which deformable segments are replaced by perfectly flexible and inextensible cables. As an illustrative example, the case of a planar robot with a punctual platform, actuated by three or more cables, is considered. Cables are assumed to be affected by sagging in the plane of movement.

In Sect. 2, the model of a single cable and its winder, undergoing transverse deformation in a plane is considered. Based on Lagrange approach, a dynamic model is derived. In Sect. 3, the model of a planar robot with three or more cables is considered. The DAE model is developed and then reduced. In Sect. 4, some simulation results are presented and discussed. The model derived with Maple and the simulation with Matlab-Simulink are available online[1].

2 Single Cable Modeling

In this section, we focus on the an elementary constitutive element of the planar robot depicted in Fig. 2, namely, one single cable winded at one side and submitted at the other side to an external force.

2.1 Single Cable Modeling

Up to four deformation fields can be considered when modeling a deformable beam under Euler-Bernoulli assumption [12]. Herein, the cable subjected to sagging is considered as a perfectly flexible and inextensible 1-dimensional body. In the current study, the only deformation field of interest is the transverse deformation in the plane of motion. The final geometry of the cable will be given as the composition of three steps: unwinding, shaping and rotation.

Let us consider a single cable $\#k$ operated by a winder $\#k$. The cable is tangent to the winder at point W_k and has an end-point denoted P_k. The unwinded portion of the cable is the planar curve between W_k and P_k of length l_k. Let $\mathscr{F}_b = (O_b, \mathbf{x_b}, \mathbf{y_b})$ and $\mathscr{F}_k = (W_k, \mathbf{x_k}, \mathbf{y_k})$ denote respectively the fixed global reference frame and the local reference frame attached to the winder $\#k$. The position of W_k and the orientation of the cable at W_k are defined by (x_{W_k}, y_{W_k}) and φ_k respectively as indicated in Fig. 1.

---

[1] http://icube-avr.unistra.fr/fr/index.php/Planar_cable_robot_with_non_straight_cables.

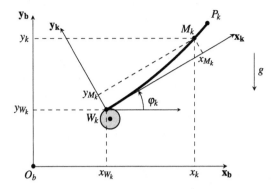

Fig. 1. General configuration of a single cable.

From an initial configuration where the cable is straight along the $\mathbf{x_k}$ direction, let us now consider a small displacement of the cable that alters the cable shape but preserves its point of tangency W_k on the winder. In this elementary displacement, a point of coordinates $(x, 0)$ with $x \in [0, l_k]$ is moved to the point M_k of coordinates $(x_{M_k} = x + \delta x_{M_k}, y_{M_k} = \delta y_{M_k})$ in the local frame $\mathscr{F}_k$. Finally, the coordinates (x_k, y_k) of the point M_k expressed in the global frame $\mathscr{F}_b$ can be obtained using an homogeneous transformation as

$$\begin{bmatrix} x_k \\ y_k \\ 1 \end{bmatrix} = \begin{bmatrix} \cos\varphi_k & -\sin\varphi_k & x_{W_k} \\ \sin\varphi_k & \cos\varphi_k & y_{W_k} \\ 0 & 0 & 1 \end{bmatrix} \begin{bmatrix} x_{M_k} \\ y_{M_k} \\ 1 \end{bmatrix}. \tag{1}$$

Notice that if the cable is inextensible, the small displacement variables are linked by

$$\left(\frac{\partial \delta x_{M_k}}{\partial x} + 1\right)^2 + \left(\frac{\partial \delta y_{M_k}}{\partial x}\right)^2 = 1 \tag{2}$$

and assuming that $\left|\frac{\partial \delta y_{M_k}}{\partial x}\right| \ll 1$, Eq. (2) yields

$$\delta x_{M_k}(x, t) = -\frac{1}{2} \int_0^x \left(\frac{\partial \delta y_{M_k}(u, t)}{\partial u}\right)^2 du. \tag{3}$$

The small displacement δy_{M_k} along the $\mathbf{y_k}$ direction is assumed to be the sum of a number of contributions that can be written, with a given basis $\Phi_k(x)$ truncated at the order N, as

$$\delta y_{M_k}(x, t) = \sum_{j=1}^N \Phi_j(x) V_{jk}(t) \tag{4}$$

where V_{jk} is the generalized coordinate for mode Φ_j. In the sequel, we choose to work with a polynomial basis of the form $\Phi_j(x) = x^{j+1}$. In this assumed mode approach, other basis could have been used, as for example the set of modal deformations described in [9].

Upon substitution of the coordinates (x_{M_k}, y_{M_k}) into Eq. (1), the position of a point M_k in the global reference frame can be readily calculated as analytic functions, namely, $x_k(\tilde{q}_k)$ and $y_k(\tilde{q}_k)$ with $\tilde{q}_k = \begin{bmatrix} x & V_{1k} & \ldots & V_{Nk} & \varphi_k \end{bmatrix}^T$. At this point, when x is set to the unwinded length of cable l_k, the model of the single cable can be parameterized by the generalized coordinate vector $q_k = \begin{bmatrix} l_k & V_{1k} & \ldots & V_{Nk} & \varphi_k \end{bmatrix}^T$ containing $N+2$ independent parameters q_{k_i}.

2.2 Cable Dynamic Model

Based on the parameterization presented in the previous subsection, a dynamic model of a single cable is now introduced as a basic example of the approach. The details of the three-cable robot are not given in the paper but are available online (see the link given at the first page).

The cable is winded at one side with a fixed winder actuated by a torque τ_k and is subject to the gravitational acceleration $-g\mathbf{y_b}$. The other end P_k is submitted to an arbitrary force $\mathbf{F_k}$ which coordinates in $\mathscr{F}_b$ are (F_{x_k}, F_{y_k}). The cylindric winder is of radius R and inertia J_0. The cable has a linear density ρ and a total length l_t. Accounting for the wounded portion of the cable, the actual inertia is $J_k = J_0 + \rho(l_t - l_k)R^2$. Furthermore, the winder angular position θ_k is related to the unwinded length of cable l_k by $l_k = -R\theta_k$. The gravitational potential energy of the single cable writes

$$V_k = \int_0^{l_k} \rho\, g\, y_k(\tilde{q}_k)\, \mathrm{d}x. \tag{5}$$

The kinetic energy of the single cable and its rotating winder writes

$$T_k = \frac{1}{2}\frac{J_k}{R^2} \dot{l}_k^2 + \frac{1}{2}\int_0^{l_k} \rho\left(\dot{x}_k(\tilde{q}_k)^2 + \dot{y}_k(\tilde{q}_k)^2\right) \mathrm{d}x \tag{6}$$

in which the velocity terms $\dot{x}_k(q_k)$ and $\dot{y}_k(q_k)$ can be calculated as $\sum_{i=1}^{N+2} \frac{\partial x_k}{\partial q_{k_i}} \dot{q}_{k_i}$ and $\sum_{i=1}^{N+2} \frac{\partial y_k}{\partial q_{k_i}} \dot{q}_{k_i}$. The kinetic energy can then be written under its quadratic form $T_k = \frac{1}{2}\dot{q}_k^T M_k(q_k)\dot{q}_k$ where $M_k(q_k)$ refers to the kinetic energy matrix. The Lagrange's equations of motion can be written as

$$\frac{\mathrm{d}}{\mathrm{d}t}\frac{\partial T_k}{\partial \dot{q}_k} - \frac{\partial T_k}{\partial q_k} = \Gamma_k\, Q_k - \frac{\partial V_k}{\partial q_k} \tag{7}$$

where $\Gamma_k = \begin{bmatrix} F_{x_k} & F_{y_k} & \tau_k \end{bmatrix}$ corresponds to the actions applied on the system and Q_k a matrix of partial velocity terms, relative to the generalized coordinates and determined from the virtual-work principle as:

$$Q_k = \begin{bmatrix} \frac{\partial x_{P_k}}{\partial q_{k_1}} & \frac{\partial x_{P_k}}{\partial q_{k_2}} & \cdots & \frac{\partial y_{P_k}}{\partial q_{k_{N+2}}} \\ \frac{\partial y_{P_k}}{\partial q_{k_1}} & \frac{\partial y_{P_k}}{\partial q_{k_2}} & \cdots & \frac{\partial x_{P_k}}{\partial q_{k_{N+2}}} \\ -\frac{1}{R} & 0 & \cdots & 0 \end{bmatrix} \tag{8}$$

where x_{P_k} and y_{P_k} denote the position functions of the point P_k at which the effort $\mathbf{F_k}$ is applied. The entries of the $1 \times (N+2)$ line matrix $\Gamma_k \, Q_k$ correspond to the generalized forces acting on the cable.

Denoting $p_k = \dfrac{\partial T_k}{\partial \dot{q}_k} = \dot{q}_k^T M_k$, the line matrix of generalized momentum, the model can be rewritten under the following state-space representation:

$$\dot{p}_k = C_k + \Gamma_k \, Q_k - G_k \tag{9}$$
$$\dot{q}_k = M_k^{-1} p_k^T \tag{10}$$

where

$$C_k = \frac{\partial T_k}{\partial q_k} = \begin{bmatrix} \frac{1}{2}\dot{q}_k^T \frac{\partial M_k}{\partial q_{k_1}} \dot{q}_k & \cdots & \frac{1}{2}\dot{q}_k^T \frac{\partial M_k}{\partial q_{k_{N+2}}} \dot{q}_k \end{bmatrix} \tag{11}$$

$$G_k = \frac{\partial V_k}{\partial q_k} = \begin{bmatrix} \frac{\partial V_k}{\partial q_{k_1}} & \cdots & \frac{\partial V_k}{\partial q_{k_{N+2}}} \end{bmatrix} \tag{12}$$

3 Planar Robot with n cables

A planar cable robot operated by several cables is now considered as presented in Fig. 2. Its platform is considered as a punctual mass m located at point P of coordinates (x_P, y_P) in the global reference frame. The number of cables in this example is three but the presented method is applicable to any number of cables.

3.1 Dynamic Model

The generalized coordinate vector q for the system includes the two parameters of the mobile platform and the n sets of parameters relative to each cable. The column vector q can be written symbolically as

$$q = \begin{bmatrix} x_P & y_P & q_1^T & \cdots & q_n^T \end{bmatrix}^T \tag{13}$$

which corresponds to $n(N+2)+2$ non independent parameters.

The total kinetic energy is calculated as the sum of contributions of each cable plus the platform of mass m, yielding to $T = \frac{1}{2}\dot{q}^T M \dot{q}$ with $M = \text{diag}(M_0, M_1, \ldots, M_n)$ where $M_0 = \text{diag}(m, m)$ is the kinetic inertia matrix of the platform and M_k, $k = 1, \ldots, n$ denotes the kinetic inertia matrix for cable $\#k$.

With the selected generalized coordinate vector q and gathering the terms $\Gamma_k Q_k$ corresponding to each cable, the generalized force vector acting on the system writes

$$\Gamma Q = \begin{bmatrix} F_{x_P} & F_{y_P} & \Gamma_1 Q_1 & \cdots & \Gamma_n Q_n \end{bmatrix} \tag{14}$$

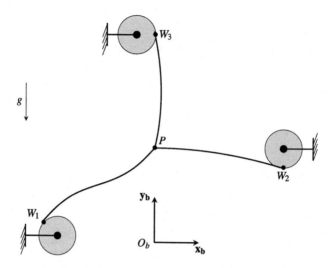

Fig. 2. Schematics of a planar cable robot with 3 cables.

where F_{x_P} and F_{y_P} are the components, in the global reference frame, of an effort $\mathbf{F_P}$ acting on the moving platform at point P. The gravitational potential energy for the whole system can be calculated as $V = \sum_{k=1}^{n} V_k - mgy_P$. In the sequel, we assume that $\mathbf{F_P} = \mathbf{0}$.

The coincidence of the positions of the platform with the cable ends provide $h = 2n$ geometric (holonomic) constraints of the form $h_r(q) = 0$, $r = 1, \ldots, h$:

$$h_{2k-1} = x_{P_k}(q_k) - x_P \tag{15}$$
$$h_{2k} = y_{P_k}(q_k) - y_P \tag{16}$$

with $k = 1, \ldots, n$.

As the $n(N+2)+2$ parameters are related by the h geometric constraints (15) and (16), the dynamic behavior of the system can be obtained using Lagrange's equations with h multipliers [6]. Upon differentiation with respect to time, the constraint relations can be written $A(q)\dot{q} = 0$ where A is the Jacobian of the constraints with respect to the generalized coordinate vector q whose entries write $A_{rk}(q) = \frac{\partial h_r(q)}{\partial q_k}$.

Using $\lambda = \begin{bmatrix} \lambda_1 \ldots \lambda_h \end{bmatrix}^\mathrm{T}$ as the column vector of the Lagrange multipliers, the Lagrange's equations can be written as:

$$\frac{\mathrm{d}}{\mathrm{d}t} \frac{\partial T}{\partial \dot{q}} - \frac{\partial T}{\partial q} = \varGamma Q - G + \lambda^\mathrm{T} A \tag{17}$$

with $G = \frac{\partial V}{\partial q}$. Given that the generalized momentum matrix $p = \frac{\partial T}{\partial \dot{q}} = \dot{q}^\mathrm{T} M$ and after differentiation of the geometric constraints, the differential-algebraic equations of the system can be obtained as

$$\begin{bmatrix} M & -A^{\mathrm{T}} \\ A & 0 \end{bmatrix} \begin{bmatrix} \ddot{q} \\ \lambda \end{bmatrix} = \begin{bmatrix} C^{\mathrm{T}} + (\Gamma Q)^{\mathrm{T}} - G^{\mathrm{T}} - \dot{M}\dot{q} \\ -\dot{A}\dot{q} \end{bmatrix}. \tag{18}$$

Since the equations set (18) is linear with respect to $\ddot{q}$ and λ, solving for $\ddot{q}$ can be done directly by inversion of the matrix $\begin{bmatrix} M & -A^{\mathrm{T}} \\ A & 0 \end{bmatrix}$ either online, numerically or offline, using a computer algebra system.

4 Simulation Results

A system composed of three cables and three winders evenly distributed on a circle with a 10 m diameter has been tested with the following set of parameters: $l_t = 5$ m, $\rho = 0.2$ kg/m, $R = 0.1$ m, $J_0 = 2.5 \cdot 10^{-3}$ kg·m² and $m = 1$ kg. The cable models have been set with one mode ($N = 1$).

A controller has been implemented in order to have the platform follow a desired trajectory (x^*, y^*). A number of approaches are available in the literature for cable robot control [3,8,13]. Herein, a simplistic approach is used, assuming that both position and speed of the platform are available.

The controller has been established on the kinetic model $\dot{\theta} = J(q_0)\dot{q}_0$ that connects the vector of the angular velocities $\dot{\theta}$ to the velocity of the platform $\dot{q}_0$ through the Jacobian matrix $J(q_0)$, assuming straight cables. The control signals (i.e. the motor torques) are computed as

$$u = u_0 \begin{bmatrix} 1 \\ 1 \\ 1 \end{bmatrix} + J^{\mathrm{T}\dagger}(q_r) \begin{bmatrix} u_x \\ u_y \end{bmatrix} \tag{19}$$

where u_0 ensures a positive tension in the cables; $J^{\mathrm{T}\dagger}$ is the pseudo-inverse of the transpose of J; u_x and u_y are the control actions in the (x, y) plane, computed with a proportional-derivative (PD) control law given in the Laplace domain:

$$u_x(s) = K(s)\left(x^*(s) - x(s)\right) \tag{20}$$
$$u_y(s) = K(s)\left(y^*(s) - y(s)\right) \tag{21}$$

where s denote the Laplace variable and $u_x(s)$ is the signal u_x in the Laplace domain. The same PD controllers with filtering are used for both x and y directions:

$$K(s) = K_p + K_d \frac{\omega_f s}{\omega_f + s} \tag{22}$$

where the coefficients have been chosen as following: the proportional gain is $K_p = 400$ N; the derivative gain is $K_d = 100$ N.s; the filtering frequency is $\omega_f = 100$ rad/s.

The robot being initialized at the center of the workspace without sagging, the reference remains at the center during 2 s before moving by 1 m along the $\mathbf{x_b}$ direction, then following a square of 2 m side length centered in the workspace at a constant speed of 1 m/s and finally coming back to the origin. The reference signals and the actual trajectory can be seen in Fig. 3. The reference trajectory is tracked with some oscillations. One can check in Fig. 4 that the tensions remain positive during operation. In Fig. 5, the actual trajectory is presented in the $(\mathbf{x_b}, \mathbf{y_b})$ plane and the geometry of the cables is plotted for three positions in order to see how sagging evolves dynamically at a fast pace. The modal coordinates V_{11}, V_{12} and V_{13} are presented in Fig. 6. One can see how they vary in term of amplitude and frequency. Notice that the sagging at rest observed at $t = 2$ s is reduced compared to the variations observed during dynamic operation.

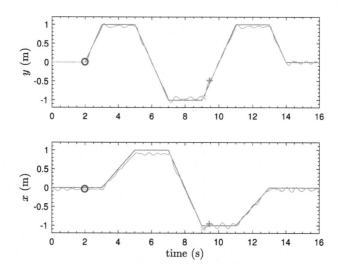

Fig. 3. Trajectory of the effector with respect to time: reference and actual position.

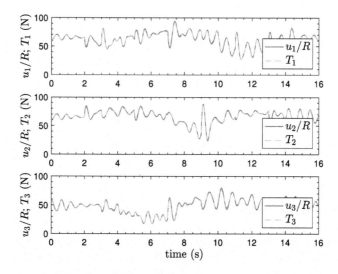

Fig. 4. Evolution of the cable tensions T_k and of the control signals u_k.

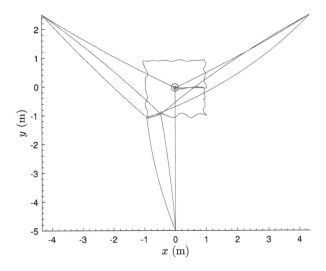

Fig. 5. Trajectory of the effector in the x-y plan and geometry of the cables at $t = 2$ s; 9.1 s and 9.45 s.

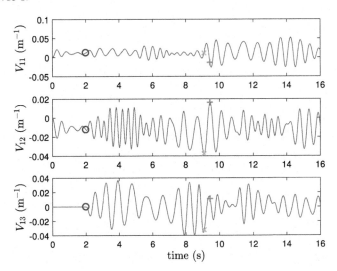

Fig. 6. Evolution of the deformation variables V_{1k} with respect to time.

In order to highlight the effect of the cable dynamics on the trajectories, the trajectories obtained for two different values of the linear density of the cables are given in Fig. 7. For a low linear density ($\rho = 0.02$ kg/m), the reference is quite well tracked whereas the dynamic behavior of the cables observed for $\rho = 0.2$ kg/m significantly degrades the system behavior.

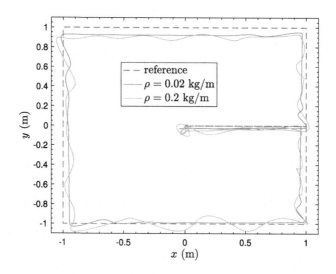

Fig. 7. Trajectories of the effector in the x-y plane for different values of ρ.

5 Conclusion

In this paper, an original approach has been proposed to account for the cable movements for very dynamic operations of CDPR. Using the assumed deformation method, a dynamic model is derived using the Lagrange's equations of motion for constrained systems. The method has been implemented in the case of a planar CDPR with three cables. Simulation results have shown the effect of the cable movements on the system behavior.

The next steps to further assess the method's efficiency will include comparisons of the obtained simulation results with experimental data as well as with other available approaches. Another perspective will consist in extending the model to account for the cable elongation in the planar case but also in the more challenging case of 3D setups.

References

1. Ansell, A.: The dynamic element method for analysis of frame and cable type structures. Eng. Struct. **27**(13), 1906–1915 (2005)
2. Arsenault, M.: Stiffness analysis of a planar 2-dof cable-suspended mechanism while considering cable mass. In: First International Conference on Cable-Driven Parallel Robots, Stuttgart, Germany (2012)
3. Chellal, R., Cuvillon, L., Laroche, E.: Model identification and vision-based H_∞ position control of 6-dof cable-driven. Int. J. Control **90**(4), 684–701 (2017)
4. Du, J., Ding, W., Bao, H.: Cable vibration analysis for large workspace cable-driven parallel manipulators. In: First International Conference on Cable-Driven Parallel Robots, Stuttgart, Germany (2012)

5. Gosselin, C.: Global planning of dynamically feasible trajectories for three-dof spatial cable-suspended parallel robots. In: First International Conference on Cable-Driven Parallel Robots, Stuttgart, Germany (2012)
6. Greenwood, D.T.: Advanced Dynamics. Cambridge University Press, Cambridge (2003)
7. Irvine, H.M.: Cable Structures. MIT Press, Cambridge (1981)
8. Khosravi, M.A., Taghirad, H.D.: Dynamic modeling and control of parallel robots with elastic cables: singular perturbation approach. IEEE Trans. Rob. **30**(3), 694–704 (2014). doi:10.1109/TRO.2014.2298057
9. De Luca, A., Siciliano, B.: Closed-form dynamic model of planar multilink lightweight robots. IEEE Trans. Syst. Man Cybern. **21**(4), 826–839 (1991)
10. De Luca, A., Siciliano, B.: Inversion-based nonlinear control of robot arms with flexible links. AIAA J. Guidance Control Dyn. **16**(6), 1169–1176 (1993)
11. Nguyen, T.T., Laroche, E., Cuvillon, L., Gangloff, J., Piccin, O.: Identification du modèle phénoménologique d'un robot parallèle à câbles. J. Européen des Systémes Automatisès **46**(6–7), 673–689 (2012)
12. Shi, P., McPhee, J., Heppler, G.: A deformation field for Euler-Bernouilli beams with application to flexible multibody dynamics. Multibody Syst. Dyn. **5**(1), 79–104 (2001)
13. Weber, X., Cuvillon, L., Gangloff, J.: Active vibration canceling of a cable-driven parallel robot in modal space. In: IEEE International Conference on Robotics and Automation, Seattle, WA, USA, pp. 1599–1604 (2015)
14. Yao, R., Li, H., Zhang, X.: A modeling method of the cable driven parallel manipulator for fast. In: First International Conference on Cable-Driven Parallel Robots, Stuttgart, Germany (2012)
15. Yuan, H., Courteille, E., Deblaise, D.: Static and dynamic stiffness analyses of cable-driven parallel robots with non-negligible cable mass and elasticity. Mech. Mach. Theory **85**, 64–81 (2015)

Manipulator Deflection for Optimum Tension of Cable-Driven Robots with Parameter Variations

Leila Notash^(✉)

Department of Mechanical and Materials Engineering,
Queen's University, Kingston, ON, Canada
leila.notash@queensu.ca

Abstract. Analytical formulation for positive cable tension is presented when the variations in parameters/data of parallel robot manipulators result in the interval form of the Jacobian matrix and external wrench. Solutions for cable tension vector, including the minimum 2-norm non-negative solution, and the pertinent deflection of manipulator due to cable stiffness and change in geometric parameters, are discussed. Example manipulators are simulated to examine the methods.

1 Introduction

In cable/wire-driven parallel manipulators, the motion of platform is controlled by cables/wires. Cable parallel manipulators have been used as robotic cranes, cable supported moving aerial cameras, and so on. Two example cable parallel robots are depicted in Fig. 1. Cables can pull but not push, i.e., their inputs are unidirectional and irreversible. Hence, for fully constrained cable robots, the number of cables/actuators are larger than the degrees of freedom (DOF) of manipulator. This results in many solutions for the cable tension vector for a given platform wrench. The minimum 2-norm tension vector is one of these solutions and is obtained using the generalized inverse (GI) of the transposed Jacobian matrix, which could result in negative tension for cables even if the platform is within the wrench closure workspace.

The workspace of cable parallel manipulators, with no variations in parameters/data, has been investigated extensively, e.g., [1–3]. A methodology for realizing a closed-form, minimum norm and continuous non-negative cable tension, in the presence of uncertainty and error, was presented in [4]. When the bounded range of variations in parameters/data are given, the relation between the cable tension and platform wrench becomes an interval expression. Solution set of an interval linear system generally is not an interval vector. The interval vector formed using the bounds of solution set (smallest "box" that includes the solution) is the "interval hull", e.g., [5, 6].

The solution set of general interval linear systems could be identified considering each orthant of the solution space, e.g., [5, 7]. For non-negative tension, solution lies in the first orthant. Because in each orthant the solution of interval linear systems is

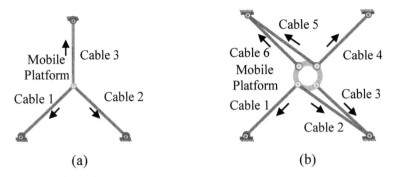

Fig. 1. Parallel manipulators (a) 2 DOF, three cables; (b) 3 DOF, six cables.

convex [8], the solution for non-negative tension is a convex set. In this article, analytical methods for identifying the boundary of solution set for positive tension and the pertinent manipulator stiffness and deflection are discussed. Results are verified by a discrete method.

2 Formulation of Wrench and Stiffness

For cable-driven robot manipulators, the $m \times n$ transposed Jacobian matrix $\mathbf{J}^T$ linearly relates the $n \times 1$ vector of cable forces $\boldsymbol{\tau} = [\tau_1 \cdots \tau_n]^T$ to the $m \times 1$ ($m \leq 6$) vector of platform wrench $\mathbf{F} = [\mathbf{f}^T; \mathbf{m}^T]^T$ as

$$\mathbf{F} = \mathbf{J}^T \boldsymbol{\tau} = [\mathbf{J}_1^T \ \mathbf{J}_2^T \ \cdots \ \mathbf{J}_i^T \ \cdots \ \mathbf{J}_{n-1}^T \ \mathbf{J}_n^T] \boldsymbol{\tau} = \sum_{j=1}^{n} \mathbf{J}_j^T \tau_j \quad (1)$$

The platform deflection $\delta \mathbf{p}$ can be formulated in terms of its stiffness matrix $\mathbf{K}$ and wrench $\mathbf{F}$, using $\delta \mathbf{F} = \mathbf{K} \delta \mathbf{p}$, by differentiating Eq. (1)

$$\delta \mathbf{F} = \delta \mathbf{J}^T \boldsymbol{\tau} + \mathbf{J}^T \delta \boldsymbol{\tau} = (\mathbf{K}_p + \mathbf{K}_c) \delta \mathbf{p} \quad (2)$$

The first term on the right-hand side of Eq. (2), $\delta \mathbf{J}^T \boldsymbol{\tau} = \sum_{j=1}^{n} \delta \mathbf{J}_j^T \tau_j$, depends on the layout and kinematic parameters of manipulators. This term results in stiffness matrix $\mathbf{K}_p$, which is due to change in the geometric (kinematic) parameters of manipulator because of actuator forces/torques (for non-negative τ_j).

With a simple linear spring model for cable stiffness $\delta \boldsymbol{\tau} = \mathbf{K}_q \delta \mathbf{l}$, the differential form of twist is related to the differential change in cable lengths $\delta \mathbf{l} = [\delta l_1 \ldots \delta l_n]^T$ by the Jacobian matrix as $\delta \mathbf{l} = \mathbf{J} \delta \mathbf{p}$. Then, $\mathbf{J}^T \delta \boldsymbol{\tau} = \mathbf{J}^T \mathbf{K}_q \mathbf{J} \delta \mathbf{p}$ and the manipulator stiffness due to cable stiffness is $\mathbf{K}_c = \mathbf{J}^T \mathbf{K}_q \mathbf{J}$, with $\mathbf{K}_c$ being dominant compared to $\mathbf{K}_p$. The diagonal matrix $\mathbf{K}_q$ is in terms of the stiffness of cable j, k_j, which is a function of cable length $l_{oj} + l_j$, $k_j = (E_j A_{cj})/(l_{oj} + l_j)$, where E_j and A_{cj} are respectively the equivalent modulus of elasticity and cross-sectional area of cable j. For a non-zero cable offset l_{oj}, a finite value for cable stiffness k_j is guaranteed when a cable attachment point on platform approaches its anchor, i.e., for $l_j \approx 0$.

Representing the variations in parameters/data as intervals, the Jacobian **J** and stiffness **K** become interval matrices. For example, interval $\mathbf{K} = [\underline{\mathbf{K}}, \overline{\mathbf{K}}]$ is in terms of real (point) lower and upper bound matrices $\underline{\mathbf{K}}, \overline{\mathbf{K}}$, and is regular if it does not include a singular matrix.

2.1 Solution Sets for Cable Tension and Platform Deflection

The non-negative solution set of interval system $\mathbf{F} = \mathbf{J}^T \boldsymbol{\tau}$ is the union of all non-negative solutions of the real $\hat{\mathbf{J}}^T \hat{\boldsymbol{\tau}} = \hat{\mathbf{F}}$ for any real matrix $\hat{\mathbf{J}}^T \in \mathbf{J}^T = [\underline{\mathbf{J}}^T, \overline{\mathbf{J}}^T]$ and real wrench $\hat{\mathbf{F}} \in \mathbf{F} = [\underline{\mathbf{F}}, \overline{\mathbf{F}}]$, i.e., the set of real $\hat{\boldsymbol{\tau}}$

$$\boldsymbol{\tau}: \left\{ \hat{\boldsymbol{\tau}} \mid \hat{\boldsymbol{\tau}} \succ 0, \hat{\mathbf{J}}^T \hat{\boldsymbol{\tau}} = \hat{\mathbf{F}}, \hat{\mathbf{J}}^T \in \mathbf{J}^T, \hat{\mathbf{F}} \in \mathbf{F} \right\} \tag{3}$$

where $\hat{\boldsymbol{\tau}} \succ 0$ specifies that each entry of $\hat{\boldsymbol{\tau}}$ is non-negative. There is at least one non-negative solution for $\mathbf{J}^T \boldsymbol{\tau} = \mathbf{F}$ if the two linear inequalities $\underline{\mathbf{J}}^T \hat{\boldsymbol{\tau}} \leq \overline{\mathbf{F}}$ and $-\overline{\mathbf{J}}^T \hat{\boldsymbol{\tau}} \leq -\underline{\mathbf{F}}$ have a non-negative solution for $\hat{\boldsymbol{\tau}}$ [9]. If each $\hat{\mathbf{J}}^T \hat{\boldsymbol{\tau}} = \hat{\mathbf{F}}$ has a nonnegative solution $\mathbf{J}^T \boldsymbol{\tau} = \mathbf{F}$ is strongly nonnegative solvable.

When the pose is in the wrench closure work space it is possible to maintain positive tension. However, the particular (minimum 2-norm) solution $\boldsymbol{\tau} = \mathbf{J}^{\#T} \mathbf{F}$ (using the Moore-Penrose GI of $\mathbf{J}^T$) may result in negative tension for k cables. Then, the non-negative solution set for $\boldsymbol{\tau}$ could be identified utilizing the following closed-form method. The reduced Jacobian matrix $\mathbf{J}_r^T$ is formed by removing the columns of $\mathbf{J}^T$ corresponding to cables with negative tension in the particular solution. The negative tension τ_p of these k cables are set to non-negative values τ_c (real or interval values; calculated or assigned as discussed in [4]). Then, the overall solution $\tau_{tot\,r}$ for the cables with positive tension in particular solution is identified using

$$\mathbf{J}_r^T \boldsymbol{\tau}_{tot\,r} = \mathbf{F} - \mathbf{F}_{\tau_{bal}} = \mathbf{F} - \sum_k \mathbf{J}_j^T \tau_{cj} \tag{4}$$

which results in the minimum 2-norm non-negative solution.

Because of the convexity of solution set (and its subsets) in each orthant, the solution is formulated using the intersection of the corresponding closed half-spaces. The solution set for non-negative cable tension lies in the first orthant ($\tau_j \geq 0$ for $j = 1, \cdots, n$), with the pertinent closed half-spaces formulated using the bounds of the entries of the interval linear system [7] as

$$\underline{\mathbf{J}}_{i1}^T \hat{\tau}_1 + \cdots + \underline{\mathbf{J}}_{in}^T \hat{\tau}_n \leq \overline{F}_i \quad , \quad \overline{\mathbf{J}}_{i1}^T \hat{\tau}_1 + \cdots + \overline{\mathbf{J}}_{in}^T \hat{\tau}_n \geq \underline{F}_i \tag{5}$$

for $i = 1, \cdots, m$. For n-dimensional solution space, the solution is formulated as the intersection of $2m + n$ half-spaces, including n closed half-spaces that signify the first orthant. For example, for $m = 2$ and $n = 3$, the intersections of seven closed half-spaces characterize the solution.

The solution set for platform deflection (and its subsets), utilizing the minimum 2-norm non-negative tension vector and applying $\delta \mathbf{F} = \mathbf{K}\, \delta\mathbf{p}$, could span over all orthants. Thus, using a similar procedure, in each orthant, the intersection of pertinent closed half-spaces is formulated, refer to [10]. The union of intersections is the solution set for platform deflection.

2.2 Subsets of Non-negative Solution

The subsets of solution set include tolerance τ_{tol}, control τ_{con} and algebraic solutions. The tolerance solution set of $\mathbf{F} = \mathbf{J}^T \tau$ includes all non-negative real vectors $\hat{\tau}$ for real $\hat{\mathbf{J}}^T \in \mathbf{J}^T$ for which the wrench remains within the required lower and upper bounds, $\hat{\mathbf{J}}^T \hat{\tau} = \hat{\mathbf{F}} \in \mathbf{F} = [\underline{\mathbf{F}}, \overline{\mathbf{F}}]$, i.e., $\mathbf{J}^T\hat{\tau} \subseteq \mathbf{F}$. The control set includes all non-negative $\hat{\tau}$ for which a $\hat{\mathbf{J}}^T \in \mathbf{J}^T$ exists such that $\hat{\mathbf{J}}^T \hat{\tau} = \hat{\mathbf{F}} \in \mathbf{F}$, i.e., $\mathbf{F} \subseteq \mathbf{J}^T \hat{\tau}$. The algebraic set is the intersection of tolerance and control solutions, i.e., the set of $\hat{\tau}$ that results in equality $\mathbf{J}^T \hat{\tau} = \mathbf{F}$. For deflection, $\delta\mathbf{p}_{tol}$ and $\delta\mathbf{p}_{con}$ satisfy $\mathbf{K}\,\delta\mathbf{p} \subseteq \delta\mathbf{F}$ and $\delta\mathbf{F} \subseteq \mathbf{K}\,\delta\mathbf{p}$, respectively.

The tolerance and control solution sets for non-negative cable tension are also formulated as the intersection of $2m + n$ closed half-spaces. In the first orthant, to characterize τ_{tol}, the closed half-spaces are

$$\left(\overline{\mathbf{J}}_{i1}^T \hat{\tau}_1 + \cdots + \overline{\mathbf{J}}_{in}^T \hat{\tau}_n\right) \leq \overline{F}_i \quad , \quad -\left(\underline{\mathbf{J}}_{i1}^T \hat{\tau}_1 + \cdots + \underline{\mathbf{J}}_{in}^T \hat{\tau}_n\right) \leq -\underline{F}_i \qquad (6)$$

for $i = 1, \cdots, m$. For real $\mathbf{F}$, $\underline{\mathbf{F}} = \overline{\mathbf{F}}$ and the inequalities reduce to $\left(\overline{\mathbf{J}}_{i1}^T \hat{\tau}_1 + \cdots + \overline{\mathbf{J}}_{in}^T \hat{\tau}_n\right) = \left(\underline{\mathbf{J}}_{i1}^T \hat{\tau}_1 + \cdots + \underline{\mathbf{J}}_{in}^T \hat{\tau}_n\right) = F_i$, i.e., $\mathbf{J}_c^T \hat{\tau} = \mathbf{F}$ and $\Delta \mathbf{J}^T |\hat{\tau}| = \mathbf{0}$ [10].

The control solution set τ_{con} is characterized by

$$\left(\overline{\mathbf{J}}_{i1}^T \hat{\tau}_1 + \cdots + \overline{\mathbf{J}}_{in}^T \hat{\tau}_n\right) \geq \overline{F}_i \quad , \quad -\left(\underline{\mathbf{J}}_{i1}^T \hat{\tau}_1 + \cdots + \underline{\mathbf{J}}_{in}^T \hat{\tau}_n\right) \geq -\underline{F}_i \qquad (7)$$

for $i = 1, \cdots, m$; and for real $\mathbf{F}$ by $\left(\underline{\mathbf{J}}_{i1}^T \hat{\tau}_1 + \cdots + \underline{\mathbf{J}}_{in}^T \hat{\tau}_n\right) \leq F_i \leq \left(\overline{\mathbf{J}}_{i1}^T \hat{\tau}_1 + \cdots + \overline{\mathbf{J}}_{in}^T \hat{\tau}_n\right)$.

The algebraic solution set is formulated using

$$\left(\overline{\mathbf{J}}_{i1}^T \hat{\tau}_1 + \cdots + \overline{\mathbf{J}}_{in}^T \hat{\tau}_n\right) = \overline{F}_i \quad , \quad -\left(\underline{\mathbf{J}}_{i1}^T \hat{\tau}_1 + \cdots + \underline{\mathbf{J}}_{in}^T \hat{\tau}_n\right) = -\underline{F}_i \qquad (8)$$

which could be rearranged as $\mathbf{J}_c^T \hat{\tau} = \mathbf{F}_c$ and $\Delta \mathbf{J}^T |\hat{\tau}| = \Delta \mathbf{F}$. Thus, for real $\mathbf{F}$, a non-trivial solution $\hat{\tau}$ for the algebraic (and tolerance) solution exist if $|\hat{\tau}| = \left|\mathbf{J}_c^{\#T}\mathbf{F}\right|$ is in the null space of $\Delta \mathbf{J}^T$.

3 Case Study

A planar parallel manipulator with three cables and point mass platform is examined. Plane of motion is the *x*-*y* plane. The platform pose in terms of the coordinates of the cable base attachments A_j, vector **a**, and cable orientations α_j is $\mathbf{p} = [a_{jx} + l_j \cos \alpha_j \; a_{jy} + l_j \sin \alpha_j]^T$ for $j = 1, \cdots, 3$. Then,

$$\mathbf{J}^T = [\mathbf{J}_1^T \; \mathbf{J}_2^T \; \mathbf{J}_3^T] = \begin{bmatrix} \cos \alpha_1 & \cos \alpha_2 & \cos \alpha_3 \\ \sin \alpha_1 & \sin \alpha_2 & \sin \alpha_3 \end{bmatrix} \quad (9)$$

The first term on the right-hand side of Eq. (2) is a function of $\delta \boldsymbol{\alpha}$

$$\delta \mathbf{J}^T \boldsymbol{\tau} = \sum_{j=1}^{n=3} \delta \mathbf{J}_j^T \tau_j = \sum_{j=1}^{n=3} \left(\frac{\partial \mathbf{J}_j^T}{\partial \alpha_j} \delta \alpha_j \right) \tau_j \quad (10)$$

and $\delta \boldsymbol{\alpha}$ is linearly related to $\delta \mathbf{p}$. Thus,

$$\delta \mathbf{J}_j^T = \begin{bmatrix} 0 & \cdots & -\sin \alpha_j & \cdots & 0 \\ 0 & \cdots & \cos \alpha_j & \cdots & 0 \end{bmatrix} \begin{bmatrix} \frac{\sin \alpha_1}{l_1} & -\frac{\cos \alpha_1}{l_1} \\ \vdots & \vdots \\ \frac{\sin \alpha_3}{l_3} & -\frac{\cos \alpha_3}{l_3} \end{bmatrix} \begin{bmatrix} \delta p_x \\ \delta p_y \end{bmatrix} \quad (11)$$

The 2×2 stiffness matrix $\mathbf{K} = \mathbf{K}_p + \mathbf{K}_c$ of manipulator becomes a symmetric matrix as the formulation of the Jacobian and stiffness matrices are with respect to a reference frame that has its origin at the operation point **p** (point of application of external force) and the same orientation as that of the fixed frame. Thus, the entries of **K** are:

$$\begin{aligned} k_{11} &= \sum_3 \frac{\tau_j \sin^2 \alpha_j}{l_j} + \sum_3 \frac{E_j A_{cj} \cos^2 \alpha_j}{l_{oj} + l_j} \\ k_{12} &= k_{21} = -\sum_3 \frac{\tau_j \cos \alpha_j \sin \alpha_j}{l_j} + \sum_3 \frac{E_j A_{cj} \cos \alpha_j \sin \alpha_j}{l_{oj} + l_j} \\ k_{22} &= \sum_3 \frac{\tau_j \cos^2 \alpha_j}{l_j} + \sum_3 \frac{E_j A_{cj} \sin^2 \alpha_j}{l_{oj} + l_j} \end{aligned} \quad (12)$$

Nominal coordinates of cable attachments A_j are (−1, −1), (1, −1), (0, 1); with 1.5 mm wire rope diameter and $E = 57.3$ GPa, (7 × 7 AISI 316 Stainless Steel).

3.1 Example 1 - Positive Particular Solution

The interval $\mathbf{J}^T$ for the 2 DOF manipulator of Fig. 1(a), with three cables, is

$$\mathbf{J}^T = \begin{bmatrix} [-0.6075, -0.5926] & [0.2362, 0.2489] & [-1.0409, -0.9605] \\ [-0.8086, -0.7915] & [-0.9808, -0.9596] & [-0.0205, 0.0205] \end{bmatrix} \quad (13)$$

which corresponds to the platform pose of $\mathbf{p} = [0.5\ 1]^T$ meters in the wrench closure workspace, and an interval radius of 5 mm for the coordinates of cable base attachments.

For wrench $\mathbf{F} = [[-46.6919, -42.8121]\ [-46.4096, -43.8042]]^T$ Newtons, $\tau_p = [[32.4646, 39.2004][13.4608, 20.4276][23.6549, 31.0288]]^T$ is the enclosure for the minimum 2-norm (particular) solution using the INTLAB [11]. The problem is also studied for real $\mathbf{F} = [-44.7239\ -45.1043]^T$ with enclosure $\tau_p = [[33.8193, 37.8240]\ [14.9810, 18.9198][25.2946, 29.3490]]^T$. Thus, particular solution lies in the first octant for both interval and real $\mathbf{F}$.

The solution sets for interval and real $\mathbf{F}$, using the intersections of half-spaces shown as rays, are depicted in Figs. 2(a) and 3(a) respectively. The pertinent enclosure vectors are used for the span of axes in the plots of this section, unless otherwise noted. The results are verified by the discrete method, with each $\hat{\tau}_p$ color-coded for its norm. Entries of the calculated enclosure vector τ_p are discretized within their bounds. Each generated real $\hat{\tau}_p$ belongs to the solution set when the calculated interval wrench $\mathbf{F}_{check} = \mathbf{J}^T \hat{\tau}_p$ and the given wrench $\mathbf{F}$ have non-empty intersection, i.e., $\mathbf{F}_{check} \cap \mathbf{F} \neq \varnothing$. The minimum 2-norm least-square solution sets for all combinations of the lower and upper bounds of the interval entries of $\mathbf{J}^T$ and $\mathbf{F}$, referred to as "minimum norm solution", are shown in Figs. 2(b) and 3(b).

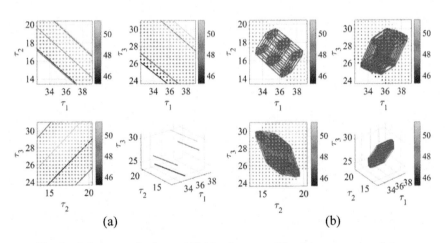

Fig. 2. Solution set for interval $\mathbf{F}$ (a) using rays; (b) minimum norm solution.

The nonempty tolerance and control solutions (τ_{tol} for $\mathbf{J}^T \hat{\tau} \subseteq \mathbf{F}$; τ_{con} for $\mathbf{F} \subseteq \mathbf{J}^T \hat{\tau}$) for interval and real $\mathbf{F}$ are depicted in Figs. 4(a) and (b) respectively. For interval (real) $\mathbf{F}$, the control (tolerance) and algebraic solutions are empty sets (for real $\mathbf{F}$, $\Delta \mathbf{J}^T |\mathbf{J}_c^{\#T}\mathbf{F}| \neq \mathbf{0}$ here) within the corresponding enclosure bounds; $\tau_{con}(\tau_{tol})$ is nonempty beyond the enclosure bounds for interval (real) $\mathbf{F}$.

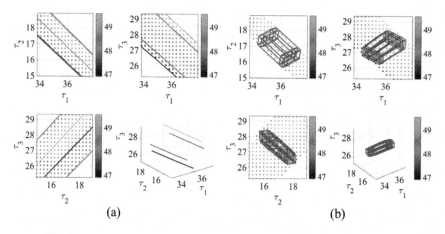

Fig. 3. Solution set for real **F** (a) using rays; (b) minimum norm solution.

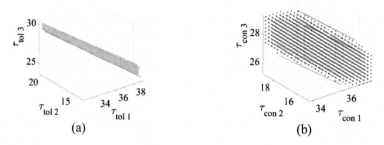

Fig. 4. (a) Tolerance solution for interval **F**; (b) control solution for real **F**.

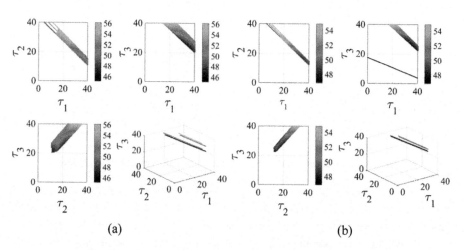

Fig. 5. Solution set for $0 \leq \tau_j \leq 40$ (a) interval **F**; (b) real **F**.

In Figs. 5(a) and (b), the solution sets for interval and real **F** are displayed for the tension range of $\tau_{min} = 0 \leq \tau_j \leq \tau_{max} = 40$ N, respectively. As illustrated (and verified by discrete method), investigating intersections of the seven closed half-spaces in the first octant, tension values are bounded. That is, for the upper limit of 40 N, the lowest value of tension is for cable 2; about 11.5 N for interval **F** (12.5 N for real **F**). For τ_2 equal to (over) 40 N, the lower bound (both bounds) of τ_1 becomes negative.

The stiffness matrix for interval **F**, in kN/m, is

$$\mathbf{K} = \mathbf{K}_p + \mathbf{K}_c = \begin{bmatrix} [172.5256, 176.4375] & [5.1088, 11.1663] \\ [5.1088, 11.1663] & [70.8496, 72.8206] \end{bmatrix} \quad (14)$$

with $1/\kappa = [0.3763, 0.4383]^T \leq (\|\Delta\mathbf{p}\|/\|\mathbf{p}\|)/(\|\Delta\mathbf{F}\|/\|\mathbf{F}\|) \leq [2.2814, 2.6571]^T = \kappa$ as the bounds of the ratio of input and output relative change (in terms of the 2-norm condition number κ of **K**). The solution set for platform deflection, with bounds of $\delta\mathbf{p} = [[-0.2555, -0.2013][-0.6429, -0.5612]]^T$ in mm, and the lines of the pertinent empty tolerance and control solution sets are depicted in Fig. 6. Thus, within the enclosure, there is no $\delta\mathbf{p}_{tol}$ to satisfy $\mathbf{K}\,\delta\mathbf{p} \subseteq \delta\mathbf{F}$, and no $\delta\mathbf{p}_{con}$ to meet $\delta\mathbf{F} \subseteq \mathbf{K}\,\delta\mathbf{p}$. For $\tau_2 = 39$ N, the pertinent plots are similar to Fig. 6 with $\delta\mathbf{p} = [[-0.2555, -0.2013][-0.6426, -0.5607]]^T$. For this example, $\mathbf{K}_c$ is four orders of magnitude larger than $\mathbf{K}_p$.

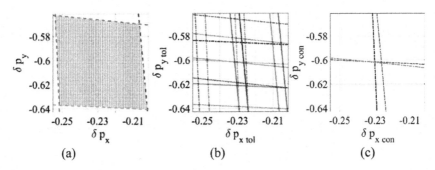

Fig. 6. Manipulator deflection (a) solution set; (b) lines of tolerance solution set; (c) rays of control solution set.

3.2 Example 2 - Negative Particular Solution

For real $\mathbf{F} = [0\ -10]^T$ Newtons at $\mathbf{p} = [0.5\ -0.5]^T$ meters and

$$\mathbf{J}^T = \begin{bmatrix} [-0.9628, -0.9348] & [0.6793, 0.7360] & [-0.3252, -0.3074] \\ [-0.3252, -0.3074] & [-0.7360, -0.6793] & [0.9348, 0.9628] \end{bmatrix} \quad (15)$$

$\tau_p = [[4.3253, 5.1630][3.1195, 3.9507][-6.6958, -5.9472]]^T$ verified enclosure includes negative value for the tension of cable 3.

Using the intersection of closed half-spaces in the first octant for the allowable range of tension $\tau_{min} = 0 \leq \tau_j \leq \tau_{max} = 30$ N for $j = 1, \cdots, 3$, the solution set for real **F** is displayed in Fig. 7. The plots indicate that by setting the tension of cable 3 to zero or a positive threshold the platform pose could be fully controlled.

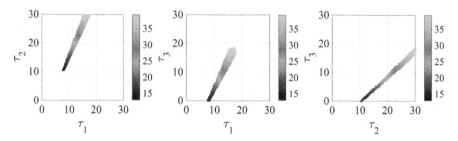

Fig. 7. Solution set using discrete method for $0 \leq \tau_j \leq 30$.

Next, the minimum norm non-negative solution set is calculated. Matrix $\mathbf{J}_r^T$ is formed by removing the third column of $\mathbf{J}^T$ ($\mathbf{J}_3^T$, related to negative tension of cable 3), and the tension of cable 3 is set to $\tau_3 = \tau_{c3} = 1$ N. Then, using $\mathbf{J}_r^T \boldsymbol{\tau}_{tot\,r} = \mathbf{F} - \mathbf{F}_{\tau_{bal}}$, the overall solution set is identified for cables 1 and 2, where $\mathbf{F}_{\tau_{bal}} = \mathbf{J}^T [[0,0]\ [0,0]\ \tau_{c3}]^T = \mathbf{J}_3^T \tau_{c3}$. The result is verified by discretizing the enclosure for $\boldsymbol{\tau}_{tot}$. Each vector $\hat{\boldsymbol{\tau}}_{tot}$ belongs to the solution set when $\mathbf{F}_{check} = \mathbf{J}^T \hat{\boldsymbol{\tau}}_{tot}$ and **F** have non-empty intersection.

The overall tensions of cables 1 and 2 are depicted in Fig. 8(a) and verified in Fig. 8(b) in green using the discrete method, along with the rays of empty tolerance solution set. The vertices, which are the minimum norm solutions for all combinations of the bounds of interval entries, are marked as well. As depicted in Fig. 8(b), some rays for $\boldsymbol{\tau}_{tol}$ form the boundary of solution set for $\boldsymbol{\tau}$ in the first orthant. The bounds of the overall tension vector is

$$\boldsymbol{\tau} = [[7.7355, 9.1240][11.1315, 12.3464][1.0000, 1.0000]]^T \tag{16}$$

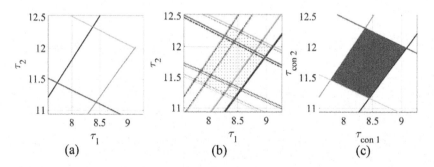

Fig. 8. Tension adjustment for real **F** (a) overall solution $\boldsymbol{\tau}$; (b) rays of tolerance solution; (c) control solution set. (Color figure online)

The control solution, displayed in Fig. 8(c), is obtained using the reduced Jacobian matrix $\mathbf{J}_r^T$ and wrench $\mathbf{F} - \mathbf{F}_{\tau_{bal}}$. As demonstrated, for real $\mathbf{F}$, the non-negative solution τ_{con} is almost the solution set.

At this pose, with real $\mathbf{F}$ and the pertinent stiffness matrix in kN/m

$$\mathbf{K} = \begin{bmatrix} [118.1253, 126.3293] & [-65.2153, -55.8640] \\ [-65.2153, -55.8640] & [118.2406, 126.4468] \end{bmatrix} \quad (17)$$

$[0.2764, 0.4053]^T \leq (\|\Delta\mathbf{p}\|/\|\mathbf{p}\|)/(\|\Delta\mathbf{F}\|/\|\mathbf{F}\|) \leq [2.4679, 3.6174]^T$. The bounds of solution set for platform deflection is $\delta\mathbf{p} = [[-0.0672, -0.0401][-0.1217, -0.0949]]^T$ mm. The solution and pertinent tolerance (empty) and control ($\delta\mathbf{p}_{con}$ to meet $\delta\mathbf{F} \subseteq \mathbf{K}\,\delta\mathbf{p}$) sets are shown in Fig. 9. It is noteworthy that for $\mathbf{F} = [0 -10]^T$ and $\tau_{c3} = 1000$ N, $\mathbf{K}_c$ is two orders of magnitude larger than $\mathbf{K}_p$ (compared to five orders of magnitude for original $\tau_{c3} = 1$) with pertinent plots similar to Fig. 9 and $\delta\mathbf{p} = [[-0.0639, -0.0380] [-0.1184, -0.0928]]^T$ mm.

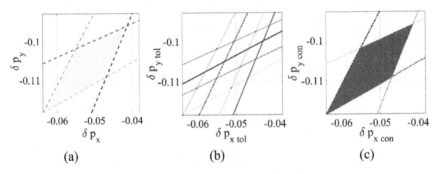

Fig. 9. Tension adjustment for real $\mathbf{F}$ (a) platform deflection; (b) lines of tolerance solution; (c) control solution set.

4 Conclusions

The effect of variations in parameters/data of cable-driven robots on the cable tension and platform stiffness was examined. The solution set for non-negative cable tension was investigated using the closed half-spaces in the first orthant, as well as for the minimum 2-norm, and was utilized to formulate the platform deflection analytically. It was presented that for fully constrained cable robots, the solution set for cable tension may be bounded for a given pose and wrench, and much smaller than the maximum allowable tension limit. In addition, it was demonstrated that the stiffness produced due to change in geometric parameters, because of wire/actuator forces/torques, has minor effect on the manipulator deflection as compared to the stiffness of cables. The results were verified by the discrete method for the cases with bounded and unbounded solution sets.

References

1. Stump, E., Kumar, V.: Workspace of cable-actuated parallel manipulators. ASME J. Mech. Des. **128**, 159–167 (2006)
2. Bosscher, P., Riechel, A.T., Ebert-Uphoff, I.: Wrench-feasible workspace generation for cable-driven robots. IEEE Trans. Rob. **22**(5), 890–902 (2006)
3. McColl, D., Notash, L.: Workspace formulation of planar wire-actuated parallel manipulators. Robotica **29**(4), 607–617 (2011)
4. Notash, L.: On the solution set for positive wire tension with uncertainty in wire-actuated parallel manipulators. J. Mech. Robot. **8**(4), 044506 (2016)
5. Hanson, E.: Global Optimization Using Interval Analysis. Marcel Dekker, New York (1992)
6. Moore, R., Kearfott, R.B., Cloud, M.J.: Introduction to Interval Analysis. SIAM, Philadelphia (2009). ISBN 978-0-898716-69-6
7. Notash, L.: Analytical methods for solution sets of interval wrench. In: Proceedings of the ASME IDETC/CIE (2015). doi:10.1115/DETC2015-47575
8. Oettli, W.: On the solution set of a linear system with inaccurate coefficients. SIAM J. Numer. Anal. B **2**(1), 115–118 (1965)
9. Fiedler, M., Nedoma, J., Ramik, J., Rohn, J., Zimmermann, K.: Linear Optimization Problems with Inexact Data. Springer, New York (2006)
10. Notash, L.: Wrench accuracy for parallel manipulators and interval dependency. J. Mech. Robot. **9**(1), 011008 (2017)
11. Rump, S.M.: INTLAB - INTerval LABoratory. In: Csendes, T. (ed.) Developments in Reliable Computing, pp. 77–104. Kluwer Academic (1999)

Sensitivity Analysis of the Elasto-Geometrical Model of Cable-Driven Parallel Robots

Sana Baklouti[1]([✉]), Stéphane Caro[2], and Eric Courteille[1]

[1] Université Bretagne-Loire, INSA-LGCGM-EA 3913,
20, avenue des Buttes de Cöesmes, 35043 Rennes, France
{sana.baklouti,eric.courteille}@insa-rennes.fr
[2] CNRS, Laboratoire des Sciences du Numérique de Nantes,
UMR CNRS n6004, 1, rue de la Noë, 44321 Nantes, France
stephane.caro@ls2n.fr

Abstract. This paper deals with the sensitivity analysis of the elasto-geometrical model of Cable-Driven Parallel Robots (CDPRs) to their geometric and mechanical uncertainties. This sensitivity analysis is crucial in order to come up with a robust model-based control of CDPRs. Here, 62 geometrical and mechanical error sources are considered to investigate their effect onto the static deflection of the moving-platform (MP) under an external load. A reconfigurable CDPR, named "CAROCA", is analyzed as a case of study to highlight the main uncertainties affecting the static deflection of its MP.

1 Introduction

In recent years, there has been an increasing number of research works on the subject of Cable-Driven Parallel Robots (CDPRs). The latter are very promising for engineering applications due to peculiar characteristics such as large workspace, simple structure and large payload capacity. For instance, CDPRs have been used in many applications like rehabilitation [1], pick-and-place [2], sandblasting and painting [3,4] operations.

Many spatial prototypes are equipped with eight cables for six Degrees of Freedom (DOF) such as the CAROCA prototype, which is the subject of this paper.

To customize CDPRs to their applications and enhance their performances, it is necessary to model, identify and compensate all the sources of errors that affect their accuracy.

Improving accuracy is still possible once the robot is operational through a suitable control scheme. Numerous control schemes were proposed to enhance the CDPRs precision on static tasks or on trajectory tracking [5–7]. The control can be either off-line through external sensing in the feedback signal [2], or on-line control based on a reference model [8].

This paper focuses on the sensitivity analysis of the CDPR MP static deflection to uncertain geometrical and mechanical parameters. As an illustrative example, a suspended configuration of the reconfigurable CAROCA prototype,

Fig. 1. CAROCA prototype: a reconfigurable CDPR (Courtesy of IRT Jules Verne, Nantes)

shown in Fig. 1, is studied. First, the manipulator under study is described. Then, its elasto-geometrical model is written while considering cable mass and elasticity in order to express the static deflection of the MP subjected to an external load. An exhaustive list of geometrical and mechanical uncertainties is given. Finally, the sensitivity of the MP static deflection to these uncertainties is analyzed.

2 Parametrization of the CAROCA Prototype

The reconfigurable CAROCA prototype illustrated in Fig. 1 was developed at IRT Jules Verne for industrial operations in cluttered environment such as painting and sandblasting large structures [3,4]. This prototype is reconfigurable because its pulleys can be displaced in a discrete manner on its frame. The size of the latter is 7 m long, 4 m wide and 3 m high. The rotation-resistant steel cables Carl Stahl Technocables Ref 1692 of the CAROCA prototype are 4 mm diameter. Each cable consists of 18 strands twisted around a steel core. Each strand is made up of 7 steel wires. The cable breaking force is 10.29 kN.

As shown in Fig. 2, the Cartesian coordinate vectors of anchor points A_i and exit points B_i are denoted $\mathbf{a}_i$ and $\mathbf{b}_i$. Vector $\mathbf{p}$ represents the Cartesian coordinates of the MP geometric center, P, expressed in $\mathscr{F}_b = \{O, x_b, y_b, z_b\}$. The Cartesian coordinates of A_i (B_i, resp.) expressed in the MP frame $\mathscr{F}_p = \{P, x_p, y_p, z_p\}$ (in the base frame $\mathscr{F}_b$, resp.) are given in Table 1. The cable frame $\mathscr{F}_i = \{B_i, x_i, y_i, z_i\}$ is associated to the ith cable, where axes z_i and z_b are parallel.

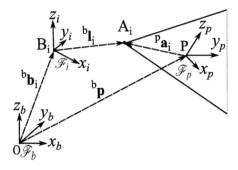

Fig. 2. The ith closed-loop of a CDPR

Table 1. Cartesian coordinates of anchor points A_i (exit points B_i, resp.) expressed in $\mathscr{F}_p$ (in $\mathscr{F}_b$, resp.)

	x (m)	y (m)	z (m)		x (m)	y (m)	z (m)
B_1	−3.5	2	3.5	A_1	0.2	0.15	0.125
B_2	3.5	2	3.5	A_2	−0.2	0.15	−0.125
B_3	−3.5	2	3.5	A_3	−0.2	−0.15	−0.125
B_4	3.5	2	3.5	A_4	0.2	−0.15	0.125
B_5	−3.5	−2	3.5	A_5	−0.2	0.15	0.125
B_6	3.5	−2	3.5	A_6	0.2	0.15	−0.125
B_7	−3.5	−2	3.5	A_7	0.2	−0.15	−0.125
B_8	3.5	−2	3.5	A_8	−0.2	−0.15	0.125

3 Elasto-Geometric Modeling

In this section, both sag-introduced and axial stiffness of cables are considered in the elasto-geometrical modeling of CDPR. The inverse elasto-geometrical model and the direct elasto-geometrical model of CDPR are presented. Then, the variations in static deflection due to external loading is defined as a sensitivity index.

3.1 Inverse Elasto-Geometric Modeling (IEGM)

The IEGM of a CDPR aims at calculating the unstrained cable length for a given pose of its MP. If both cable mass and elasticity are considered, the inverse kinematics of the CDPR and its static equilibrium equations should be solved simultaneously. The IEGM is based on geometric closed loop equations, cable sagging relationships and static equilibrium equations.

The geometric closed-loop equations take the form:

$$^b\mathbf{p} = {}^b\mathbf{b}_i + {}^b\mathbf{l}_i - {}^b\mathbf{R}_p{}^p\mathbf{a}_i, \qquad (1)$$

where $^b\mathbf{R}_p$ is the rotation matrix from $\mathscr{F}_b$ to $\mathscr{F}_p$ and $\mathbf{l}_i$ is the cable length vector.

The cable sagging relationships between the forces ${}^i\mathbf{f}_i = [{}^if_{xi}, 0, {}^if_{zi}]$ applied at the end point A_i of the ith cable and the coordinates vector ${}^i\mathbf{a}_i = [{}^ix_{Ai}, 0, {}^iz_{Ai}]$ of the same point resulting from the sagging cable model [9] are expressed in $\mathscr{F}_i$ as follows:

$${}^ix_{Ai} = \frac{{}^if_{xi}L_{usi}}{ES} + \frac{|{}^if_{xi}|}{\rho g}[\sinh^{-1}(\frac{{}^if_{zi}}{f_{xi}^{C_i}}) - \sinh^{-1}(\frac{{}^if_{zi} - \rho g L_{usi}}{{}^if_{xi}})], \quad (2a)$$

$${}^iz_{Ai} = \frac{{}^if_{xi}L_{usi}}{ES} - \frac{\rho g L_{usi}^2}{2ES} + \frac{1}{\rho g}[\sqrt{{}^if_{xi}{}^2 + {}^if_{zi}{}^2} - \sqrt{{}^if_{xi}{}^2 + ({}^if_{zi} - \rho g L_{usi})^2}], \quad (2b)$$

where L_{usi} is the unstrained length of ith cable, g is the acceleration due to gravity, S is the cable cross sectional area, ρ denotes the cable linear mass and E the cable modulus of elasticity.

The static equilibrium equations of the MP are expressed as:

$$\mathbf{W}\mathbf{t} + \mathbf{w}_{ex} = 0, \quad (3)$$

where $\mathbf{W}$ is the wrench matrix, $\mathbf{w}_{ex}$ is the external wrench vector and $\mathbf{t}$ is the 8-dimensional cable tension vector. Those tensions are computed based on the tension distribution algorithm described in [10].

3.2 Direct Elasto-Geometrical Model (DEGM)

The direct elasto-geometrical model (DEGM) aims to determine the pose of the mobile platform for a given set of unstrained cable lengths. The constraints of the DEGM are the same as the IEGM, i.e., Eqs. (1) to (3). If the effect of cable weight on the static cable profile is non-negligible, the direct kinematic model of CDPRs will be coupled with the static equilibrium of the MP. For a 6 DOFs CDPR with 8 driving cables, there are 22 equations and 22 unknowns. In this paper, the non-linear Matlab function *"lsqnonlin"* is used to solve the DEGM.

3.3 Static Deflection

If the compliant displacement of the MP under the external load is small, the static deflection of the MP can be calculated by its static Cartesian stiffness matrix [11]. However, once the cable mass is considered, the sag-introduced stiffness should be taken into account. Here, the small compliant displacement assumption is no longer valid, mainly for heavy or/and long cables with light mobile platform. Consequently, the static deflection can not be calculated through the Cartesian stiffness matrix. In this paper, the IEGM and DEGM are used to define and calculate the static deflection of the MP under an external load. The CDPR stiffness is characterized by the static deflection of the MP. Note that only the positioning static deflection of the MP is considered in order to avoid the homogenization problem [12].

As this paper deals with the sensitivity of the CDPR accuracy to all geometrical and mechanical errors, the elastic deformations of the CDPR is involved.

This problem is solved by deriving the static deflection of the CDPR obtained by the subtraction of the poses calculated with and without an external payload. For a desired pose of the MP, the IEGM gives a set of unstrained cable lengths $\mathbf{L_{us}}$. This set is used by the DEGM to calculate first, the pose of the MP under its own weight. Then, the pose of the MP is calculated when an external load (mass addition) is applied. Therefore, the static deflection of the MP is expressed as:

$$\mathbf{dp}_{j,k} = \mathbf{p}_{j,k} - \mathbf{p}_{j,1}, \tag{4}$$

where $\mathbf{p}_{j,1}$ is the pose of the MP considering only its own weight for the jth pose configuration and $\mathbf{p}_{j,k}$ is the pose of the MP for the set of the jth pose and k^{th} load configuration.

4 Error Modeling

This section aims to define the error model of the elasto-geometrical CDPR model. Two types of errors are considered: geometrical errors and mechanical errors.

4.1 Geometrical Errors

The geometrical errors of the CDPR are described by $\delta\mathbf{b}_i$, the variation in vector $\mathbf{b}_i$, $\delta\mathbf{a}_i$, the variation in vector $\mathbf{a}_i$, and $\delta\mathbf{g}$, the uncertainty vector of the gravity center position; So, 51 uncertainties. The geometric errors can be divided into base frame geometrical errors and MP geometrical errors and mainly due to manufacturing errors.

4.1.1 Base Frame Geometrical Errors

The base frame geometrical errors are described by vectors $\delta\mathbf{b}_i$, (i = 1..8). As the point B_i is considered as part of its correspondent pulley, it is influenced by the elasticity of the pulley mounting and its assembly tolerance. $\mathbf{b}_i$ is particularly influenced by pulleys tolerances and reconfigurability impact.

4.1.2 Moving-Platform Geometrical Errors

The MP geometrical errors are described by vectors $\delta\mathbf{a}_i$, (i = 1..8), and $\delta\mathbf{g}$. The gravity center of the MP is often supposed to coincide with its geometrical center P. This hypothesis means that the moments generated by an inaccurate knowledge of the gravity center position or by its potential displacement are neglected. The Cartesian coordinate vector of the geometric center G does not change in frame $\mathscr{F}_p$, but strongly depends on the real coordinates of exit points A_i that are related to uncertainties in mechanical welding of the hooks and in MP assembly.

4.2 Mechanical Errors

The mechanical errors of the CDPR are described by the uncertainty in the MP mass (δm) and the uncertainty on the cables mechanical parameters ($\delta \rho$ and δE). Besides, uncertainties in the cables tension δt affect the error model. As a result, 11 mechanical error sources are taken into account.

4.2.1 End-Effector Mass

As the MP is a mechanically welded structure, there may be some differences between the MP mass and inertia matrix given by the CAD software and the real ones. The MP mass and inertia may also vary in operation In this paper, MP mass uncertainty δm is about $\pm 10\%$ the nominal mass.

4.2.2 Cables Parameters

Linear mass: The linear mass ρ of CAROCA cables is equal to $0.1015\,\text{kg/m}$. The uncertainty of this parameter can be calculated from the measurement procedure as: $\delta \rho = \dfrac{m_c\,\delta L + L\,\delta m_c}{L^2}$, where m_c is the measured cable mass for a cable length L. δL and δm_c are respectively the measurement errors of the cable length and mass.

Modulus of elasticity: This paper uses experimental hysteresis loop to discuss the modulus of elasticity uncertainty. Figure 3 shows the measured hysteresis loop of the 4 mm cable where the unloading path does not correspond to the loading path. The area in the center of the hysteresis loop is the energy dissipated due to internal friction in the cable. It depicts a non-linear correlation in the lower area between load and elongation.

Based on experimental data presented in Fig. 3, Table 2 presents the modulus of elasticity of a steel wire cable for different operating margins, when the cable is in loading or unloading phase. This modulus is calculated as follows:

$$E_{p-q} = L_c \frac{F_{q\%} - F_{p\%}}{S(x_q - x_p)}, \tag{5}$$

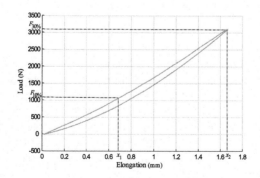

Fig. 3. Load-elongation diagram of a steel wire cable measured in steady state conditions at the rate of $0.05\,\text{mm/s}$

where S is the metallic cross-sectional area, i.e. the value obtained from the sum of the metallic cross-sectional areas of the individual wires in the rope based on their nominal diameters. x_p and x_q are the elongations at forces equivalent to $p\%$ and $q\%$ ($F_{p\%}$ and $F_{q\%}$), respectively, of the nominal breaking force of the cable measured during the loading path (Fig. 3). L_c is the measured initial cable length.

Table 2. Modulus of elasticity while loading or unloading phase

Modulus of elasticity (GPa)	E_{1-5}	E_{5-10}	E_{5-20}	E_{5-30}	E_{10-15}	E_{10-20}	E_{10-30}	E_{20-30}
Loading	72.5	83.2	92.7	97.2	94.8	98.3	102.2	104.9
Unloading	59.1	82.3	96.2	106.5	100.1	105.1	115	126.8

For a given range of loads (Table 2), the uncertainty on the modulus of elasticity depends only on the corresponding elongations and tensions measurements. In this case, the absolute uncertainty associated with applied force and resulting elongation measurements from the test bench outputs is estimated to be ± 1 N and ± 0.03 mm, respectively; so, an uncertainty of ± 2 GPa can be applied to the calculation of the modulus of elasticity.

According to the International Standard ISO 12076, the modulus of elasticity of a steel wire cable is E_{10-30}. However, the CDPR cables do not work always between $F_{10\%}$ and $F_{30\%}$ in real life and the cables can be in loading or unloading phase. The mechanical behavior of cables depends on MP dynamics, which affects the variations in cable elongations and tensions. From Table 2, it is apparent that the elasticity moduli of cables change with the operating point changes. For the same applied force, the modulus of elasticity for loaded and unloaded cables are not the same. While the range of the MP loading is unknown, a large range of uncertainties on the modulus of elasticity should be defined as a function of the cable tensions.

4.2.3 Tension Distribution

Two cases of uncertainties of force determination can be defined depending on the control scheme:

The first case is when the control scheme gives a tension set-point to the actuators resulting from the force distribution algorithm. If there is no feedback about the tensions measures, the range of uncertainty is relatively high. Generally, the effort of compensation does not consider dry and viscous friction in cable drum and pulleys. This non-compensation leads to static errors and delay [13] that degrade the CDPR control performance. That leads to a large range of uncertainties in tensions. As the benefit of tension distribution algorithm used is less important in case of a suspended configuration of CDPR than the fully-constrained one [14], a range of ± 15 N is defined.

The second case is when the tensions are measured. If measurement signals are very noisy, amplitude peaks of the correction signal may lead to a failure of the force distribution. Such a failure may also occur due to variations in the MP and pulleys parameters. Here, the deviation is defined based on the measurement tool precision. However, it remains lower than the deviation of the first case by at least 50%.

5 Sensitivity Analysis

Due to the non-linearities of the elasto-geometrical model, explicit sensitivity matrix and coefficients [15,16] cannot be computed. Therefore, the sensitivity of the elasto-geometrical model of the CDPR to geometrical and mechanical errors is evaluated statistically. Here, MATLAB has been coupled with mode-FRONTIER, a process integration and optimization software platform [17] for the analysis.

The RMS (Root Mean Square) of the static deflection of CAROCA MP is studied. The nominal mass of the MP and the additional mass are equal to 180 kg and 50 kg, respectively.

5.1 Influence of Mechanical Errors

In this section, all the uncertain parameters of the elasto-geometrical CAROCA model are defined with uniformly distributed deviations. The uncertainty range and discretization step are given in Table 3. In this basis, 2000 SOBOL quasi-random observations are created.

Table 3. Uncertainties and steps used to design the error model

Parameter	m (kg)	ρ (kg/m)	E (GPa)	$\mathbf{a}_i$ (m)	$\mathbf{b}_i$ (m)	δt_i (N)
Uncertainty range	±18	±0.01015	±18	±0.015	±0.03	±15
Step	0.05	$3*10^{-5}$	0.05	0.0006	0.0012	0.1

In this configuration, the operating point of the MP is supposed to be unknown. A large variation range of the modulus of elasticity is considered. The additional mass corresponds to a variation in cable tensions from 574 N to 730 N, which corresponds to a modulus of elasticity of 84.64 GPa. Thus, while the operating point of the MP is unknown, an uncertainty of ±18 GPa is defined with regard to the measured modulus of elasticity $E = 102$ GPa.

Figure 4a displays the distribution fitting of the static deflection RMS. It shows that the RMS distribution follows a quasi-uniform law whose mean μ_1 is equal to 1.34 mm. The RMS of the static deflection of the MP is bounded between a minimum value RMS_{min} equal to 1.12 mm and a maximum value RMS_{max} equal to 1.63 mm; a variation of 0.51 mm under all uncertainties, which presents 38% of the nominal value of the static deflection.

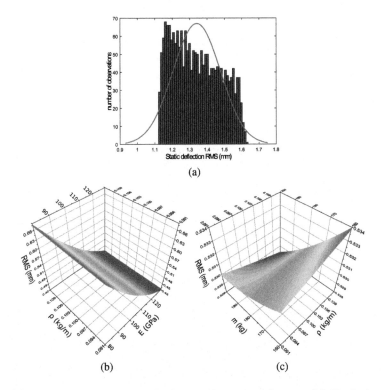

Fig. 4. (a) Distribution of the RMS of the MP static deflection (b) Evolution of the RMS under a simultaneous variations of E and ρ (c) Evolution of the RMS under a simultaneous variations of m and ρ

Figure 4b depicts the RMS of the MP static deflection as a function of variations in E and ρ simultaneously, whose values vary respectively from 0.09135 to 0.11165 kg/m and from 84.2 to 120.2 GPa. The static deflection is very sensitive to cables mechanical behavior. The RMS varies from 0.42 mm to 0.67 mm due to the uncertainties of these two parameters only. As a matter of fact, the higher the cable modulus of elasticity, the smaller the RMS of the MP static deflection. Conversely, the smaller the linear mass of the cable, the smaller the RMS of the MP static deflection. Accordingly, the higher the sag-introduced stiffness, the higher the MP static deflection. Besides, the higher the axial stiffness of the cable, the lower the MP static deflection.

Figure 4c illustrates the RMS of the MP static deflection as a function of variations in ρ and m, whose value varies from 162 kg to 198 kg. The RMS varies from 0.52 mm to 0.53 mm due to the uncertainties of these two parameters only. The MP mass affects the mechanical behavior of cables: the heavier the MP, the larger the axial stiffness, the smaller the MP static deflection. Therefore, a fine identification of m and ρ is very important to establish a good CDPR model.

Comparing to the results plotted in Fig. 4b, it is clear that E affects the RMS of the MP static deflection more than m and ρ. As a conclusion, the integration of cables hysteresis effects on the error model is necessary and improves force algorithms and the identification of the robot geometrical parameters [16].

5.2 Influence of Geometrical Errors

In this section, the cable tension set-points during MP operation are supposed to be known; so, the modulus of elasticity can be calculated around the operating point and the confidence interval is reduced to ± 2 GPa. The uncertainty range and the discretization step are provided in Table 4.

Table 4. Uncertainties and steps used to design the error model

Parameter	m (kg)	ρ (kg/m)	E (GPa)	a_i (m)	b_i (m)	δt_i (N)
Uncertainty range	± 18	± 0.01015	± 2	± 0.015	± 0.03	± 15
Step	0.05	$3*10^{-5}$	0.05	0.0006	0.0012	0.1

Figure 5a displays the distribution fitting of the MP static deflection RMS. It shows that the RMS distribution follows a normal law whose mean μ_2 is equal to 1.32 mm and its standard deviation σ_2 is equal to 0.01 mm. This deviation is relatively small, which allows to say that the calibration through static deflection is not obvious. The RMS of the static deflection of the MP is bounded between a minimum value RMS_{min} equal to 1.28 mm and a maximum value RMS_{max} equal to 1.39 mm; a variation of 0.11 mm under all uncertainties. The modulus of elasticity affects the static compliant of the MP, which imposes to always consider E error while designing a CDPR model.

The bar charts plotted in Fig. 5b and c present, respectively, the effects of the uncertainties in a_i and b_i, (i = 1..8), to the static deflection of the CAROCA for symmetric (0 m, 0 m, 1.75 m) and non-symmetric (3.2 m, 1.7 m, 3 m) robot configurations. These effects are determined based on t-student index of each uncertain parameter. This index is a statistical tool that can estimate the relationships between outputs and uncertain inputs. The t-Student test compares the difference between the means of two samples of designs taken randomly in the design space:

- M_+ is the mean of the n_+ values for an objective S in the upper part of domain of the input variable,
- M_- is the mean of the n_- values for an objective S in the lower part of domain of the input variable.

The t-Student is defined as $t = \dfrac{|M_- - M_+|}{\sqrt{\dfrac{V_g^2}{n_-} + \dfrac{V_g^2}{n_+}}}$, where V_g is the general variance [18].

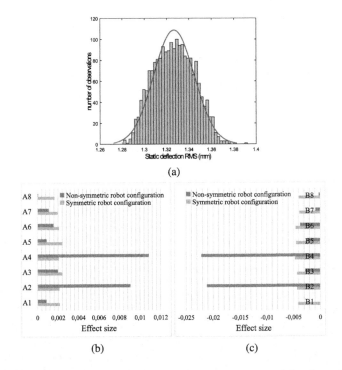

Fig. 5. (a) Distribution of the RMS of the MP static deflection (b) Effect of uncertainties in $\mathbf{a}_i$ (c) Effect of uncertainties in $\mathbf{b}_i$

When the MP is in a symmetric configuration, all attachment points have nearly the same effect size. However, when it is located close to points B_2 and B_4, the effect size of their uncertainties becomes high. Moreover, the effect of the corresponding mobile points (A_2 and A_4) increases. It means that the closer the MP to a given point, the higher the effect of the variations in the Cartesian coordinates of the corresponding exit point of the MP onto its static deflection. That can be explained by the fact that when some cables are longer than others and become slack for a non-symmetric position, the sag effect increases. Consequently, a good identification of geometrical parameters is highly required. In order to minimize these uncertainties, a good calibration leads to a better error model.

6 Conclusion

This paper dealt with the sensitivity analysis of the elasto-geometrical model of CDPRs to mechanical and geometrical uncertainties. The CAROCA prototype was used as a case of study. The validity and identifiability of the proposed model are verified for the purpose of CDPR model-based control. That revealed the importance of integrating cables hysteresis effect into the error modeling to enhance the knowledge about cables mechanical behavior, especially when there is no feedback about tension measurement. It appears that the effect of geometrical errors onto the static deflection of the moving-platform is significant too. Some calibration [19,20] and self-calibration [21,22] approaches were proposed to enhance the CDPR performances. More efficient strategies for CDPR calibration will be performed while considering more sources of errors in a future work.

References

1. Merlet, J.P.: MARIONET, a family of modular wire-driven parallel robots. In: Lenarcic, J., Stanisic, M. (eds.) Advances in Robot Kinematics: Motion in Man and Machine, pp. 53–61. Springer, Dordrecht (2010)
2. Dallej, T., Gouttefarde, M., Andreff, N., Michelin, M., Martinet, P.: Towards vision-based control of cable-driven parallel robots. In: 2011 IEEE/RSJ International Conference on Intelligent Robots and Systems (IROS), pp. 2855–2860. IEEE, September 2011
3. Gagliardini, L., Caro, S., Gouttefarde, M., Girin, A.: A reconfiguration strategy for reconfigurable cable-driven parallel robots. In: 2015 IEEE International Conference on Robotics and Automation (ICRA), pp. 1613–1620. IEEE, May 2015
4. Gagliardini, L., Caro, S., Gouttefarde, M., Girin, A.: Discrete reconfiguration planning for cable-driven parallel robots. Mech. Mach. Theory **100**, 313–337 (2016)
5. Jamshidifar, H., Fidan, B., Gungor, G., Khajepour, A.: Adaptive vibration control of a flexible cable driven parallel robot. IFAC-PapersOnLine **48**(3), 1302–1307 (2015)
6. Fang, S., Franitza, D., Torlo, M., Bekes, F., Hiller, M.: Motion control of a tendon-based parallel manipulator using optimal tension distribution. IEEE/ASME Trans. Mechatron. **9**(3), 561–568 (2004)
7. Zi, B., Duan, B.Y., Du, J.L., Bao, H.: Dynamic modeling and active control of a cable-suspended parallel robot. Mechatronics **18**(1), 1–12 (2008)
8. Pott, A., Mütherich, H., Kraus, W., Schmidt, V., Miermeister, P., Verl, A.: IPAnema: a family of cable-driven parallel robots for industrial applications. In: Bruckmann, T., Pott, A. (eds.) Cable-Driven Parallel Robots, pp. 119–134. Springer, Heidelberg (2013)
9. Irvine, H.M.: Cable Structures. Dover Publications, New York (1992)
10. Mikelsons, L., Bruckmann, T., Hiller, M., Schramm, D.: A real-time capable force calculation algorithm for redundant tendon-based parallel manipulators. In: IEEE International Conference on Robotics and Automation, 2008, ICRA 2008, pp. 3869–3874. IEEE, May 2008
11. Carbone, G.: Stiffness analysis and experimental validation of robotic systems. Front. Mech. Eng. **6**(2), 182–196 (2011)
12. Nguyen, D.Q., Gouttefarde, M.: Stiffness matrix of 6-DOF cable-driven parallel robots and its homogenization. In: Lenarčič, J., Khatib, O. (eds.) Advances in Robot Kinematics, pp. 181–191. Springer International Publishing, Cham (2014)
13. De Wit, C.C., Seront, V.: Robust adaptive friction compensation. In: 1990 IEEE International Conference on Robotics and Automation, 1990, Proceedings, pp. 1383–1388. IEEE, May 1990

14. Lamaury, J.: Contribution a la commande des robots parallles a câblesà redondance d'actionnement. Doctoral dissertation. Université Montpellier II-Sciences et Techniques du Languedoc (2013)
15. Zi, B., Ding, H., Wu, X., Kecskeméthy, A.: Error modeling and sensitivity analysis of a hybrid-driven based cable parallel manipulator. Precis. Eng. **38**(1), 197–211 (2014)
16. Miermeister, P., Kraus, W., Lan, T., Pott, A.: An elastic cable model for cable-driven parallel robots including hysteresis effects. In: Pott, A., Bruckmann, T. (eds.) Cable-Driven Parallel Robots, pp. 17–28. Springer International Publishing, Bern (2015)
17. modeFRONTIER. www.esteco.com
18. Courteille, E., Deblaise, D., Maurine, P.: Design optimization of a delta-like parallel robot through global stiffness performance evaluation. In: IEEE/RSJ International Conference on Intelligent Robots and Systems, 2009, IROS 2009, pp. 5159–5166. IEEE, October 2009
19. dit Sandretto, J.A., Trombettoni, G., Daney, D., Chabert, G.: Certified calibration of a cable-driven robot using interval contractor programming. In: Thomas, F., Perez Gracia, A. (eds.) Computational Kinematics, pp. 209–217. Springer, Amsterdam (2014)
20. Joshi, S.A., Surianarayan, A.: Calibration of a 6-DOF cable robot using two inclinometers. In: Performance Metrics for Intelligent Systems, pp. 3660–3665 (2003)
21. Miermeister, P., Pott, A., Verl, A.: Auto-calibration method for overconstrained cable-driven parallel robots. In: 7th German Conference on Robotics; Proceedings of ROBOTIK 2012, pp. 1–6. VDE, May 2012
22. Borgstrom, P.H., Jordan, B.L., Borgstrom, B.J., Stealey, M.J., Sukhatme, G.S., Batalin, M.A., Kaiser, W.J.: Nims-pl: a cable-driven robot with self-calibration capabilities. IEEE Trans. Rob. **25**(5), 1005–1015 (2009)

CASPR-ROS: A Generalised Cable Robot Software in ROS for Hardware

Jonathan Eden[1,2(✉)], Chen Song[2], Ying Tan[1], Denny Oetomo[1], and Darwin Lau[2]

[1] The University of Melbourne, Melbourne, Australia
jpeden@student.unimelb.edu.au, {yingt,doetomo}@unimelb.edu.au
[2] The Chinese University of Hong Kong, Sha Tin, Hong Kong
chensong@link.cuhk.edu.hk, darwinlau@mae.cuhk.edu.hk

Abstract. In this paper, the software platform CASPR-ROS is introduced to extend the author's recently developed simulation platform CASPR. To the authors' knowledge, no single software framework exists to implement different types of analyses onto different hardware platforms. This new platform therefore takes the advantages of CASPR, including its generalised CDPR model and library of different analysis tools, and combines them with the modular and flexible hardware interfacing of ROS. Using CASPR-ROS, hardware based experiments can be performed on arbitrarily CDPR types and structures, for a wide range of analyses, including kinematics, dynamics and control. The case studies demonstrate the potential to perform experiments on CDPRs, directly compare algorithms and conveniently add new models and analyses. Two robots are considered, a spatial cable robot actuated by PoCaBot units and an anthropomorphic arm actuated by MYO-muscle units.

1 Introduction

Cable-driven parallel robots (*CDPRs*) are a class of mechanisms in which actuation is transmitted through cables in place of rigid links. CDPRs possess a range applications including payload manipulation [1–3], motion simulation [4], exoskeletons [5] and musculoskeletal robots [6–9]. An important feature of cable actuation is that cables can only transmit forces in tension. This creates many unique challenges in CDPR modelling [6], design [10], inverse dynamics [11–13], forward kinematics [14–16] and motion control [17,18].

Most CDPR algorithms can be applied onto different classes of CDPR. However, CDPR research typically either validates algorithms in simulation or through implementation only on the research group's robots. This inhibits CDPR development and research in a number of ways: (1) The evaluation of new techniques often neglects the effect of CDPR structures such that the impact of varying attachments/degrees of freedom/cables may be unknown. (2) There are no benchmarking algorithms for performance comparison. (3) There is a significant cost in implementing new results on hardware, where researchers often re-implement existing models and algorithms.

To address these concerns, different software platforms have been developed for the study of CDPRs. In [19], a MATLAB/Simulink control simulation software interfaced to the dynamics simulator XDE was presented. This platform was developed for single link CDPRs, where the addition of other CDPRs is not simple. For planar and spatial CDPRs, the ARACHNIS [20] and WireCenter [3] software platforms were developed. These platforms were not designed for algorithm benchmarking and additional algorithms cannot be added. Recently, CASPR [21], a MATLAB based simulation platform for the study of CDPRs, was developed. This platform addresses the previous issues by allowing the analysis of arbitrary CDPRs with the possibility to accommodate different algorithms.

CASPR is primarily a simulation platform and does not favour online hardware control and analysis due to its object oriented MATLAB implementation. The extension of CASPR for hardware implementation would allow the operation of arbitrary CDPR hardware to benefit from the flexibility, robustness and extendibility of CASPR. ROS represents one existing means of interfacing robotics hardware which provides a modular and well supported interface for extension and integration [22].

In this paper, CASPR-ROS is introduced as a software platform for CDPR hardware implementation. This platform implements the generalised and object-oriented principles of CASPR into ROS to take advantage of ROS's flexible and modular hardware interfacing capabilities. Through the addition of a new extendible hardware interfacing layer, it is shown that hardware implementations can be performed onto arbitrary CDPRs using arbitrary hardware units. The convenience and efficiency of benchmarking and online implementation in CASPR-ROS is then demonstrated through experimental results obtained on different hardware platforms.

2 Background

2.1 System Model

Consider the general single (SCDR) and multi-link (MCDR) CDPRs depicted in Fig. 1. The n degree of freedom robot configuration (*joint space*) is represented by the pose vector $\mathbf{q} \in \mathbb{R}^n$. The m cable actuation can be described by the cable length and force (*cable space*) vectors $\mathbf{l} = [l_1 \ \ldots \ l_m] \in \mathbb{R}^m$ and $\mathbf{f} = [f_1 \ \ldots \ f_m] \in \mathbb{R}^m$, respectively, where $l_i, f_i \geq 0$ denote the length and force of cable i.

The kinodynamic equations for the CDPRs depicted in Fig. 1 are given by

$$\dot{\mathbf{l}} = L(\mathbf{q})\dot{\mathbf{q}} \tag{1}$$

$$M(\mathbf{q})\ddot{\mathbf{q}} + \mathbf{C}(\dot{\mathbf{q}}, \mathbf{q}) + \mathbf{G}(\mathbf{q}) + \mathbf{w}_e = -L^T(\mathbf{q})\mathbf{f}$$

$$\mathbf{0} \leq \mathbf{f}_{min}(\mathbf{q}) \leq \mathbf{f} \leq \mathbf{f}_{max}(\mathbf{q}), \tag{2}$$

where $L \in \mathbb{R}^{n \times m}$ is the cable Jacobian matrix, $M \in \mathbb{R}^{n \times n}$ is inertia matrix and $\mathbf{C}, \mathbf{G}, \mathbf{w}_e \in \mathbb{R}^n$ are the Coriolis/centrifugal vector, the gravitational vector and the external wrench, respectively. The vectors $\mathbf{f}_{min}, \mathbf{f}_{max} \in \mathbb{R}^m$ are the minimum and maximum cable force bounds. They are constant for ideal cables and pose dependent for spring, variable stiffness [10,23] and muscle inspired cables [24,25].

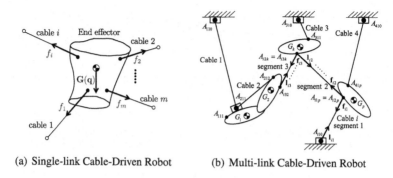

Fig. 1. General single-link and multi-link CDPRS

2.2 Models in CASPR

CASPR is a software platform for the simulation and analysis of CDPRs [21]. CASPR models CDPRs with fundamental Eqs. (1) and (2) using the cable routing matrix based model [6]. This model can represent SCDRs and MCDRs provided that it is given the inertia and joint properties of each link as well as the actuation and attachment specifications of each cable. These specifications are provided in an easily reconfigurable manner through the use of XML scripts[1].

The model representation used by CASPR provides a generic form from which any new CDPR can be added. As such, CASPR-ROS makes use of this same representation allowing for new models to be easily added in addition to the existing CASPR supported models including: NIST RoboCrane [1], CoGiRo [2], IPAnema family [3], the MyoArm [9], and CAREX [5]. To use these models on hardware it is necessary that the algorithms are made suitably computationally efficient and that interfaces are provided to connect the computational component of CASPR with the hardware.

2.3 Analysis in CASPR

CASPR supports CDPR analysis using an inheritance based object oriented approach. For each analysis problem, an abstract based class is created to represent the problem. New algorithms are then generated by inheriting the base class and implementing abstract methods which map the input to the appropriate outputs. Using this approach, different algorithms for a range of problems including inverse dynamics, forward kinematics, control and workspace analysis are supported in CASPR [21].

In hardware implementation, the resolution of joint space wrench into cable forces, conversion of cable lengths into joint space pose and the tracking of reference trajectories must be considered. CASPR-ROS therefore uses the generalised inheritance based paradigm of CASPR [21] for the inverse dynamics,

---

[1] A detailed explanation of CASPR models and the use of XML scripts is provided in [21].

forward kinematics and control problems. CASPR-ROS currently contains the following algorithms:

- Inverse Dynamics - The computationally efficient closed form method [11] and the quadratic programming method using the qpOASES solver [26].
- Forward Kinematics - The Jacobian pseudo-inverse method [14] and the non-linear least squares method [15].
- Control - The computed torque [17] and Lyapunov based static [18] controllers.

3 Interfacing Hardware in CASPR-ROS

3.1 Integrating CASPR with ROS

Using the reconfigurable model representation and inheritance based object oriented paradigm discussed in Sect. 2, CASPR-ROS users can apply a range of CDPR analysis techniques onto different CDPR models. To connect this generic CASPR-ROS computational module to hardware it is necessary for the module to be interfaced with hardware specific sensors and actuators. ROS is chosen for this connection due to its widespread usage, existing hardware support capabilities and open source nature. ROS messaging is then used to translate between the cable space variables and the actuator and sensor information.

Figure 2 shows two different ROS-based hardware communication schemes supported in CASPR-ROS: the centralised method and the distributed method. It can be seen that the centralised method (Fig. 2(a)) connects hardware to CASPR-ROS through a single experiment node. This node facilitates all possible CASPR-ROS operations including forward kinematics, trajectory generation, control and inverse dynamics. In addition, the node is also responsible for translating the generic cable space information into hardware specific feedback and setpoint ROS messages. As a result, the centralised method allows for a more direct process of porting CASPR code into ROS. However, the method prevents common operations, such as forward kinematics, from being continuously operated while the hardware is active and limits algorithm changes to only occur in between experiments. This approach is therefore best used for single experiments that are not repeated such as calibration.

In contrast, the distributed method (Fig. 2(b)) distributes the common CASPR-ROS operations across a number of ROS nodes thereby allowing for the nodes to be run independently if desired. As a result, this method can allow for nodes and analysis algorithms to be changed within single experiments and is best used for repetitive operation of experiments. In addition to having ROS publishers and subscribers associated with the feedback and setpoint messages, the distributed method also requires CASPR-ROS messages to be defined for communication between CASPR-ROS nodes. The following messages are provided:

- `joint_kinematics` - Contains the kinematic vectors $\mathbf{q}$, $\dot{\mathbf{q}}$ and $\ddot{\mathbf{q}}$.
- `master_command` - Contains the current operation and timing information.
- `model_update_command` - Provides the command variables $\mathbf{q}_{cmd}$, $\dot{\mathbf{q}}_{cmd}$, $\ddot{\mathbf{q}}_{cmd}$ and $\mathbf{w}_{cmd}$ for updating the equation of motion (2).

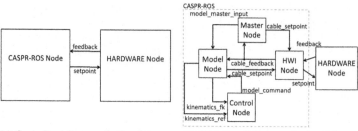

(a) Centralised Communication Scheme   (b) Distributed Communication Scheme

Fig. 2. ROS based hardware communication methods used in CASPR-ROS

3.2 Hardware Interfaces

Common to all CASPR-ROS communication is the need to translate between generic cable space variables and hardware specific ROS messages. CASPR-ROS does this by providing an abstract hardware interfacing class. This class therefore contains all of the basic ROS messaging publisher and subscriber objects in addition to providing the rules for cable space-hardware translation. In this manner, the CASPR-ROS computational core remains generic and need not consider hardware specific requirements such as filtering and relative/absolute data conversion.

The abstract class `HardwareInterfaceBase` therefore sits between the hardware and the CASPR-ROS computational core. Implementations of this class are responsible for translating the hardware specific contents of the feedback message into the associated cable space variables and in writing the actuator specific commands given knowledge of the command cable space variables. This is achieved by implementing the abstract methods `updateFeedback(..)`, `publishForceSetpoint(..)`, `publishLengthSetpoint(..)` and `publishVelocitySetpoint(..)`. To show CASPR-ROS's use of hardware interfaces, the `FlexrayInterface` and `PoCaBotInterface` classes have been constructed for MYO-muscle modules [27] and PoCaBot units[2].

4 Experiments in CASPR-ROS

4.1 Adding New Experiments

Experiments in CASPR-ROS represent executables which define the operation of a CDPR using the analysis techniques and hardware interfaces discussed in Sects. 2 and 3. New experiments can be added into CASPR-ROS through the addition of new master ROS nodes. A single master node is therefore responsible for generating a reference in addition to possibly specifying the CDPR model, hardware interface and analysis algorithm when the centralised scheme is used.

---

[2] PoCaBot unit specifications can be found at https://github.com/darwinlau/CASPR/wiki.

Like the modelling and hardware interface classes described in [21] and Sect. 3, respectively, CASPR-ROS master nodes use an inheritance based object oriented design principle. The abstract classes `ScriptBase` and `MasterNodeBase` are classes which comprise of a single `mainLoop` function which is to be implemented by all new centralised and distributed experiments, respectively. Code Sample 1 illustrates the `mainLoop` function used in CASPR-ROS.

Code Sample 1. mainLoop function used in `ScriptBase` class.

```
// Variable Initialisation
bool is_initialised = 0, first_time = 1;
// General Operation
while((ros::ok()) && (!terminating_condition(t))){
    // Only proceed once the hardware is ready
    if(hardware_interface.hardwareReady()){
        // Check the initialisation status of the experiment
        if(!is_initialised){
            // Run the initialisation procedure
            is_initialised = initialising_function(first_time);
            if(first_time){ first_time = false; }
        } else {
            // Run the main procedure
            main_function(t); t += SYSTEM_PERIOD;
        }
    }
    // ROS management
    ros::spinOnce(); loopRate.sleep();
}
// Terminate the script
terminating_function();
```

It can be seen that the `mainLoop` function represents a single function that defines the behaviour of the robot throughout operation. This behaviour is defined for each particular experiment through the implementations of four abstract methods: `terminating_condition(..)` which defines the terminating condition for the main function, `initialising_function(..)` which initialises the hardware, `main_function(..)` which defines the desired general CDPR behaviour and `terminating_function(..)` which safely terminates the experiment.

4.2 Operating Procedure

To run experiments in CASPR-ROS the following procedure is required: (1) Configure the model parameters and trajectories using CASPR XML scripting. (2) Configure the experiment settings using the roslaunch files associated with the desired experiments. (3) Run the relevant ROS nodes using the appropriate ROS launch files.

5 Experimental Results

Two case studies are presented using the CDPRs depicted in Fig. 3. These studies show the application of CASPR-ROS on arbitrary CDPRs and CASPR-ROS's potential application for benchmarking different analysis algorithms on hardware[3].

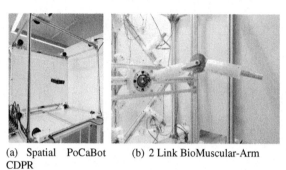

(a) Spatial PoCaBot CDPR (b) 2 Link BioMuscular-Arm

Fig. 3. Case study CDPRs

5.1 Case Study: Spatial Cable Robots Using PoCaBot Units

This case study shows the use of CASPR-ROS in the online length control of the spatial CDPR driven by PoCaBot units (depicted in Fig. 3(a)) with the units attached on the corners of a $84 \times 54 \times 80$ cm frame. Figure 4 depicts the performance of the system for each of the trajectories depicted in Fig. 5, where $\begin{bmatrix} q_1 & q_2 & q_3 & q_4 & q_5 & q_6 \end{bmatrix}^T = \begin{bmatrix} x & y & z & \alpha & \beta & \gamma \end{bmatrix}^T$ and the orientation $\begin{bmatrix} \alpha & \beta & \gamma \end{bmatrix}^T$ is represented by the XYZ Euler angle convention. It can be seen from Fig. 4 that the robot tracks the desired lengths with only a small lag and tracking error. In addition it can be seen from Fig. 4 that the obtained length feedback is provided within the PoCaBot operating frequency of 20 Hz (for 8 motors) indicating the capability of CASPR-ROS to be configured for this online constraint.

From this case study it can be seen that online kinematic length control can be achieved using CASPR-ROS. Furthermore, by using XML scripts and modular ROS nodes, the resulting code is flexible to changes in the experimental set-up, such as different experimental trajectories, without the need for separate experiment scripts.

5.2 Case Study: BioMuscular Arm Using MYO-muscles

This case study illustrates the use of CASPR-ROS in benchmarking two different inverse dynamics algorithms over a range of cable sets. The experiment is

---

[3] The case study specifications can be found in the folder data/model_config/models at the repository https://github.com/darwinlau/CASPR. Case Studies 1 and 2 are contained in the folders PoCaBot_spatial and BM_arm, respectively.

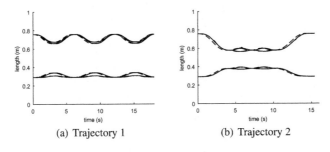

(a) Trajectory 1

(b) Trajectory 2

Fig. 4. Cable length command (Dashed) and feedback (Solid)

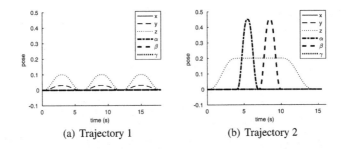

(a) Trajectory 1

(b) Trajectory 2

Fig. 5. Reference joint space trajectory

performed using the 2 link BioMuscular Arm (*BM-Arm*) with Myomuscle units, depicted in Fig. 3(a). The performance of the closed form [11] and minimum force norm quadratic program (QP) based inverse dynamics algorithms are compared by tracking the reference joint space trajectories (shown with dashed lines) in Fig. 6 with the cable sets CS1 and CS2, where $\begin{bmatrix} q_1 & q_2 & q_3 & q_4 \end{bmatrix}^T = \begin{bmatrix} \alpha & \beta & \gamma & \theta \end{bmatrix}^T$. To ensure accurate tracking, a computed torque controller is also implemented.

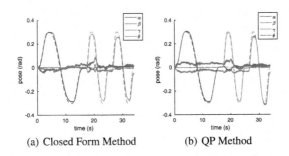

(a) Closed Form Method

(b) QP Method

Fig. 6. Joint space command (dashed) and forward kinematics (solid) - CS1

Figure 7 shows the cable force solutions for each algorithm using cable set CS1. It can be seen that the QP solver (due to its solving objectives) requires typically lower forces and that in both cases the closed loop control has resulted in oscillating cable forces. The resulting tracking performance (obtained using the pseudo-inverse forward kinematics method) of each inverse dynamics solver is shown in Fig. 6. In this case, the closed form solver results in a slightly larger lag and steady state error particularly in the twist axis β and revolute θ axes. This is likely the result of the discretised MYO-muscle sensor resolution, where the resolution is larger over smaller force values.

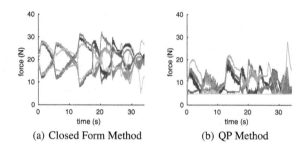

Fig. 7. Closed loop inverse dynamics solutions for CS1

Figures 8 and 9 depict the cable forces and tracking performance for cable set CS2, respectively. It can be seen that the resulting solutions for both methods are different to that observed using cable set CS1, however the tracking performance is quite similar. The relative performance of the two solvers is however similar in which the use of lower cable forces leads to more reliable tracking performance.

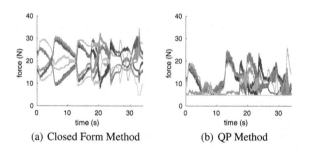

Fig. 8. Closed loop inverse dynamics solutions for CS2

Table 1 shows the computational time used in solving the inverse dynamics, forward kinematics and control for each experiment. It can be seen that the closed form method is on average slightly faster and possesses a lower maximum time. It is also noted that the period of operation was less than that required by the 150 Hz frequency of the BM-Arm in all cases for both algorithms.

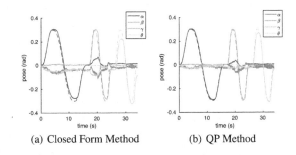

(a) Closed Form Method (b) QP Method

Fig. 9. Joint space command (dashed) and forward kinematics (solid) - CS2

Table 1. Computational time specifications

	Closed form - CS1	QP - CS1	Closed form - CS2	QP - CS2
Maximum time (ms)	5.20	4.79	2.12	5.14
Average time (ms)	1.33	1.41	1.27	1.38

From this case study the use of CASPR-ROS in comparing different analysis techniques can be observed. This case study also displays the flexibility of CASPR-ROS to consider arbitrary cables sets without the need for system model derivation.

6 Conclusion

CASPR-ROS was presented as a tool for the hardware implementation of algorithms onto arbitrary CDPRs. The platform aims to address the lack of a comprehensive CDPR hardware implementation software by integrating the hardware connectivity of ROS with the generic and flexible qualities of CASPR. The modular design of CASPR-ROS makes it convenient to develop new models, analysis algorithms, hardware interfaces and executable scripts. The presented case studies illustrate the flexibility of using CASPR-ROS on different hardware platforms. Future work for CASPR-ROS will look to increase the types of analyses provided and to broaden the cable models considered to include sagging cables and other actuator dynamics.

Acknowledgements. The work was supported by a grant from the Germany/Hong Kong Joint Research Scheme sponsored by the Research Grants Council of Hong Kong and the German Academic Exchange Service of Germany (Reference No. G-CUHK410/16). Acknowledgements to the CUHK T-Stone Robotics Institute for supporting this work.

References

1. Albus, J., Bostelman, R., Dagalakis, N.: The NIST robocrane. J. Robot. Syst. **10**(5), 709–724 (1993)
2. Lamaury, J., Gouttefarde, M.: Control of a large redundantly actuated cable-suspended parallel robot. In: Proceedings of the IEEE International Conference on Robotics and Automation, pp. 4659–4664 (2013)
3. Pott, A., Mütherich, H., Kraus, W., Schmidt, V., et al.: IPAnema: a family of cable-driven parallel robots for industrial applications. In: Cable-Driven Parallel Robots. Mechanisms and Machine Science Series, vol. 12, pp. 119–134 (2012)
4. Miermeister, P., Lächele, M., Boss, R., Masone, C., et al.: The CableRobot simulator large scale motion platform based on cable robot technology. In: Proceedings of IEEE/RSJ International Conference on Intelligent Robots and Systems, pp. 3024–3029 (2016)
5. Mao, Y., Agrawal, S.K.: Design of a cable-driven arm exoskeleton (CAREX) for neural rehabilitation. IEEE Trans. Robot. **28**(4), 922–931 (2012)
6. Lau, D., Oetomo, D., Halgamuge, S.K.: Generalized modeling of multilink cable-driven manipulators with arbitrary routing using the cable-routing matrix. IEEE Trans. Robot. **29**(5), 1102–1113 (2013)
7. Kozuki, Y., Mizoguchi, H., Asano, T., Osada, M., et al.: Design methodology for thorax and shoulder of human mimetic musculoskeletal humanoid kenshiro - a thorax with rib like surface. In: Proceedings of the IEEE/RSJ International Conference on Intelligent and Robotic Systems, pp. 3687–3692 (2012)
8. Lau, D., Eden, J., Halgamuge, S.K., Oetomo, D.: Cable function analysis for the musculoskeletal static workspace of a human shoulder. In: Cable-Driven Parallel Robots. Mechanisms and Machine Science Series, vol. 32, pp. 263–274 (2015)
9. Richter, C., Jentzsch, S., Garrido, J.A., Ros, E., et al.: Scalability in neural control of musculoskeletal robots. IEEE Robot. Autom. Mag. (2016)
10. Yeo, S., Yang, G., Lim, W.: Design and analysis of cable-driven manipulators with variable stiffness. Mech. Mach. Theory **69**, 230–244 (2013)
11. Pott, A.: An improved force distribution algorithm for over-constrained cable-driven parallel robots. In: Computational Kinematics, pp. 139–146 (2014)
12. Müller, K., Reichert, C., Bruckmann, T.: Analysis of a real-time capable cable force computation method. In: Cable-Driven Parallel Robots. Mechanisms and Machine Science Series, vol. 32, pp. 227–238 (2015)
13. Lau, D., Oetomo, D., Halgamuge, S.K.: Inverse dynamics of multilink cable-driven manipulators with the consideration of joint interaction forces and moments. IEEE Trans. Robot. **31**(2), 479–488 (2015)
14. Bruckmann, T., Mikelsons, L., Brandt, T., Hiller, M., et al.: Wire robots part I: kinematics, analysis & design. In: Parallel Manipulators New Developments. ARS Robotic Books Series (2008)
15. Pott, A., Schmidt, V.: On the forward kinematics of cable-driven parallel robots. In: Proceedings of the IEEE/RSJ International Conference on Intelligent Robots and Systems, pp. 3182–3187 (2015)
16. Merlet, J.P.: A generic numerical continuation scheme for solving the direct kinematics of cable-driven parallel robot with deformable cables. In: Proceedings of the IEEE/RSJ International Conference on Intelligent Robots and Systems, pp. 4337–4343 (2016)
17. Williams II, R.L., Xin, M., Bosscher, P.: Contour-crafting-cartesian-cable robot system: dynamics and controller design. In: DETC 2008: 32nd Annual Mechanisms and Robotics Conference, Vol. 2, Parts A & B, pp. 39–45 (2009)

18. Alp, A.B., Agrawal, S.K.: Cable suspended robots: design, planning and control. In: Proceedings of the IEEE International Conference on Robotics and Automation, pp. 4275–4280 (2002)
19. Michelin, M., Baradat, C., Nguyen, D.Q., Gouttefarde, M.: Simulation and control with XDE and matlab/simulink of a cable-driven parallel robot (CoGiRo). In: Cable-Driven Parallel Robots. Mechanisms and Machine Science Series, vol. 32, pp. 71–83 (2015)
20. Ruiz, A.L.C., Caro, S., Cardou, P., Guay, F.: ARACHNIS: analysis of robots actuated by cables with handy and neat interface software. In: Cable-Driven Parallel Robots. Mechanisms and Machine Science Series, vol. 32, pp. 293–305 (2015)
21. Lau, D., Eden, J., Tan, Y., Oetomo, D.: CASPR: A comprehensive cable-robot analysis and simulation platform for the research of cable-driven parallel robots. In: Proceedings of the IEEE/RSJ International Conferrence on Intelligent Robots and Systems, pp. 3004–3011 (2016)
22. Quigley, M., Faust, J., Foote, T., Leibs, J.: ROS: an open-source robot operating system. In: ICRA Workshop on Open Source Software, vol. 3, p. 5 (2009)
23. Yang, K., Yang, G., Wang, J., Zheng, T., et al.: Design analysis of a 3-DOF cable-driven variable-stiffness joint module. In: IEEE International Conference on Robotics and Biomimetics, pp. 529–534. IEEE (2015)
24. Zajac, F.E.: Muscle and tendon: properties, models, scaling, and application to biomechanics and motor control. Crit. Rev. Biomed. Eng. **17**(4), 359–411 (1989)
25. Lau, D., Eden, J., Oetomo, D., Halgamuge, S.: Musculoskeletal static workspace of the human shoulder as a cable-driven robot. IEEE/ASME Trans. Mechatron. **20**(2), 978–984 (2015)
26. Ferreau, H.J., Kirches, C., Potschka, A., Bock, H.G., et al.: qpOASES: a parametric active-set algorithm for quadratic programming. Math. Program. Comput. **6**(4), 327–363 (2014)
27. Marques, H.G., Maufroy, C., Lenz, A., Dalamagkidis, K., et al.: MYOROBOTICS: a modular toolkit for legged locomotion research using musculoskeletal designs. In: Proceedings of the 6th International Symposium on Adaptive Motion of Animals and Machines, AMAM 2013 (2013)

A Polymer Cable Creep Modeling for a Cable-Driven Parallel Robot in a Heavy Payload Application

Jinlong Piao[1,2], XueJun Jin[1,2], Eunpyo Choi[1,2], Jong-Oh Park[1,2(✉)], Chang-Sei Kim[1,2], and Jinwoo Jung[2(✉)]

[1] School of Mechanical Engineering,
Chonnam National University, Gwangju, South Korea
piaojinlong622@163.com, harkjoon27@gmail.com,
eunpyochoi@chonnam.ac.kr, {jop,ckim}@jnu.ac.kr
[2] Medical Microrobot Center, Robot Research Initiative,
Chonnam National University, Gwangju, South Korea
{eunpyochoi,jwjung}@chonnam.ac.kr

Abstract. A polymer cable driven parallel robot can be an effective system in many fields due to its fast dynamics, high payload capability and large workspace. However, creep behavior of polymer cables may yield a posture control problem, especially in high payload pick and place application. The aim of this paper is to predict creep behavior of polymer cables by using different mathematical models for loading and unloading motion. In this paper, we propose a five-element model of the polymer cable that is made with series combination of a linear spring and two Voigt models, to portray experimental creep in simulation. Ultimately, the cable creep can be represented by payloads and cable length estimated according to the changes of actual payloads and cable lengths in static condition.

Keywords: Cable suspended parallel robot · Polymer cable · Creep · Parameter identification

1 Introduction

A cable-driven parallel robot (CDPR) is a parallel robot whose end-effector is controlled by winding and unwinding flexible cables. Unlike a robotic system with rigid body links, the use of flexible light cable or rope as actuator scan significantly reduce actuator weight, thus CDPR has the advantages in the application of heavy materials handling and larger workspace with low cost [1–4]. Although the CDPR as a modified version of robotic cranes has a potential of effective applications, viscoelastic property of cable yields accuracy problems while position or tracking control.

The cable modeling has been motivated by several researchers to develop cable models to overcome issues in using the flexible cable as an actuator of robotic system. Kraus et al. use a computationally efficient and real-time applicable model to identify payload, estimate the end-effector position and compensate position error. Their stiffness model can reduce position errors came by cable elasticity [5]. Also hysteresis behavior of cable driven system has been studied by Miermeister et al. who developed an improved cable model including the hysteresis effect during cable force computation. In order to avoid a complex distributed parameter model, they used a black box model to represent the cable [6]. M. Miyasaka et al. developed the hysteresis model for longitudinally loaded cables based on the Bouc-Wen hysteresis model. Also, the model is capable of emulating thin *and thick stainless* steels used for the RAVEN II system and its parameters were optimized using the genetic algorithm [7]. In addition, the elasticity and uncertainties of a CDPR can be neutralized by robust control. M.H. Korayem compensated the flexibility and uncertainties of cable suspended robot using sliding mode control [8]. Those approaches has improved tracking performance of CDPR obviously. However, there is considerable cable creeping behavior in specific applications, such as high precision part assembly and high payload pick and place application. Thus, using a creep model can improve performance of CDPR. In this paper, we propose a five-element model to properly describe significant experimental creep.

This paper is organized as follow. First, the development of a high payload cable robot is briefly described. Second, a five-element model is introduced for describing experimental creep. Third, experimental results for showing creep behavior are discussed for building a creep model. Fourth, via surface fitting process, the parameters estimation results for the suggested model based on the experimental data by different payloads and cable lengths will be shown. Finally, conclusion and future work will be discussed.

2 CNU Cable Robot System and Creep Experiments

2.1 CNU Cable Robot System

The high payload CDPR, shown in Fig. 1, is developed for the purpose of heavy part assembly and pick and place application. The size of base frame is $4\,m \times 4\,m \times 4\,m$ and eight pulleys are fixed near the corners of top frame. Configurations of cable connection were designed for a wider workspace [9]. Maximum payload that the robot can manipulate is 65 kg and the goal of this robot is to have capability of handling more than 200 kg.

As a light flexible actuator of the high payload cable robot system, polyethylene Dyneema® cable, LIROS D-Pro 01505-0600, are used. The weight of cable is 1 kg/100 m with a diameter of 6 mm. Polyethylene cable has an advantage of light-weight compared to steel cable in industrial application. However, it may cause a concern of viscoelastic behavior such as creep in loading and unloading motion.

2.2 Cable Creep Experiment

When a polymer cable is subjected to load and unload a heavy object, it exhibits time dependent elongation characteristic called creep behavior. There are several factors influencing creep behavior of a polymer cable, such as type of material property, weight of load, length of cable and temperature. Although temperature is one of the important factors that can accelerate cable creep behavior, we only consider the load

Fig. 1. High payload CDPR

weight and cable length in our experiment by assuming that our system is operated in a room temperature. In order to measure cable creep behavior in experiments, creep tests for the actual CDPR cable are conducted. There are various kinds of creep testing equipment that are most commonly used in experiments to create a creep curve [10]. However, it is difficult to realize different loading and unloading behavior happened while heavy load handling for the actual cable driven robot system. Tensile testing machine can express strain-stress curve of materials [11], but it is also not easy to test long length of cables due to the hardware limitations. Then, we used actual winch-cable-pulley system and a crane for loading and unloading weight blocks to measure cable creeping behavior because it can handle high weight and long cable lengths.

The experimental setup for the creep test is shown in Fig. 2. We employed two different experiments for different loads and lengths shown in Figs. 2 and 3 respectively. For a fixed cable length, different payloads are applied one by one and repeated for other length cable with the same payload conditions, and all the measurements are acquired simultaneously.

As shown in Fig. 2, the cable starts from the fixed winch and pass through the pulleys to the hook in the testing mass. For loading, the crane moves down and for unloading, the crane moves up to the initial position. The displacement sensor, an optical tracking system (OTS) from NDI with a root mean square (RMS) of 0.3 mm resolution is used and its sensor tool with four ball markers are fixed at the cable.

In Fig. 2, we have nine weight blocks and the total sum of loads can be more than 100 kg including middle bar. And shown in Fig. 3, we have conducted cable length tests for three different cable lengths, 5.4 m, 7.4 m and 8.4 m, considering our current system configuration and operating ranges.

Fig. 2. Experimental setup with nine weight blocks

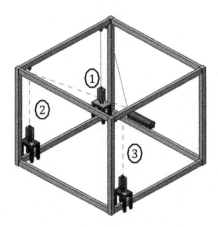

Fig. 3. Experimental setup with three different cable lengths

3 Creeping Modeling and Parameters Estimation

3.1 Cable Creep Modeling

A well-known very simple model that can describe creep behavior is a Voigt model. It consists of a Hookean spring and a Newtonian dashpot, but it does not fully explain complex experimental creep behavior in real system. In general, Burgers model can be used for describing complicated creep behavior [12]. This model is the serial combination of a Maxwell model and a Voigt model. However, by using Burgers model, it is also difficult to express long term creep behavior of polymer cable during unloading period.

In order to properly emulate long term creep behavior, we propose a five-element model based on Burgers model whose time-response can emulate three different time dependent behavior of a polymer cable. The model is the serial combination of a linear spring and two Voigt models, as shown in Fig. 4. According to the superposition principle the creep of cable is the sum of instantaneous elastic response, transient creep and long term creep [13]. Also, the creep behavior of loading and unloading can be different and therefore we use different model parameters indicated by subscripts Li and ULi. From loading $t = 0$ to unloading $t = t_r$, creeping behavior can be modeled as

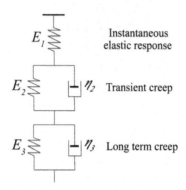

Fig. 4. A five-element model for cable creeping.

$$\varepsilon(t) = \frac{\sigma}{E_{L1}} + \frac{\sigma}{E_{L2}}\left(1 - e^{-\frac{E_{L2}t}{\eta_{L2}}}\right) + \frac{\sigma}{E_{L3}}\left(1 - e^{-\frac{E_{L3}t}{\eta_{L3}}}\right), \quad t = 0 \text{ to } t_r \quad (1)$$

where σ is the applied stress and ε is the strain of the cable. The model has five parameters in steady creep which are elastic modulus E_1, E_2, and E_3 for elastic property, transient creep, model and long-term creep effect, respectively and viscosity η_2 and η_3 for transient creep and long-term creep phenomena.

If the stress σ is removed when unloading the payload at $t = t_r$, $-\sigma$ can be added to (1). In the beginning of unloading, the negative stress applied to the cable increases up to $-\sigma_0$ until unloading completes ($t_r \leq t < t_2$) and remains as constant value $-\sigma_0$ ($t_2 \leq t$). Thus, strain recovery can be mathematically modeled as

$$\begin{aligned}\varepsilon(t) &= \frac{\sigma_0}{E_{UL1}} + \frac{\sigma_0}{E_{UL2}}\left(1 - e^{-\frac{E_{UL2}t}{\eta_{UL2}}}\right) + \frac{\sigma_0}{E_{UL3}}\left(1 - e^{-\frac{E_{UL3}t}{\eta_{UL2}}}\right) \\ &+ \left(\frac{\sigma}{E_{UL1}} + \frac{\sigma}{E_{UL2}}\left(1 - e^{-\frac{E_{UL2}(t-t_r)}{\eta_{UL2}}}\right) + \frac{\sigma}{E_{UL3}}\left(1 - e^{-\frac{E_{UL3}(t-t_r)}{\eta_{UL2}}}\right)\right), \quad \begin{aligned}t &= t_r \text{ to } t_2, \\ \sigma &= 0 \text{ to } -\sigma_0\end{aligned}\end{aligned} \quad (2)$$

$$\varepsilon(t) = \frac{\sigma_0}{E_{UL2}}(e^{\frac{E_{UL2}t_r}{\eta_{UL2}}} - 1)e^{-\frac{E_{UL2}t}{\eta_{UL2}}} + \frac{\sigma_0}{E_{UL3}}(e^{\frac{E_{UL3}t_r}{\eta_{UL3}}} - 1)e^{-\frac{E_{UL3}t}{\eta_{UL3}}}, \quad t = t_2 \text{ to ending time} \quad (3)$$

3.2 Individual Model Parameters Estimation: Polynomial Fitting

Figures 5 and 6 show the experimental creep responses and the predictions of the five element model under a series of payloads and cable lengths, respectively. Experimental results show that creep behavior becomes more obvious with higher payload and longer cable length. Experiments also show that the creep behavior of loading and unloading is different. The creep behavior of unloading period is more obvious than that of loading period. In order to portray experimental creep, the five element model is used and the parameters of creep model are estimated by minimizing root mean square error (RMSE) between experimental data and model's output via brutal model parameter searching. Estimated parameters show that five parameters are changing in terms of different payloads and different cable lengths. Also, parameters are different with loading and unloading periods. In order to minimize the complexity of the model, we fixed less changed parameters when fitting experimental data and estimated the rest of parameters again as shown in Tables 1 and 2. Figures 5 and 6 show that the predictions of our model match with experiment data which the creeping behavior is different with different loads and cable lengths, also different when loading and unloading. These differences can be caused by nonlinear property of polymer cable. Also, loading weight has an influence on the cross sectional area of cable and changes the cable properties which can be illustrated using (4)

$$E = \frac{F/L_0}{A_0 \Delta L} \qquad (4)$$

where E is elastic modulus, F is the force exerted on the cable, A_0 is the actual cross-sectional area through which the force is applied, ΔL is the amount by which the length of the cable changes and L_0 is the original length of cable.

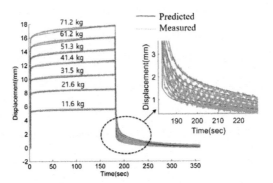

Fig. 5. Experimental creep responses and the predictions of the five element model under a series of payloads

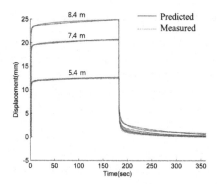

Fig. 6. Experimental creep responses and their predictions under a series of cable lengths

3.3 Dual Model Parameters Estimation: Surface Fitting

According to the experiments and estimated parameter, model parameters are influenced by loads and cable lengths. Thus, model parameters can be expressed as a function of mass (m) and cable length (l). Mass and length are utilized for the function instead of stress and strain because mass and cable length can be directly measured using force sensor and encoder equipped in our CDPR.

Table 1. Loading parameters

Parameters		Values
Elastic modulus (N/m^2)	E_{L1}	$E_{L1}(m, l)$
	E_{L2}	9.5×10^9
	E_{L3}	5.0×10^{10}
Viscosity ($N \cdot s/m^2$)	η_{L2}	1.0×10^{10}
	η_{L3}	1.0×10^{13}

Table 2. Unloading parameters

Parameters		Values
Elastic modulus (N/m^2)	E_{UL1}	$E_{UL1}(m, l)$
	E_{UL2}	$3.5 \times 10^{10} (N/m^2)$
	E_{UL3}	$E_{UL3}(m, l)$
Viscosity ($N \cdot s/m^2$)	η_{UL2}	$1.1 \times 10^{11} (N \cdot s/m^2)$
	η_{UL3}	$\eta_{UL3}(m, l)$

Table 1 and 2 show different parameters under loading and unloading respectively. The loading parameters E_{L2}, E_{L3}, η_{L2}, η_{L3} have constant values and E_{L1} is a function of mass (m) and cable length (l). The unloading parameter E_{UL2}, η_{UL2} are constant values and E_{UL1}, E_{UL3}, η_{UL3} are also a function of mass (m) and cable length (l).

Table 3. Polynomial model parameters

	E_{L1}	E_{UL1}	E_{UL3}	η_{UL3}
p_{00}	-5.036×10^{10}	-1.342×10^{9}	-8.301×10^{10}	2.803×10^{12}
p_{10}	3.07×10^{9}	5.933×10^{8}	-1.293×10^{9}	-1.993×10^{11}
p_{01}	1.81×10^{10}	2.052×10^{9}	2.799×10^{10}	-7.127×10^{11}
p_{20}	2.519×10^{7}	1.897×10^{4}	-9.385×10^{6}	2.543×10^{9}
p_{11}	-9.892×10^{8}	-1.438×10^{8}	8.516×10^{8}	5.164×10^{10}
p_{02}	-1.404×10^{9}	-1.841×10^{8}	-2.206×10^{9}	8.817×10^{10}
p_{30}	-3.985×10^{4}	-8.251×10^{3}	-2.668×10^{4}	-2.286×10^{7}
p_{21}	-2.358×10^{6}	8.45×10^{4}	4.106×10^{5}	5.996×10^{7}
p_{12}	7.51×10^{7}	9.708×10^{6}	-6.894×10^{7}	-4.337×10^{9}

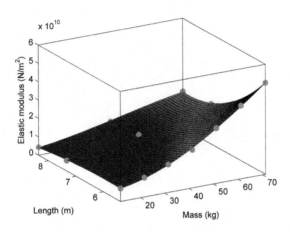

Fig. 7. Surface fitting for E_{L1} with loading condition

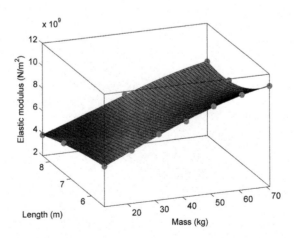

Fig. 8. Surface fitting for E_{UL1} with unloading condition

The change of parameters can be described using polynomial function with two variables, mass and cable length, as in (5).

$$P(l,m) = p_{00} + p_{10}m + p_{01}l + p_{20}m^2 + p_{11}ml \\ + \ldots p_{02}l^2 + p_{30}m^3 + p_{21}m^2l + p_{12}ml^2 \quad (5)$$

where $P(l,m) = E_{L1}, E_{UL1}, E_{UL3}, \eta_{UL3}$ and all values are listed in Table 3.

In order to examine the trend of dual parameters variation, i.e., mass and cable length, surface fitting is established by using (5). Figure 7 shows the loading elastic modulus E_{L1} which increases with high payload and short cable length. Figures 8, 9 and 10 shows the unloading elastic modulus E_{UL1}, E_{UL3} and viscosity η_{UL3}. In Figs. 8 and 10, E_{UL1} and η_{UL3} have the similar trend with E_{L1}. Although experimentally

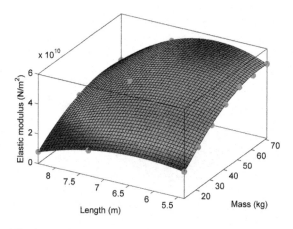

Fig. 9. Surface fitting for E_{UL3} with unloading condition

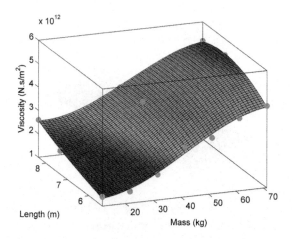

Fig. 10. Surface fitting for η_{UL3} with unloading condition

estimated E_{UL3} increases with high payload and short cable length, its polynomial function does not properly represent the trend of E_{UL3} in Fig. 9. It is expected that more experimental data will improve the function of the surface fitting.

4 Conclusion and Future Works

In this paper, a data based five-element model is proposed for describing creep behavior of polymer cables when loading and unloading. Model parameters are estimated by searching a minimum RMSE of a measured signal and an estimated model comparison. It is found that the parameters are considerably different during loading and unloading period with different payloads and different cable lengths due to the nonlinear properties of polymer cable. Polynomial functions for model parameters are built for fitting the experimental data in a function of mass and cable length. The comparison results between model and experiments in time domain shows reasonable good possibilities of the suggested parametric model utilization. Surface model including both mass and cable length parameters shows that the elastic modulus E_{L1}, E_{UL1} and viscosity η_{UL3} increase with high payload and short cable length. However, the estimation E_{UL3} is needed to be improved by using more experimental data.

In the future, our model will be expanded to cover a wide range of excitations and other factors will be considered as well as the effect of temperature. In addition to the model expansion, the effectiveness of a parameter estimation by minimizing RMSE between our model and experimental data will be quantitatively evaluated by accompanying other simple fitting functions. Also, a model based position control will be implemented for our 8 cable-driven parallel robot system and the expanded cable model will be included as part of dynamic model of our CDPR to compensate errors induced by cable creeping.

Acknowledgments. Research supported by Leading Foreign Research Institute Recruitment Program through the National Research Foundation of Korea (NRF) funded by the Ministry of Science, ICT and Future Planning (MSIP) (No. 2012K1A4A3026740).

References

1. Kawamura, S., Kino, H., Won, C.: High-speed manipulation by using parallel wire-driven robots. Robotics **18**(3), 13–21 (2000)
2. Maeda, K., Tadokoro, S., Takamori, T., Hiller, M., Verhoeven, R.: On design of a redundant wire-driven parallel robot WARP manipulator. In: IEEE International Conference on Robotics and Automation, pp. 895–900 (1999)
3. Kawamura, S., Choe, W., Tanaka, S., Pandian, S.: Development of an ultrahigh speed robot FALCON using wire drive system. In: 1995 Proceedings of the IEEE International Conference on Robotics and Automation, vol. 1, pp. 215–220 (1995)
4. Dallej, T., Gouttefarde, M., Andreff, N., Dahmouche, R., Martinet, P.: Vision-based modeling and control of large-dimension cable-driven parallel robots. In: 2012 IEEE/RSJ International Conference on Intelligent Robots and Systems (IROS), pp. 1581–1586 (2012)

5. Kraus, W., Schmidt, V., Rajendra, P., Pott, A.: Load identification and compensation for a cable-driven parallel robot. In: Proceedings of International Conference on Robotics and Automation, Karlsruhe, Germany, pp. 2485–2490 (2013)
6. Miermeister, P., Kraus, W., Lan, T., Pott, A.: An Elastic Cable Model for Cable-Driven Parallel Robots Including Hysteresis Effects, pp. 17–28. Springer International Publishing, Switzerland (2015)
7. Miyasaka, M., Haghighipanah, M., Li, Y., Hannaford, B.: Hysteresis model of longitudinally loaded cable for cable driven robots and identification of the parameters. In: Proceedings of IEEE International Conference on Robotics and Automation, Stockholm, Sweden, pp. 4051–4057 (2016)
8. Korayem, M.H., Taherifar, M., Tourajizadeh, H.: Compensating the flexibility uncertainties of a cable suspended robot using SMC approach. Robotica **33**(3), 578–598 (2014). Cambridge University Press
9. Seon, J., Park, S., Ko, S.Y., Park, J.-O.: Cable configuration analysis to increase the rotational range of suspended 6-DOF cable driven parallel robots. In: Proceedings of the 16th International Conference on Control, Automation and Systems, Gyeongju, Korea, pp. 1047–1052 (2016)
10. Momoh, J.J., Shuaib-babata, L.Y., Adelegan, G.O.: Modification and performance evaluation of a low cost electro-mechanically operated creep testing machine. Leonardo J. Sci. **16**, 83–94 (2010)
11. Czichos, H.: Springer Handbook of Materials Measurement Methods, pp. 303–304. Springer, Berlin (2006)
12. Findley, W.N., Davis, F.A.: Creep and Relaxation of Nonlinear Viscoelastic Materials, pp. 58–62. Courier Corporation (2013)
13. Lurzhenko, M., Mamunya, Y., Boiteux, G., Lebedev, E.: Creep/stress relaxation of novel hybrid organic-inorganic polymer systems synthesized by joint polymerization of organic and inorganic oligomers. Macromol. Symp. **341**(1), 51–56 (2014)

Bending Fatigue Strength and Lifetime of Fiber Ropes

Martin Wehr[1], Andreas Pott[2,3](✉), and Karl-Heinz Wehking[1]

[1] Institute for Mechanical Handling and Logistics,
University of Stuttgart, Stuttgart, Germany
`martin.wehr@ift.uni-stuttgart.de`
[2] Institute for Control Engineering and Manufacturing Units,
University of Stuttgart, Stuttgart, Germany
`andreas.pott@isw.uni-stuttgart.de`
[3] Fraunhofer IPA, Stuttgart, Germany

Abstract. Modern fiber ropes have several distinctive properties which predestine them amongst others for high dynamic applications in robotics. Beside their great breaking load due to their high tensile streghth, the extremely low density and weight are the most important advantages over steel wire ropes. For steel wire ropes, it is generally known that their lifetime drops when raising the dynamic stress on running or static ropes. The long-time behavior of high-dynamically stressed fiber ropes is totally unexplored up to now. This lack impedes the breakthrough of fiber ropes and causes a safety gap, which has to be closed. This paper describes the research on modern fiber ropes regarding their lifetime in normal and high dynamic applications. The derived results are interpretered with respect to application in robotics.

Keywords: Cable-driven parallel robots · Cable wear · Experimental testing · Lifetime

1 Introduction

Ropes are used in many different applications. In the field of mechanical handling, they are used for elevators, cranes, and hoisting devices. A comparably new field of application are cable robots. Interestingly, in the field of mechanical handling, the transmission elements are called *ropes* where in the field of robotics they are referred to as *cables*. In this paper, we stick to the term *rope* as we approach the subject from the perspective of rope technology. However, all results relate to the transmission element of cable-driven robots. Also in this paper, the terms *wire* for steel ropes and *fiber* for synthetic ropes relate to the smallest subelements of which the rope is made of.

Safety, availability, and economy of rope applications are mainly influenced by the lifetime of the applied ropes [20]. In contrast to other components, ropes have a great axial stiffness, which results from the stiffness of the single wires or fibers [6]. A rope is called *running* if it is guided over sheaves (pulleys) and coiled onto drums. Ropes in running applications are not fatigue endurable but have a limited lifetime [23] which is measured in the number of bending cycles. The parallel arrangement of the single elements results in a redundant structure which prevents the complete system from a sudden failure. The number of broken wires or fibers on a defined reference length characterizes the condition of the used rope and offers the possibility to rate the degree of wear using inspection methods.

Running ropes, how they are used for example in cable robots, are characterized by their axial movement pattern. On this occasion, they run over sheaves and are winded on drums which leads beside the tensile stress to a bending stress in the ropes. These alternating stresses in combination with wear processes on the surface and in the ropes lead to a limited lifetime of the single wires or fibers and thus of the rope as a whole.

In this context of cable-driven parallel robots, running ropes have to be regarded as wear parts like other machine elements, for example V-belts, brake blocks, or bearings. For safety reasons, the ropes of a robot must not break in operation to prevent excessive damage. Thus, the rope as wear part has to be exchanged in time before reaching a critical situation. Therefore, the condition of the rope has to be monitored and classified using a non-destructive test method or the average lifetime of the rope in the specific application has to be well-known. Additionally, the rope drive and the rope itself have to be designed in the way that the lifetime of the rope reaches a manageable maximum.

Many research institutes and companies around the world work on different ways in order to develop a non-destructive test method for fiber ropes. Today, there is no established method to measure the condition of fiber ropes and to predict the remaining lifetime. Some simple methods are based on the outer appearance of the rope [14], but these methods are very subjective and do not give an exactly result. Additionally, the inner condition of the rope stays unconsidered, which is a huge lack, especially for mantled ropes. Other approaches are based on resistive strings, measuring of geometrical parameters, or magnetic stray field methods [8]. Steel wire ropes can be examined completely using the magnetizability of the steel wires in combination with stray field measurements (Magnetic Rope Testing – MRT) [4].

Existing high dynamic rope drives, for example the roller coaster Space Mountain in Euro Disney Land Paris reach a number of 120,000 operations only by shifting the wire rope on the drum. As a consequence of a cycle time of 36 s, the used high performance steel wire rope has to be assessed by non-destructive testing in short terms and has to be changed on average every three months [1].

These days, there is only a small number of applications in which established steel wire ropes could be replaced by modern fiber ropes. One example is an aramid rope for elevators, which was invented and applied by the Swiss elevator manufacturer Schindler AG [17,22]. In 2016, the Teufelberger Seil GmbH presented their new High-Modulus-High-Tensity-fiber rope *soLITE*, which was invented in cooperation with the crane manufacturer Liebherr especially for mobile cranes [19]. Since 2014, the American Manitowoc Crane Group operates Grove RT770E mobile cranes with KTM100 fiber ropes produced by Samson rope [15]. All these single examples show that different parties try to bring fiber ropes into established steel wire rope markets. Up to now, this goal could only be reached in a few specific applications and not global.

In the field of cable-driven parallel robots, different types of fiber ropes have been applied in a number of demonstrator systems [2,10,13]. Qualitative observations indicate an acceptable lifetime if the cables are guided over pulleys. However, no quantities measures or experimentally grounded data are available in the literature.

In this paper, an attempt is made to present latest findings in the field of fiber rope testing to the application in cable robots by providing data in performance numbers that can be applied in cable robot design and analysis.

2 State of the Art in Rope Testing

Ropes are not fatigue endurable and can be operated safely only in a limited period of time [23]. Due to a huge number of local contact points between wires or fibers, the combination of wear, notches, and frictional heat result in a complex damage behavior, which limits the lifetime of ropes [12,16,21]. Because of this complex damage behavior the lifetime of ropes cannot be estimated analytically until now. The number of bending cycles over sheaves under defined parameters still has to be determined experimentally [6].

In this experimental research, cycling bending-over-sheaves-tests (BOS) on bending-over-sheaves-machines performed by the author at the University of Stuttgart are established. In the following section, the machine type Stuttgart is mentioned. A test rope is wrapped over a bigger drive sheave in the upper part of the machine and over a smaller test sheave in the lower part. This relation between the sheave diameters ensures that the rope will break on the smaller test sheave. The test sheave is mounted vertically moveable and connected to a weight-cage, filled with a defined number of steel plates. The constant rotational movement of the AC engine gets transferred into an oscillating by a thrust rod (Fig. 1, [7]). Figure 1 shows the resulting movement profile of the test sheave in standard BOS machines.

The important parameters for BOS-tests are:

- D/d-ratio of the diameter of the sheave D and the diameter of the rope d,
- diameter-related tension force S/d^2
- geometry of the test sheave and
- bending length l.

The D/d-ratio describes the ratio between the diameter of the test sheave D and the nominal rope diameter d. It is well-known that the lifetime of ropes depends largely on D/d where longer lifetime is achieved for larger D/d-ratio. For steel wire ropes, $D/d = 25$ can be regarded as a minimum while the D/d-ratio can go down to 10 or even less when using fiber ropes. In technical applications, the actual D/d-ratio is a compromise between technical or economic feasibility and acceptable lifetime of the ropes.

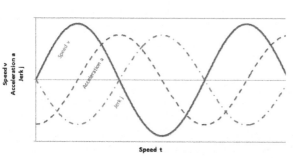

Fig. 1. Movement profile of standard BOS machines

Due to scaling reasons, the tension force of the rope gets related to the square of the nominal rope diameter. This resulting diameter-related tension force S/d carries the unit [N/mm^2] which characterizes a normalized tension in the rope. This force or tension has to be regarded in reference to the nominal breaking load of the rope. Building the quotient between nominal breaking load and the present load leads to the so-called safety factor. The lifetime of running ropes increases dramatically when high diameter-related forces or high safety factors can be realized.

The bending length l is the length of the part of the rope which gets bent over sheaves during a bending test. Its influence on the lifetime is statistically. The possibility of defects in a part of a rope grows with longer bending lengths but approximates to threshold value with increasing length. Feyrer used these factors to predict the lifetime N of running steel wire ropes under ideal conditions [7]. The equations read

$$\lg N = a_0 + \left(a_1 + a_3 \lg \frac{D}{d}\right)\left(\lg \frac{S}{d^2} - 0.4 \lg \frac{R_0}{1770}\right) + a_2 \lg \frac{D}{d} + \lg f_d + \lg f_l \quad (1)$$

$$f_d = \frac{0.52}{-0.48 + \left(\frac{d}{16}\right)^3} \quad (2)$$

$$f_l = \cfrac{1.54}{2.54 - \left(\cfrac{\frac{l}{d} - 2.5}{57.5}\right)^{-0.14}} \tag{3}$$

where the factors a_i, $i = 0, \ldots, 3$ of this *Feyrer-formula* [7] have to be determined for every type of rope within BOS-tests. Not ideal conditions, for example high dynamic stress, can be taken into account using correction factors f_{Ni}. The factor R_0 describes the nominal strength of the used wires.

Until now, there is no comparable formula for fiber ropes. Beside the new field of research in HM-HT fiber ropes, a reason for that is the big number of different fiber materials, rope constructions, coatings, and manufacturers on the market. Slight changes in the manufacturing process can influence the properties of the final product significantly. Additionally, there is no standardized procedure and documentation method for BOS tests on fiber ropes which makes it difficult to compare independent research results. In this paper, three materials being used for cables in the recent years are considered

- High-modulus Polyethylene (HMPE)
- Aromatic Polyamide (Aramid)
- High-modulus Polyester (TLCP).

The most famous representative of the group of Polyethylene is the brand Dyneema, which can be counted to the Ultra-High-Molecular-Weight Polyethylene (UHMW-PE). It has a huge mechanical strength of up to $4000 \, \text{N/mm}^2$ and an extremely low density of $0.97 \, \text{g/cm}$, which makes it lighter than water and therewith floatable. The surface is very smooth which leads to a friction coefficient μ between 0.08 and 0.1. Aramids are mainly known under the brands Kevlar (DuPont) and Technora (Teijin). Beside a high mechanical strength, this fiber has a negative coefficient of thermal expansion. TLCPs are liquid crystal polymers. Most famous representative is Vectran from Kuraray Co. Ltd, Japan. Figure 2 shows the most important properties of the three fiber materials Dyneema, Technora, and Vectran in comparison to steel wire ropes which illustrates the huge opportunities of modern fiber ropes.

3 Rope Testing Under High Dynamic Parameters

When moving a mass in a cable-driven parallel robot under high dynamic parameters, two different effects relating to the ropes appear:

- High-dynamic bending over sheaves
- Pulsating tension load due to acceleration of inertial masses (tension-tension stress)

In order to find and identify relevant influence parameters, it is necessary to separate different influence parameters and vary them independently.

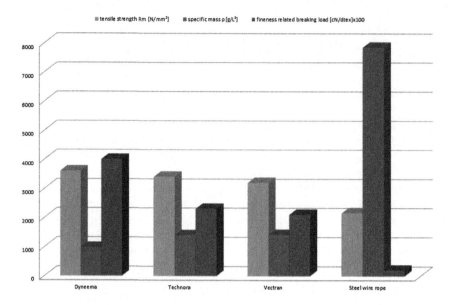

Fig. 2. Comparison of main properties of fiber ropes and steel wire ropes.

3.1 Tension-Tension Tests

In this setting, pulsating tension load can be tested in pulsating tension-tension-tests. Therefore, different tension-tension-test-machines are available at the Institute of Mechanical Handling and Logistics (IFT) of the University of Stuttgart. All machines satisfy the highest requirement class according ISO 7500-1 [9]. For testing, the sample rope has to be mounted in the test machine. Therefore, conical Epoxy resin socketing in the style of DIN EN 13411-4 [5] is favorable. DIN EN ISO 2307 recommends to mount the ropes using friction clamps [3], which can be used for break-load-tests but not for tension-tension tests. Oscillating tension forces $S(t)$ are applied in the form of a sinusoidal wave with periodic time T

$$S(t) = S_m + (S_a \sin(2\pi/T) + \delta), \qquad (4)$$

where S_m is the mean stress, S_a is the amplitude of the pulsating stress, and δ is the delay angle. Here, the two parameters load and frequency shall be varied within the present work. The finite life fatigue strength is self-defined to $0.5 \cdot 10^6$ cycles. Tension-tension tests were performed with five different test ropes, two different loads (15–30% and 30–50% of nominal breaking load) and two different frequencies (2 Hz and 4 Hz). After performing all tests, no single break of a rope could be detected within $0.5 \cdot 10^6$ load cycles. Because of this, all test items were tested in following load tests. Figure 3 shows the result of these tests exemplary for the test rope *D2* made from Dyneema.

This figure shows that preceding tension-tension stress did not drop the residual strength. On the contrary, tension-tension stress could raise the residual strength. This can be explained with extension processes of the long-chain

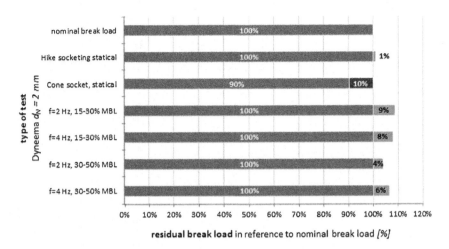

Fig. 3. Residual strength of the test rope $D2$

Dyneema fibers. A qualitative analogical result can mostly be seen for all test ropes.

3.2 High-Dynamic Bending Tests

Standard bending machines generate an oscillating speed of $v_{max} = 0.2\,\text{m/s}$. In cable-driven parallel robots, rope speeds of up to $10\,\text{m/s}$ and more are realized. Additionally, accelerations up to $100\,\text{m/s}^2$ and more are implemented. In order to reproduce these high-dynamic demands, a new and unique test machine was designed, engineered, constructed, built and brought into service in the IFT laboratory. With this new test machine, it is possible to test fiber ropes up to diameter $d = 6\,\text{mm}$ on four independent test stations. The movement profile (speed and acceleration) can be set completely unrestrained up to $v = 10\,\text{m/s}$ and $a = 100\,\text{m/s}$. Figure 4 show a drawing of the new high dynamic test machine and the movement profile.

The new high-dynamic test machine covers a base area of $2\,\text{m} \times 3.20\,\text{m}$, has a height of $6.50\,\text{m}$ and enables a free rope length of approx. $4700\,\text{mm}$. Four IPAnema 3.2 winches [13] in combination with asynchronous servo motors are used to drive and buffer the test rope. The use of cable robot for the fatigue tests ensures that the operation conditions match the targeted application. Compact winches minimize the polar moment of inertia and offer the possibility to generate high dynamic motion. Similar to established bending machines, the drive units are installed in the upper part of the machine frame while the test sheaves and tension weights are situated in the lower part. A compensation for diagonal-pull moves the test sheave synchronously to the mobile run-on and run-off points of the rope on the winch and prevents the rope from lifetime reduction by diagonal pull [11, 18]. In order to identify the influence of high-dynamic stress on the

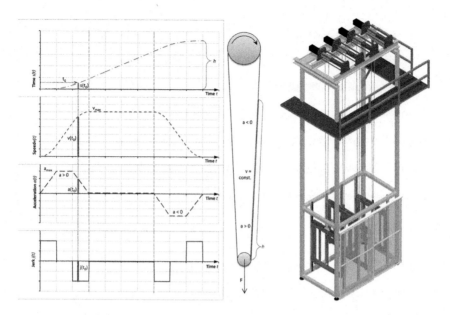

Fig. 4. Movement profile used on the high-dynamic test machine (left) and CAD rendering of the test machine (right)

lifetime of running fiber ropes, conventional low-dynamic bending tests were performed. Overall six different fiber ropes made from three different fiber materials in two different constructions and two diameters were examined.

Additionally, one steel wire rope was tested in order to get a relationship between the different materials. The results of these conventional bending tests are shown in Fig. 5.

For performing high-dynamic bending tests, a constant test frequency, load, length of constant speed, and test sheave were applied. Figure 6 shows the result of high-dynamic bending tests with varied speed and acceleration for a 12-fold braided 2 mm Dyneema fiber rope *D2*. The lifetime N is related to the corresponding lifetime of conventional, low-dynamic bending tests N_{ref}.

It appears that the lifetime drops off with increasing speed. At the same time, the lifetime tends to a global limit at approx. 30% of the reference lifetime of standard BOS-tests. This means that 70% of the lifetime gets lost by raising the dynamic of cycling BOS-tests. Due to the immunity of HM-HT-fiber ropes against tension-tension stress under dynamic load change discovered before, the reason for the decrease of lifetime in high-dynamic BOS-tests has to be found in the bending stress itself. Slippage measurements show that there is slip between the rope and the sheave which can be responsible for outer wear on the fiber rope. The appearance of the broken ropes was investigated microscopically. Here it could be found out that outer wear dominates the appearance of the ropes. Completely different properties can be found for other test ropes; for example the covered rope *M6*, see Fig. 7.

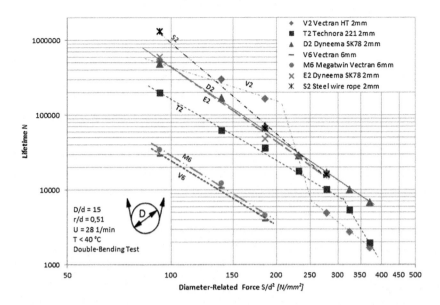

Fig. 5. Lifetime of test ropes in conventional BOS tests

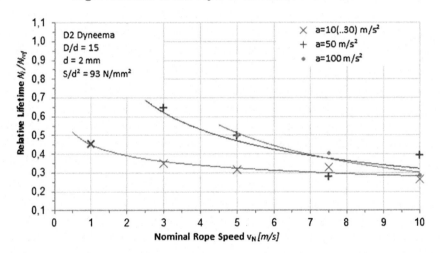

Fig. 6. Relative lifetime of the test rope $D6$ in high-dynamic BOS tests

The covered rope $M6$ shows only a marginally small decrease in lifetime of maximum 10% when raising the dynamic parameters speed and acceleration. Based on the knowledge of the outer wear process in high-dynamic BOS-tests, the cover of the rope $M6$ acts like a shelter for the inner, load-bearing HM-HT-fibers. Consequently, the leading fibers do not get damaged by outer wear, which leads to the significantly better performance in high-dynamic BOS-tests.

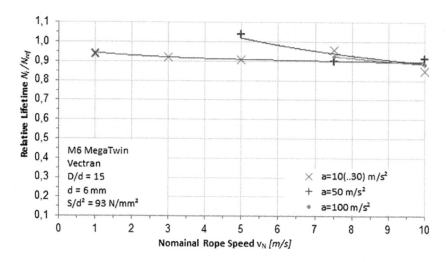

Fig. 7. Relative lifetime of the test rope $M6$ in high-dynamic BOS tests

4 Application in Cable Robot Design

The multiplicity of conducted experiments result in an extensive database, which can be used directly in real applications like cable robots in order to increase the lifetime of the wear part rope by proper usage. In this case, the lifetime-limiting parameters are determined within the design process of the cable robot. Due to the fact that the lifetime of ropes is directly proportional to the number of bending cycles, the number of bending cycles should be as small as possible for each specific rope section. Thereby, the D/d-ratio between sheave-diameter D and rope-diameter d has to be a compromise between a big bending radius $D/2$ and a low polar moment of inertia J. To accommodate this conflicting requirement, lightweight constructions with smooth-running bearings should be preferred. The surface of the sheaves, especially on the bottom of the grooves, should have a good quality in order to avoid damage from sensitive surface of the fiber ropes. To stabilize the flexible fibre rope structure, the radius of the bottom of the groove r should fit as tight as possible to the real rope diameter d_r without squeezing the rope.

During the application, the rope and the rope drive have to be maintained regularly and carefully by a competent person. The inspection intervals can be stated by performing defined bending tests in combination with the reduction factors for high-dynamic applications, factors for considering the statistical spread of BOS tests and chosen safety factors. The work aids to classify the rope condition and discard criteria for fiber ropes have begun and are anticipated for the next years. Additionally, the wear resistance of newer fiber ropes will improve although the arrival of fatigue endurable fiber cannot be expected.

5 Conclusion

In this paper, experimental research are presented on the lifetime of high-dynamical-ly stressed fiber ropes. Major criteria for a limited lifetime of running fiber ropes are presented and possibilities to improve the lifetime are presented by modifying construction parameters. Furthermore, the necessity of periodical inspections of the wear part rope for long-term applications is highlighted.

It is shown that the lifetime of ropes has to be determined experimentally what requires special bending machines. Especially for high-dynamic bending tests, a completely new test machine was designed and built at the IFT laboratory which enables bending tests with dynamic parameters correlating to the requirements of modern cable robots. Bending tests identified for the very first time the material-depending, lifetime reducing influence of high dynamic stress on modern fiber ropes.

Acknowledgements. The authors would like to thank the German Research Foundation (DFG) for financial support of the project under (WE 2187/29-1) and within the Cluster of Excellence in Simulation Technology (EXC 310/1) at the University of Stuttgart.

References

1. Arfa, A., Bellecul, A., Oplatka, G., Teissier, J.-M., Verreet, R.: Space mountain at Euro Disney. A 120 million dollar wire rope test machine (1997)
2. Bruckmann, T., Lalo, W., Nguyen, K., Salah, B.: Development of a storage retrieval machine for high racks using a wire robot. In: ASME 2012 International Design Engineering Technical Conferences & Computers and Information in Engineering Conference, p. 771, Sunday 12 August 2012
3. DIN 2011: DIN EN ISO 2307: Faserseile – Bestimmung einiger physikalischer und mechanischer Eigenschaften, 2307
4. DIN EN 2005: DIN EN 12927-8: Sicherheitsanforderungen für Seilbahnen für den Personenverkehr - Seile - Teil 8: Magnetische Seilprüfung (MRT), 12927-8
5. DIN EN 2011: DIN EN 13411-4: Endverbindungen für Drahtseile aus Stahldraht - Sicherheit - Teil 4: Vergiessen mit Metall und Kunstharz, 13411-4
6. Feyrer, K.: Wire Ropes. Tension, Endurance, Reliability. Springer, Heidelberg (2015)
7. Feyrer, K., Hemminger, R.: New rope bending fatigue machines. Constructed in the traditional way. OIPEEC Bull. **45**, 59–66 (1983)
8. Huntley, E., Grabandt, O., Gaetan, R.: Non-destructive test methods for high-performance synthetic rope. In: OIPEEC Conference 2015 Together with 5th International Stuttgart Ropedays, pp. 191–198 (2015)
9. ISO 2016: DIN EN ISO 7500: Metallische Werkstoffe – Kalibrierung und Überprüfung von statischen einachsigen Prüfmaschinen – Teil 1: Zug- und Druckprüfmaschinen – Kalibrierung und Überprüfung der Kraftmesseinrichtung. Beuth-Verlag, 7500
10. Merlet, J.-P.: MARIONET, a family of modular wire-driven parallel robots. In: Lenarcic, J., Stanisic, M. (eds.) Advances in Robot Kinematics (ARK), pp. 53–61. Springer, Dordrecht (2010)

11. Novak, G., Winter, S., Wehking, K.-H.: Geringe Masse - wenig Energie. Einsatz hochmodularer Faserseile in Regalbediengeräten. Hebezeuge Fördermittel **3**, 32–34 (2016)
12. Papailiou, O.K.: Die Seilbiegung mit einer durch die innere Reibung, die Zugkraft und die Seilkrümmung veränderlichen Biegesteifigkeit. Dissertation, Eidgenössische technische Hochschule Zürich (1995)
13. Pott, A., Mütherich, H., Kraus, W., Schmidt, V., Miermeister, P., Verl, A.: IPAnema: a family of cable-driven parallel robots for industrial applications. In: Bruckmann, T., Pott, A. (eds.) Cable-Driven Parallel Robots. Mechanisms and Machine Science, pp. 119–134. Springer, Heidelberg (2012)
14. Samson Rope Technologies: Samson Technical Bulletin. Inspection & Retirement Pocket Guide, Ferndale, WA 98248, USA (2013)
15. Samson Rope Technologies: K-100 Synthetic Crane Hoist Line. Rope Handling, Installation, Inspection, and Retirement Guidelines, Ferndale, WA 98248, USA (2016)
16. Schiffner, G.: Spannungen in laufenden Drahtseilen. Dissertation, Universität Stuttgart (1986)
17. Schindler Holding AG: Tätigkeitsbericht 2000, 73. Geschäftsjahr (2000)
18. Schönherr, S.: Einfluss der seitlichen Seilablenkung auf die Lebensdauer von Drahtseilen beim Lauf über Seilscheiben. Berichte aus dem Institut für Fördertechnik und Logistik, IFT, Universität Stuttgart (2005)
19. Teufelberger: soLITE. Das hochfeste Faserseil für Krane (2016)
20. VDI 2012: VDI 2358 - Drahtseile für Fördermittel, 2358
21. Vogel, W.: Tragmittel für Treibscheibenaufzüge. Lift-Report **29**(5), 6–16 (2003)
22. Wehking, K.-H.: Lebensdauer und Ablegereife von Aramidfaserseilen in Treibscheibenaufzügen der Schindler AG. Interner Forschungsbericht, Ebikon/Schweiz (2000)
23. Wehking, K.-H.: Laufende Seile. Bemessung und Überwachung. Kontakt & Studium 673. expert-Verlag, Renningen (2014)

Bending Cycles and Cable Properties of Polymer Fiber Cables for Fully Constrained Cable-Driven Parallel Robots

Valentin Schmidt[(✉)] and Andreas Pott

Fraunhofer IPA, Stuttgart, Germany
{valentin.schmidt,Andreas.Pott}@ipa.fraunhofer.de

Abstract. In most practical applications for cable-driven parallel robots, cable lifetime is an important issue. While there is extensive knowledge of steel cables in traditional applications such as elevators or cranes, it cannot be easily applied to cable robots. Especially new polymer based materials behave substantially different, but also the conditions for the cable change dramatically. Cable robots have more bending points and a higher variability in cable force and speed than traditional applications. This paper presents a form of bending cycle analysis which can be applied to assess cable wear. This algorithm counts the number of bends per trajectory in each cable segment. The sum gives an indication how much wear a cable receives. Experiments are conducted on a cable robot using different kinds of polymer fibers. The results show that this method is successful in predicting the point at which a cable finally breaks.

1 Introduction

A very important element of cable-driven parallel robots are the cables themselves. The properties of cables, being able to transmit a force over a long distance while remaining flexible, give rise to many of the advantages and disadvantages of cable-driven parallel robots. As a mechanical component, cables (or ropes) have a long history, with the first being made exclusively from biological materials by civilizations several thousand years ago. Today, steel is the dominant material for making cables for a variety of industries. Polymer fibers are finding more applications in recent times, as some show a superior strength to weight ratio and higher flexibility [4]. It is expected that their market share increases in the upcoming years, as manufacturers of lifts, cranes, and other hoisting devices make use of these advantages. In all applications, estimating the lifetime of the cables is notoriously difficult and poses a challenge for the adoption of polymer cables. From previous works with steel cables, we know that the lifetime related to several key factors such as bending radius, cable force, and whether it is moving continuously or not. These are summarized in the Freyer Formula which bundles the criteria and vast amounts of historical data to estimate cable lifetimes. But even this knowledge is not easily applicable for cable robots as higher speeds and more complex wear profiles make the application unique. Some recent investigations are looking into these factors such as the abrasive wear [7] or cable fretting [11]. The equivalent does not exist for polymer fibers [10]. There are some recent bending cycle experiments [9], but the amount of data is much less than that for steel cables.

Interestingly, some investigations have been made for deep sea applications [1, 2]. Here, polymer fibers are used extensively in very large mooring applications. As such, wear is investigated extensively including factors such as tension and torsion fatigue [6]. The associated risks and high costs of failure have spurred a lot of research in this application [2].

Even in the case of steel cables, cable-driven parallel robots pose very different stresses and strains on the cables than previous applications. Generally, the number of pulleys is greater, loads are more varying, and directional changes more frequent. This means conventional methods of predicting cable lifetimes (such as the Freyer Formula) have yet to be validated through experiments [10].

In previous works, it is attempted to use bending cycle experiments in order to obtain an estimate for the behavior of polymer fiber ropes. These showed that the lifetime is highly dependent on the cable force, somewhat dependent on the bending diameters, and to some extent also on the cable speed [9]. An expansion on this is to investigate the behavior of polymer fiber ropes in the actual cable robot itself. This includes assessing the bending cycles for a trajectory and monitoring cable lifetime.

This paper tries to give a brief overview of possible polymer fiber materials and their properties, presents a method of analyzing bending cycles for cable robot trajectories, and gives some experimental results to show possible ways of dealing with the cable lifetime issue in the application of cable-driven parallel robots. Breaking strength is still a clear indicator of cable lifetime (when ignoring susceptibility of degradation by ultraviolet light). The breaking strength of the cable provides a basis for sizing cables. Currently, high safety factors (>10) are taken for granted as the lifetime is hard to predict. The experimental results show that the bending cycles are significant in determining cable lifetime. Additionally, monitoring the elastic properties of cables continuously may help in predicting cable rupture. The polymer creep curve can perhaps be monitored to identify a cable near breaking due to its elastic properties [8].

2 Overview of Cables and Cable Robot Properties

In order to explain the motivation and notation conventions used, some preliminaries are introduced here.

A cable-driven parallel robot with four or eight cables indexed by i is primarily defined through vectors which indicate cable attachment points on the moving platform $\mathbf{b}_i$ and the stationary frame $\mathbf{a}_i$. The pose of the platform is defined through a position vector $\mathbf{r}$ and a rotation matrix $\mathbf{R}$. This gives rise to the inverse kinematics where the cable length l of cable i, is determined as

$$l_i = \|\mathbf{a}_i - (\mathbf{r} + \mathbf{b}_i \mathbf{R})\|_2. \tag{1}$$

Static analysis using this relation is sufficient to determine the necessary control properties. For pulley kinematics, the point a_i is defined by the cable point entering the last pulley not in the direction of the platform (as this changes with position), but the one in the direction of the winch.

The cable material is not the only property significant to the cables. Major factors of wear include material dependent ones such as UV-radiation and temperature. Further production parameters are also relevant such as the meshing of individual strands, and coating materials. Others factors influenced by the cable robot include the bending cycles and forces applied to the cables. These cause abrasion within the cable as individual strands rub against one another to cause friction. In common bending cycle experiments, the number of successful bending cycles is used as a measure for the lifetime of the cable. Thus, it seems prudent to use a similar approach for cable lifetime within cable-driven parallel robots. To the author's knowledge, there has been no algorithm so far to give the number of bending cycles per rope segment for a given trajectory. Thus, this is presented in the following section. As it is known that cable force is also an important contributing factor, the presented method also provides the ability to take the cable force into account, if it is known.

3 Algorithm for Counting Bending Cycles

As mentioned, bending cycles are an important factor in cable lifetime. However, for cable-driven parallel robots, it is not trivial to assess the bending cycles, as these are not consistent. Other applications such as hoists or cranes have simple one to one relationships between usage cycles and bending cycles.

A single bending cycle is defined by the cable going from a straight position into a bending position or vice versa. Thus, a single pulley actually has two positions where a bending cycle occurs. The first when the rope enters the pulley and is bent around at the pulley radius, and the second when the rope exits the pulley in the other direction. The number of these bends which a cable is subjected to, over a usage period, is a measure of wear. When using a cable robot, not all sections of a cable receive to the same amount of bending cycles. An algorithm (Algorithm 1), can be constructed to count these bending positions over a trajectory.

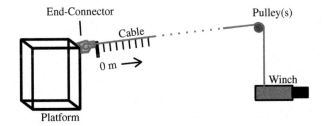

Fig. 1. Cable parameterization

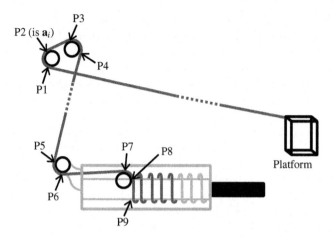

Fig. 2. Bending positions for IPAnema 3 cable-driven parallel robot

First a parameterization needs to be chosen. In this case, the attachment point $\mathbf{b}_i$ of the cable on the platform is chosen as zero, as is shown in Fig. 1. The reason for this is that it is consistent for a given robot geometry. When considering a cable robot under external force, it is at least theoretically conceivable that it moves to an infinite position. This means no definite fixed end can be defined on the winch. In this algorithm, cable wear is abstracted into a value representing the number of bending cycles for finite cable segments. The result is a one-dimensional array $\mathbf{v}_i$ in $\mathbb{N}^n$ for each cable i of size

---

Algorithm 1. Algorithm for counting the bending cycles during a trajectory

1: **Inputs:**
   $\mathbf{v}_i$: to store wear characteristics for cable i
   l_s, l_e: determined by the inverse kinematics
   $\mathbf{b}_{pos}$: bending positions
   l_y: segment size of a single segment
   f_f: optional force factor
2: **Begin**:
3: **for** every b_{pos} in $\mathbf{b}_{pos}$ **do**
4:    **if** $l_s < l_e$ **then**
5:       $i_{start} \leftarrow \frac{-b_{pos}+l_s}{l_y}$
6:       $i_{end} \leftarrow \frac{-b_{pos}+l_e}{l_y}$
7:    **else**
8:       $i_{start} \leftarrow \frac{-b_{pos}+l_e}{l_y}$
9:       $i_{end} \leftarrow \frac{-b_{pos}+l_s}{l_y}$
10:    **for** v in $\mathbf{v}\,[i_{start}, i_{end}]$ **do**
11:       $v++$ **or** $v \leftarrow v + f_f$

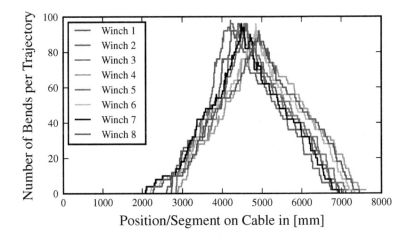

Fig. 3. Bending cycles for an example trajectory on the IPAnema 3

$$n = \frac{l_a}{l_y}, \qquad (2)$$

where l_a is the absolute cable length of the entire cable and l_y is the desired segment size to be investigated.

Now we define the bending positions of the cable robot geometry. These are presumed to be static with respect to the $\mathbf{a}_i$ points and the winches which are not expected to move with respect to the robot frame. When the pose of the robot $(\mathbf{r}, \mathbf{R})$ is known, the bending positions can easily be transferred into the cable parameterization. Bending positions are defined by the point $\mathbf{a}_i$, (in the case of pulley kinematics in itself a bending position) in a one-dimensional parameter which refers to cable length. The bending positions are stored in another one-dimensional array $\mathbf{b}_{pos}$ in $\mathbb{R}^m$ where m is the number of bending positions defined. Figure 2 shows these bending positions for the IPAnema configuration.

In order to determine a correct conversion between the cable parameterization and the bending positions a reference needs to be defined. From this position the length to the first bending position close to $\mathbf{a}_i$ can be determined. For this, a zero or reference position is chosen which is part of many controllers anyway.

For a given trajectory, we can use the inverse kinematics to calculate cable lengths at different points in the trajectory. It is assumed that the points on the trajectory are sufficiently close to one another that the starting length l_s changes linearly to ending length l_e. For each cable, the algorithm then determines which segment has moved across the bending position during this cycle and increments this accordingly. This is shown in Algorithm 1.

Essentially, each of the segments in $\mathbf{v}_i$ which have moved over any of the bending positions in $\mathbf{b}_{pos}$ are incremented regardless of which direction the cable has moved in. It is possible to incorporate other factors such as force, by using a factor f_f proportional to a measured or estimated force. This would turn $\mathbf{v}_i$

into a vector in $\mathbb{R}^n$, but this is an almost trivial change. This was not done in the initial analysis.

The result of this algorithm is shown in Fig. 3. Here we can see the cable as it is parameterized with zero length at the platform and the associated bending with each segment. Segments with the most wear are those were the number of bending cycles is high.

4 Experimental Setup

In order to test the influence of bending cycles, some experiments were performed on a cable robot using polymer fiber cables. The robot in question is the IPAnema 3 cable robot which is part of the IPAnema family of robots developed by the Fraunhofer IPA [5].

The investigations are part of a larger test where materials and other properties are also investigated. The aim was to continuously actuate the cable robot until cable rupture. This gives an insight into the bending cycle analysis. The robot had dimensions of $16.25 \times 11.30 \times 3.79$ m and a cable force range of 10 N to 3 kN. In this experiment, several cables with diameter 2.5 mm were used. The cable materials used in this experiment are outlined in Table 1.

Table 1. Cable materials used in the experimental evaluation

Material	Shorthand (Brand)	Weight [g/100m]	Breaking Strength [N]
Ultra High Molecular Weight Polyethylene	Dyneema SK78 ®	336	4070
Aramid (Tejin)	Twaron ®	355	3090
Aramid (DuPont)	Kevlar ®	309	3150
Aramid (Tejin)	Technora ®	304	3400
Thermotropic Liquid Crystalline Polymer	Vectran ®	302	3220
Polybenzobisoxazole	Zylon ®	318	5450

It can be seen that the cable breaking strength is very close to the maximum of the robot capabilities. Usually a much higher factor (often around 10) is used to ensure that the cables do not rupture. In this case, the cable force was chosen to be particularly high in order to ensure that the cables rupture quickly to shorten the overall experiment time. Even with this specification, the trials were conducted over a period of several weeks.

In an endless loop, the cable robot drove the specified trajectory shown in Fig. 4. The trajectory was chosen to have the similar wear in all cables. Unfortunately, the cable force is highly dependent on the pose of the platform and cannot be maintained in the same manner. However, force levels were ensured

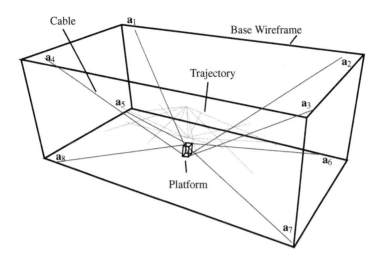

Fig. 4. Test trajectory (red) for polymer cables (blue) in IPAnema 3 geometry (black wireframe) (Color figure online)

to be consistent using a cable force control algorithm [3]. Once a cable breaks, the breaking point is recorded and compared with the theoretical rupture zone.

The experiment was divided into three tests, the first and third at a trajectory speed of 10.000 mm/min while the second was conducted at a trajectory speed of 20.000 mm/min. The distribution of cables is shown in Table 2. The cable numbers indicate in which experiment and which winch position each cable has on the frame as it is shown in Fig. 4. As the force was considerably different between four cables attached at the top frame (1 to 4), and the four cables arranged at the bottom of the frame (5 to 8), these are also given in the table. While the cable forces vary around ten percent for each cable, the force along the entire trajectory is very consistent at each repetition and have an almost identical force profile.

Table 2. Test distribution per winch for three tests

Test One			Test Two			Test Three		
C-Nr.	Material	Force	C-Nr.	Material	Force	C-Nr.	Material	Force
1.1	Dyneema ®	approx.	2.1	Kevlar ®	approx.	3.1	Dyneema ®	approx.
1.2	Technora ®	270 N	2.2	Twaron ®	270 N	3.2	Technora ®	400 N
1.3	Dyneema ®		2.3	Technora ®		3.3	Dyneema ®	
1.4	Technora ®		2.4	Zylon ®		3.4	Technora ®	
1.5	Vectran ®	approx.	2.5	Kevlar ®	approx.	3.5	Vectran ®	approx.
1.6	Zylon ®	500 N	2.6	Twaron ®	500 N	3.6	Zylon ®	600 N
1.7	Vectran ®		2.7	Technora ®		3.7	Vectran ®	
1.8	Zylon ®		2.8	Zylon ®		3.8	Zylon ®	

Using force sensors in the winch, a cable rupture can be detected by the sudden loss in cable force. The robot then can be paused and the ruptured cable replaced.

5 Experimental Results

The results of the experiment discussed in the previous section show that the bending cycle analysis closely predicts the rupture point of the cable. Table 3 shows the sixteen ruptures experienced across the three tests. The cables lasted around 500 to almost 2000 trajectory cycles, which equates to about 50,000 and 200,000 bending cycles.

Many of the ruptures occurred very close to the theoretically predicted rupture zone. The theoretical rupture zone is sometimes a range and sometimes a specific point because the bending cycle analysis indicated a defined peak of maximum wear, and sometime a wider range of segments with maximum but equally exposed to wear. This is independent of the cable material speed or force applied. There are some four outliers were the rupture position is far away (>10%) from the theoretical rupture zone. Generally, this means that the cable bending is a predominant factor even for polymer fibers.

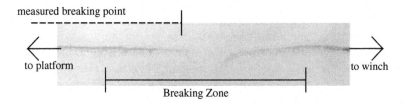

Fig. 5. Determining the breaking position on a cable

There is an inherent difficulty when measuring the rupture position. Figure 5 shows a typical cable broken under continuous bending cycles. The breaking zone is defined as the zone where the rupture cable is loosing width on either side, as several individual strands are broken. The measurement of the rupture position was done as indicated with distance from the winch. Since polymer fibre cables have more elasticity, the breaking zone is probably more pronounced than for a steel cable. The breaking zone due to bending fatigue is expected to be smaller for steel cables.

For two of the four aforementioned outliers, the cable breaking zone was very large as several very thin strands extended, both on the cable portion leading to the winch and the portion leading to the platform.

Two other conclusions can be drawn from the experiences of driving the cable robot up until cable failure. Firstly, as the cable rupture occurs on the most bent segments of the rope, it will usually occur within or near the winch. This is true for the IPAnema winch, which has five bending positions (as seen in Fig. 2). For

Table 3. Cable rupture position

Rupture Nr.	Cable Nr.	Rupture Pos. [m]	Theoretical Rupture Zone [m]	Deviation [%]	Bending Cycles
1	1.5	4.40	4.67–4.68	5.8	130
2	1.4	6.87	7.85–7.90	12.5	573
3	1.4	6.12	7.85–7.90	22.0	1009
4	2.6	3.88	4.36	11.0	2098
5	2.5	4.20	4.67–4.68	10.1	2149
6	2.2	7.27	8.03	9.5	2183
7	3.6	9.54	9.58	0.42	138
8	3.6	9.86	9.58	2.92	420
9	3.7	10.30	10.45–10.47	1.53	181
10	3.7	9.64	10.45–10.47	7.84	443
11	3.3	14.16	14.99	5.54	573
12	3.6	9.42	9.58	1.67	163
13	3.8	10.49	10.68–10.70	1.87	565
14	3.7	10.28	10.45–10.47	1.63	97
15	3.6	9.88	9.58	3.13	1185
16	3.7	10.48	10.45–10.47	0.19	1178

the operation of cable-driven parallel robots, this is positive because a cable does rarely shoot out and becomes a potential harm as it whips across the workspace. Most of the time, the cable remained stuck somewhere in the chain of pulleys. This means that it is harder to replace once ruptured. Secondly, the redundancy in cables in the eight cable robot ensured that the platform remained fairly close to the pose where the cable ruptured along the trajectory. The platform will move to a configuration within the seven actuators robot that remain when one cable is ruptured. At no point during the experimentation did the platform fall to the ground or deviate by more than a couple of centimeters from the trajectory. This also have positive implications on the safety of operation. Of course, it should be noted that the trajectory remained at the center of the workspace.

6 Conclusion

The lifetime of polymer fiber cables in the use of cable robots is still a very open issue. It was shown that previous knowledge on the significant factors gives an indication of how long a typical cable may last. However, many more investigations need to be done to complete this knowledge. Part of the difficulty is the amount of variation in polymer fiber cables. There are not only very many materials, but the material properties show large variations, even for a single material.

A bending cycle analysis was introduced in this paper. This is easy to implement and can be applied to any cable robot in the current form. It can even be applied to a controller and additionally monitoring cable forces. This should give a very accurate indication of which segment of the rope has endured the most wear. It is expected that this analysis can be applied for steel cables, but a smaller bending zone is expected due to less elasticity.

Experimental results show that the rupture zone in an example cable robot coincides closely with that predicted by the bending cycle analysis. The implication is that bending during use is one of the major contributors to cable wear during the operation in cable robots.

Acknowledgements. The research leading to these results has received funding from Federal Ministry of Economics and Technology (BMWi) under grant agreement no. KA3053503CJ3 and from the European Unions H2020 Programme (H2020/2014-2020) under grant agreement no. 732513.

References

1. Flory, J.F., Banfield, S.J.: Fiber rope myths. In: OCEANS 2011 MTS/IEEE KONA, pp. 1–8 (2011)
2. Gordon, R.B., Brown, M.G., Allen, E.M.: Mooring integrity management: A state-of-the-art review. In: Offshore Technology Conference, 05 May 2014. doi:10.4043/25134-MS
3. Kraus, W., Schmidt, V., Rajendra, P., Pott, A.: System identification and cable force control for a cable-driven parallel robot with industrial servo drives. In: IEEE International Conference on Robotics and Automation (ICRA), pp. 5921–5926 (2014). doi:10.1109/ICRA.2014.6907731
4. McKenna, H.A., Hearle, J.W.S., O'Hear, N.: Handbook of Fibre Rope Technology. CRC Press/Woodhead, Boca Raton/Cambridge (2004)
5. Pott, A., Mütherich, H., Kraus, W., Schmidt, V., Miermeister, P., Verl, A.: IPAnema: A family of cable-driven parallel robots for industrial applications. In: Bruckmann, T., Pott, A. (eds.) Cable-Driven Parallel Robots. Mechanisms and Machine Science, pp. 119–134. Springer, Heidelberg (2012)
6. Ridge, I.: Tensiontorsion fatigue behaviour of wire ropes in offshore moorings. Ocean Eng. **36**(910), 650–660 (2009). doi:10.1016/j.oceaneng.2009.03.006
7. de Silva, A., Fong, L.W.: Effect of abrasive wear on the tensile strength of steel wire rope. Eng. Fail. Anal. **9**(3), 349–358 (2002). doi:10.1016/S1350-6307(01)00012-7
8. Vlasblom, M.P., Bosman, R.L.M.: Predicting the creep lifetime of HMPE mooring rope applications. In: OCEANS 2006, pp. 1–10 (2006). doi:10.1109/OCEANS.2006.307013
9. Vogel, W., Wehking, K.H.: Neuartige Maschinenelemente in der Fördertechnik und Logistik: Hochfeste, laufende Faserseile. Logistics J. nicht-referierte Veröffentlichungen (2004). doi:10.2195/LJ_Not_Ref_d_Vogel_1020042
10. Weis, J.C., Ernst, B., Wehking, K.H.: Use of high strength fibre ropes in multi-rope kinematic robot systems. In: Bruckmann, T., Pott, A. (eds.) Cable-Driven Parallel Robots. Mechanisms and Machine Science, pp. 185–199. Springer, Heidelberg (2012). doi:10.1007/978-3-642-31988-4_12
11. Zhang, D., Ge, S., Qiang, Y.: Research on the fatigue and fracture behavior due to the fretting wear of steel wire in hoisting rope. Wear **255**(712), 1233–1237 (2003). doi:10.1016/S0043-1648(03)00161-3. 14th International Conference on Wear of Materials

Displacement and Workspace Analysis

A New Approach to the Direct Geometrico-Static Problem of Cable Suspended Robots Using Kinematic Mapping

Manfred Husty[1(✉)], Josef Schadlbauer[1], and Paul Zsombor-Murray[2]

[1] Unit Geometry and CAD, University Innsbruck, Innsbruck, Austria
{Manfred.Husty,Josef.Schadlbauer}@uibk.ac.at
[2] CIM, McGill University, Montreal, Canada
paul@cim.mcgill.ca

Abstract. The direct kinematic problem of $n-n$ ($n = 2, 3, 4, 5$) underconstrained cable manipulators has been solved previously by exploiting the line geometric equilibrium condition and using optimization techniques, heavy algebraic or numeric algebraic computation. In this paper another solution method is proposed. It uses kinematic mapping, distance constraint equations and a local plane constraint. This method can be used for all cases of underconstrained cable manipulators and it is also applicable to the case of $n-i$ equilibria of $n-$cable manipulators. Univariate polynomials are computed in examples for the $3-3$ and $5-5$ cases as well as for $n-1$ equilibria of the $5-5$ case.

1 Introduction

The direct kinematic problem of underconstrained $n-$cable robots has been addressed and solved in several papers. In all cases ($n = 2, 3, 4, 5$) the maximal number of solutions was obtained. The data according to the number of cables are obtained in these publications (see [1,5,6] and the references herein) are: (Table 1).

Adressing the problems of the above mentioned approaches that try to obtain the solutions via a univariate polynomial of the system of equations the paper [2] solves the $n-n$ underconstrained cable problem using interval analysis.

The kinematics of underconstrained cable robots is determined by two sets of equations. The first set consists of the cable length equations and in the second set are the force-moment equilibrium equations. In the case of the direct geometrico-static problem (DGP) the cable lengths are given. The equilibrium conditions are packed in the Jacobian matrix $\mathbf{J}$ whose columns consist of the Plücker coordinates of the lines that coincide with the cables and the line of action of the gravity force. $\mathbf{J}$ is therefore an $6 \times (n+1)$ matrix. Both systems contain the pose variables and it is easy to see that equilibrium is only possible when the $rank(\mathbf{J}) \leq n$. The set of equilibrium conditions is established in [6] as

Table 1. .

Number of cables	2	3	4	5
Number of solutions over $\mathbb{C}$	24	156	216	140

$$[\mathscr{L}_1, \ldots \mathscr{L}_n, \mathscr{L}_e,] \begin{bmatrix} \frac{\tau_1}{\rho_1} \\ \vdots \\ \frac{\tau_1}{\rho_1} \\ \ddot{u}Q \end{bmatrix} = 0, \tag{1}$$

where $\mathscr{L}_i$ denote the Plücker coordinates of the leg lines and $\mathscr{L}_e$ the line of action of the gravity force. τ_i represent the cable tension forces, Q, the gravity and ρ_i the cable lengths. Now, equilibrium is computed via the observation that "at equilibrium, the variation of the total potential energy of the platform due to a virtual displacement must be zero" [6]. In [4] for the $3-3$ case the minors of **J** are exploited and the solutions of the corresponding set of nonlinear equations in the sought displacement parameters are found with Groebner base algorithms, elimination techniques and numerical continuation.

In [8] a different approach was taken to obtain the solution of the DGP for the planar 2-2 case. Essentially, the workspace of the center of gravity is computed via the coupler curve of the corresponding coupler motion and the local minima of the curve are determined, which of course correspond to the minima of the constrained optimization problem cited above. Some interesting cases arise when the coupler curve has cusps.

In this paper a different algorithm is proposed to find the solutions of the DGP of all $n-n$ underconstrained cable robots. It exploits the cable length constraint varieties in the kinematic image space. But the corresponding polynomial equations are not enough to determine all pose parameters of the end effector in equilibrium. An additional observation is necessary: a local plane constraint is introduced and the linear dependency of the tangent spaces to this constraint and the distance constraints is used to obtain a set of equations that allow to solve the DGP of all types efficiently. The proposed algorithm also works to determine the poses where an $n-n$ underconstrained cable robot is in equilibrium with less than n cables.

The paper is organized as follows: In Sect. 2 a brief introduction to the kinematic image space is given. Section 3 contains the manipulator description, Sect. 4 derives the constraint equations and in Sect. 5 the solution algorithm is derived. In Sect. 6 the algorithm is exemplified for a $3-3$ cable robot. The paper finishes with a conclusion.

2 Kinematic Mapping

Euclidean rigid body displacements are often described by homogeneous 4×4 matrices **M**, that act on a point **x** located in a moving frame according to

$\mathbf{x}' = \mathbf{M}\mathbf{x}$. $\mathbf{x}'$ is the image point in a base frame, the lower right 3×3 sub matrix of $\mathbf{M}$ is a proper orthogonal matrix encoding the orientation of the moving fame with respect to the base frame and the first column of $\mathbf{M}$ contains the vector connecting the origins of both frames therefore representing the translational part of the transformation. Using Study's kinematic mapping κ (see [7,13]), the displacement given by $\mathbf{M}$ is mapped to a point $\mathbf{d} = [x_0, x_1, x_2, x_3, y_0, y_1, y_2, y_3]^T$ in a seven dimensional projective space P^7. These point coordinates are called the *Study parameters* of the displacement. They fulfill the quadratic condition

$$x_0 y_0 + x_1 y_1 + x_2 y_2 + x_3 y_3 = 0, \qquad (2)$$

which is called *Study condition*. Its zero set is the *Study quadric* $S_6^2 \subset P^7$.

In the inverse kinematic mapping a point on S_6^2 minus the *exceptional three space* $\mathscr{E}: x_0 = x_1 = x_2 = x_3 = 0$ yields the matrix

$$\mathbf{M} := \kappa^{-1}(\mathbf{d}) = \frac{1}{\Delta} \begin{bmatrix} 1 & 0 & 0 & 0 \\ t_1 & x_0^2 + x_1^2 - x_3^2 - x_2^2 & -2x_0 x_3 + 2x_2 x_1 & 2x_3 x_1 + 2x_0 x_2 \\ t_2 & 2x_2 x_1 + 2x_0 x_3 & x_0^2 + x_2^2 - x_1^2 - x_3^2 & -2x_0 x_1 + 2x_3 x_2 \\ t_3 & -2x_0 x_2 + 2x_3 x_1 & 2x_3 x_2 + 2x_0 x_1 & x_0^2 + x_3^2 - x_2^2 - x_1^2 \end{bmatrix} \qquad (3)$$

where $\Delta = x_0^2 + x_1^2 + x_2^2 + x_3^2$ and

$$\begin{aligned} t_1 &= 2x_0 y_1 - 2y_0 x_1 - 2y_2 x_3 + 2y_3 x_2, \\ t_2 &= 2x_0 y_2 - 2y_0 x_2 - 2y_3 x_1 + 2y_1 x_3, \\ t_3 &= 2x_0 y_3 - 2y_0 x_3 - 2y_1 x_2 + 2y_2 x_1. \end{aligned} \qquad (4)$$

However, as shown in [10], the Study condition (2) is not necessary for $\mathbf{M}$ to describe a rigid body displacement in image space. Thus, the range of κ^{-1} may be extended to $P^7 \setminus E$. Injectivity is lost, but the non-linear Study condition can be neglected. This fact is interesting, opens up a lot of possibilities in motion interpolation, but will not be exploited in the following.

Kinematic properties of serial chains are described with respect to (arbitrarily chosen) coordinate frames in its base and the end-effector. Possible locations of the end-effector with respect to the base correspond to algebraic varieties described by sets of polynomial equations in $\mathbb{P}^7$. Coordinate transformations in the base and the end-effector frame induce linear mappings T in $\mathbb{P}^7$ that preserve several interesting geometric objects:

1. the *Study quadric* S_6^2,
2. the *Null cone* defined by $\mathscr{N} : x_0^2 + x_1^2 + x_2^2 + x_3^2 = 0$, which is quadric in $\mathbb{P}^7$, that has only complex points with exception of its 3-dimensional vertex space $\mathscr{E} : x_0 = x_1 = x_2 = x_3 = 0$. $\mathscr{E}$ is entirely contained in S_6^2 and is called *exceptional generator space*,
3. the *exceptional quadric* $\mathscr{Y} : y_0^2 + y_1^2 + y_2^2 + y_3^2 = 0 \subset \mathscr{E}$,
4. all quadrics $\mathscr{Q} = \lambda S_6^2 + \mu \mathscr{N}$, $\lambda, \mu \in \mathbb{R}$ in the pencil spanned by the Study quadric and the Null cone.

A detailed derivation and proofs for these statements and some interesting examples can be found in [12]. The invariant objects essentially govern the kinematics of 3D-Euclidean displacements[1].

3 Manipulator Description

Figure 1 shows the model of an underconstrained $3-3$ cable robot. With respect to this figure the attachment points of the cables in the base frame are denoted with A_i, those in the moving frame are B_i. W.l.o.g. one can assume that the center of gravity in the endeffector is located in the origin of the moving frame. p, q, r are the connecting cables. In the DGP the cable lengths are assigned and the equilibrium pose of the end effector system is sought.

The coordinates of the attachment points are denoted by $A_i(a_{1i}, a_{2i}, a_{3i})$ resp. $B_i(b_{1i}, b_{2i}, b_{3i})$. In the following we assume that there is no sagging in the cables and furthermore that all $a_{ji}, b_{ji} \in \mathbb{Q}$.

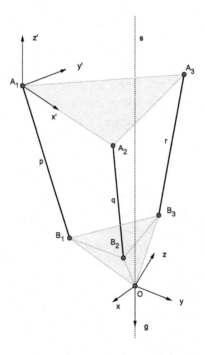

Fig. 1. 3-3 cable manipulator

---

[1] The kinematic images of planar and spherical displacements subordinate completely to this description because both cases are obtained by three dimensional sub-spaces of $\mathbb{P}^7$. The corresponding geometry of their image spaces and the algorithms to derive these geometries can be found in [3] p.393ff. resp. [11] p.60ff.

4 Constraint Equations

In most of the previous papers the rank deficiency of the matrix **J** in Eq. 1 is exploited whereas in this paper the constraint varieties corresponding to the assigned distances will be used. The distance constraint varieties were derived in [9] to solve the direct kinematics of Stewart-Gough platforms. The constraint equation for the first cable p may be written in image space coordinates

$$(w - 2(a_{11} b_{11} + a_{21} b_{21} + a_{31} b_{31})) x_0^2 + (w + 2(a_{21} b_{21} + a_{31} b_{31} - a_{11} b_{11}) x_1^2$$
$$+ (2(a_{11} b_{11} - a_{21} b_{21} + a_{31} b_{31}) + w) x_2^2 + (2(a_{11} b_{11} + a_{21} b_{21} - a_{31} b_{31}) + w) x_3^2$$
$$+ 4\,[(a_{21} b_{31} - a_{31} b_{21}) x_0 x_1 + (-a_{11} b_{31} + a_{31} b_{11}) x_0 x_2 + (a_{11} b_{21} - a_{21} b_{11}) x_0 x_3$$
$$+ (-a_{11} b_{21} - a_{21} b_{11}) x_1 x_2 + (-a_{11} b_{31} - a_{31} b_{11}) x_1 x_3 + (-a_{21} b_{31} - a_{31} b_{21}) x_2 x_3$$
$$+ (a_{11} - b_{11}) x_0 y_1 + (a_{21} - b_{21}) x_0 y_2 + (a_{31} - b_{31}) x_0 y_3 + (-a_{11} + b_{11}) x_1 y_0$$
$$+ (a_{31} + b_{31}) x_1 y_2 + (-a_{21} - b_{21}) x_1 y_3 + (-a_{21} + b_{21}) x_2 y_0 + (-a_{31} - b_{31}) x_2 y_1$$
$$+ (a_{11} + b_{11}) x_2 y_3 + (-a_{31} + b_{31}) x_3 y_0 + (a_{21} + b_{21}) x_3 y_1 + (-a_{11} - b_{11}) x_3 y_2$$
$$+ (y_0^2 + y_1^2 + y_2^2 + y_3^2)] = 0, \tag{5}$$

where $w := a_{11}^2 + a_{21}^2 + a_{31}^2 + b_{11}^2 + b_{21}^2 + b_{31}^2 - r_1^2$. This is the most general form of the distance constraint equation, which simplifies considerably when the coordinate systems are adapted to the robot. In an $n-n$ cable robot $n \leq 5$ at most 5 distance constraint equations will exist ($h_i, i = 1,\ldots 5$). Furthermore the Study condition Eq. 2 and a normalization condition (e.g. $h_n := x_0^2 + x_1^2 + x_2^2 + x_3^2 = 1$ or $x_0 = 1$) are available. These, at most $n+2$ conditions, are obviously not enough to determine the eight Study parameters. $6-n$ additional conditions are necessary. These conditions must replace the equilibrium rank conditions of the Jacobian matrix **J**. They are established by the observation that the center of gravity of the end effector must locally fulfill a plane condition and the plane has to be normal to the direction of the gravity force, therefore it can be written $z = R_z$. The unknown constant R_z determines the position of the constraint plane. As the center of gravity is the origin of the moving frame z has to be replaced with the last entry of the first column of the matrix **M** (Eq. 3) to obtain its equation in the fixed frame

$$h_p := 2(-x_0 y_3 - x_1 y_2 + x_2 y_1 + x_3 y_0) - R_z(x_0^2 + x_1^2 + x_2^2 + x_3^2) = 0. \tag{6}$$

The constraint equations (h_i, h_p), the normalization condition (h_n) and the Study condition Eq. 2 (h_s) are assembled to the system of polynomial equations

$$sys := [h_i, h_p, h_s, h_n], \qquad i = 3, 4, 5 \tag{7}$$

In practical computations it has turned out that $h_n : x_0 = 1$ has advantages concerning computation time, therefore it is used in the following. But one has to be aware that this assumption needs to test if $x_0 = 0$ allows a solution. This can be done by imposing this condition and the run the solution algorithm. Because of the condition the system becomes much simpler and a possible solution can be found fast.

Before going into the solution algorithm a further observation will simplify the system of equations. The last line in Eq. 5 shows that the coefficients of all $y_i^2, i = 0, \ldots 3$ are free of design parameters a_{kl}, b_{st}. Therefore differences of these equations are void of these squares. One can use $n-1$ such difference equations.

5 Solution Algorithm

The additionally needed equations are found by the observation that the Jacobian $\mathbf{J}_{sys}$ of the system sys has to be rank deficient. $\mathbf{J}_{sys}$ is obtained by differentiating with respect to the Study parameters x_i, y_i. Because all constraint equations are quadratic in the Study parameters the entries of $\mathbf{J}_{sys}$ are linear in these parameters. The following three cases exist

1. $n = 5 \rightarrow \mathbf{J}_{sys}$ is 8×8 and $\det \mathbf{J}_{sys}$ yields one more equation.
2. $n = 4 \rightarrow \mathbf{J}_{sys}$ is 8×7 and 7×7 minors yield additional equations.
3. $n = 3 \rightarrow \mathbf{J}_{sys}$ is 8×6 and 6×6 minors yield additional equations.

Case 1
$\det \mathbf{J}_{sys}$ is added to the system sys. This system consists now of nine equations and nine unknowns (x_i, y_i, R_z). This system is now at first inter-reduced and then immediately a Grbner basis is computed using total degree ordering. From this basis a basis with lexicographic term order providing a univariate polynomial is computed using the FGLM algorithm. As expected the univariate is of degree 140.

In this case elimination also leads to a solution: One can solve from three difference equations, h_p, the Study condition and h_n the unknowns $x_0, y_0, y_1, y_2, y_3, R_z$ linearly and only three unknowns (x_1, x_2, x_3) remain. Substitution of the solutions for x_0, y_0, y_1, y_2, y_3 into the three remaining equations (two from sys and $\det \mathbf{J}_{sys}$) yields after two elimination steps the univariate of degree 140.

Case 3
In this case sys consists only of six equations. Seven out of 28 determinants of the 6×6 minors of $\mathbf{J}_{sys}$ vanish, so one can extract 21 equations. In a next step these 21 equations are added to the the system sys. One could argue that that only three out of the 21 equations should result in a zero dimensional ideal. But it was already mentioned in [6] that an abundance of equations simplifies the computation of the Gröbner basis. Actually, there is another reason to keep all available information: in kinematics the solution set is in many cases smaller than the variety obtained by a minimal set of generators of the variety.

Inter-reduction of the 27 equations yields a relatively simple system of 26 equations out of which six are either linear or bilinear in the unknowns. A total degree Gröbner basis of this system is computed in less than 1 min. The basis consists of 195 polynomials. 10 polynomials of this basis are either linear or bilinear in the unknowns the remaining ones are at most of degree 4. The transformation of this basis into a lexicographic order needs much more time but Maple comes up with a polynomial of degree 156, as expected. The ordering of the unknowns plays an important role for the computation time of the basis. In the computed examples the ordering $plex(y_0, y_1, y_2, y_3, x_1, x_2, x_3, R_z)$ was used. Because of this ordering the univariate polynomial is in R_z. Referring to Fig. 1 a necessary (but not sufficient!) condition for a feasible equilibrium is a negative value of R_z. For these values of R_z stable equilibrium must be tested according to the methods provided e.g. in [6].

The remaining case 2 was not computed explicitly at the time of writing this paper, but it should be straight forward.

6 Example

In this section an example for the 3−3 cable case is shown. The design parameters are:

$a_{11} = 0, a_{21} = 0, a_{31} = 0, a_{12} = 7, a_{22} = 0, a_{32} = 0, a_{13} = 5, a_{23} = 7, a_{33} = 0,$
$b_{11} = -6, b_{21} = 0, b_{31} = 0, b_{12} = -3, b_{22} = 5, b_{32} = 0, b_{13} = -3, b_{23} = 4, b_{33} = 5,$
$r_1 = 7, r_2 = 11, r_3 = 13$

After running the algorithm derived in the last section one obtains the following solutions. (Only the negative R_z solutions are listed).

$R_{z_1} = -12.99999341, \quad R_{z_2} = -11.84761041, \quad R_{z_3} = -9.93666435,$
$R_{z_4} = -9.21554122, \quad R_{z_5} = -3.39366902, \quad R_{z_6} = -2.53790343,$
$R_{z_7} = -1.46030247, \quad R_{z_8} = -0.99393093, \quad R_{z_9} = -0.24655762,$

In Fig. 2 the second solution is displayed. It was shown that the four lines containing the four blue line segments, representing the three cables and the line of action of the gravity force are contained in a hyperboloid.

7 4-Cable Equilibria of 5-Cable Robots

In Sect. 3 the 5-cable equilibria of a $5-5$ cable robot were computed. The question arises if those poses of the end effector can be computed where only 4 cables in tension provide an equilibrium pose, therefore one cable being slack. Computationally this means that the determinants of all 7×7 minors of $\mathbf{J}_{sys}$ have to vanish. This yields 64 polynomials and one would think that there is no chance to compute the corresponding variety. Nevertheless, in several computed examples it turned out that the system consisting of these 64 polynomials and the starting system sys has two degrees of freedom, meaning that two of the unknowns can be chosen arbitrarily and then the variety belonging to the remaining system becomes zero dimensional. After assigning values to two unknowns the system of 64 polynomials becomes relatively simple with only 54 polynomials remaining.

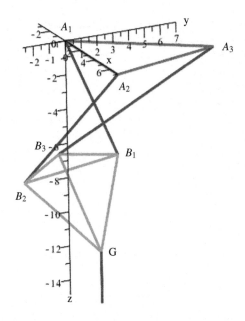

Fig. 2. Equilibrium pose of 3-3 cable system

Adding these 54 polynomials to *sys* a total degree Gröbner basis can be computed. This basis contains 58 polynomials. The FGLM algorithm converts this basis into a lexicographic basis having a univariate polynomial of degree 80 with only even powers. Back substitution and solving for the remaining unknowns showed that in the corresponding pose of the end-effector indeed only four out of the five cables provided equilibrium.

These properties could only be shown in examples, which is of course no mathematical proof. Therefore we are only able to formulate

Conjecture 1. In a $5-5$ underconstrained cable system exists a two parameter set of poses in which only four cables provide equilibrium.

8 Conclusion

In this paper a new approach for solving the DGP of underconstrained cable robots was presented. Polynomial constraint equations, comprising distance equations and a local plane equation provided enough information to compute the DGP of this type of cable manipulator. In examples to was shown that this description also allows to compute the $n-i$ equilibria of $n-n$ underconstrained cable manipulators. It has to be admitted that several statements have been made by observation of examples, which is of course no mathematical proof. Nevertheless solution algorithms have been developed. Future research will have to proof the general validity of the presented algorithms.

Acknowledgements. The authors acknowledge the support of the FWF project I 1750-N26 "Kinematic Analysis of Lower-Mobility Parallel Manipulators Using Efficient Algebraic Tools".

References

1. Abbasnejad, G., Carricato, M.: Direct geometrico-static problem of underconstrained cable-driven parallel robots with n cables. IEEE Trans. Robot. **31**(2), 468–478 (2015)
2. Berti, A., Merlet, J.P., Carricato, M.: Solving the direct geometrico-static problem of underconstrained cable-driven parallel robots by interval analysis. Int. J. Robot. Res. **35**(6), 723–739 (2016)
3. Bottema, O., Roth, B.: Theoretical Kinematics, 1st edn. North-Holland Publishing Company, Amsterdam (1979)
4. Carricato, M.: Direct geometrico-static problem of underconstrained cable-driven parallel robots with three cables. J. Mech. Robot. **5**(3), 1–10 (2013)
5. Carricato, M., Abbasnejad, G., Walter, D.: Inverse geometrico-static analysis of under-constrained cable-driven parallel robots with four cables. In: Lenarčič, J., Husty, M. (eds.) Latest Advances in Robot Kinematics, pp. 365–372. Springer, Dordrecht (2012)
6. Carricato, M., Merlet, J.P.: Stability analysis of underconstrained cable-driven parallel robots. IEEE Trans. Robot. **29**(1), 288–296 (2013). https://hal.archives-ouvertes.fr/hal-01425222
7. Husty, M., Pfurner, M., Schröcker, H.P.: Algebraic methods in mechanism analysis and synthesis. Robotica **25**(6), 661–675 (2007)
8. Husty, M., Zsombor-Murray, P.: Geometric Contributions to the Analysis of 2-2 Wire Driven Cranes, pp. 11–19. Springer, Cham (2015)
9. Husty, M.L.: An algorithm for solving the direct kinematics of general stewart-gough platforms. Mech. Mach. Theory **31**(4), 365–379 (1996)
10. Pfurner, M., Schröcker, H.-P., Husty, M.: Path planning in kinematic image space without the study condition. In: Jadran Lenarčič, J.P.M. (ed.) Proceedings of the 15th International Conference on Advances in Robot Kinematics, France (2016). https://hal.archives-ouvertes.fr/hal-01339423
11. Müller, H.R.: Sphärische Kinematik. VEB Deutscher Verlag der Wiss, Berlin (1962)
12. Rad, T.D., Scharler, D., Schröcker, H.P.: The kinematic image of RR, PR and RP Dyads. In: arXive: 1607.08119v1 [csRO] 27, July 2016
13. Study, E.: Geometrie der Dynamen. B.G. Teubner, Leipzig (1903)

Determination of the Cable Span and Cable Deflection of Cable-Driven Parallel Robots

Andreas Pott$^{(\boxtimes)}$

Institute for Control Engineering and Manufacturing Units, Fraunhofer IPA,
University of Stuttgart, Stuttgart, Germany
andreas.pott@isw.uni-stuttgart.de

Abstract. In this paper, a method is proposed to compute the so-called cable span, i.e. the space occupied by the cables when the robot is moving within its workspace. As the cables are attached to a mostly fixed point on the robot frame, the shape of the cable span is a generalized cone. We present an efficient method *polar sorting* to compute the surface of this cone. Furthermore, the found geometry of the cone is employed in the design of the cable anchor points in order to dimension its deflection capabilities and to compute a suitable orientation for the installation of the mechanical unit.

Keywords: Cable-driven parallel robots · Cable span · Collision · Deflection angles · Design · Workspace

1 Introduction

Cable-driven parallel robots possess a number of advantages such as light-weight design, huge workspace, and excellent dynamic capabilities. These features come at some costs in terms of difficult geometric design for complex tasks. For a cable robot, a number of collision problems need to be addressed. Firstly, the problem of cable-cable collision can significantly reduce the usable workspace and was extendedly studied (e.g. in [3,6,7,9]). A couple of robot design with so-called *cross-over configurations* are proposed that offer a large collision-free workspace [5,8,13]. Another problem arising from the application of cable robots is related to the possible collisions of the cables with the environment. Furthermore, the mechanical design of the cable deflection units with large deflection angles is involved and applies both to the distal end of the cable at the mobile platform and to the proximal guiding on the machine frame. The cable-environment interference as well as the design of the cable guiding are related and discussed within this paper. To the best of the authors knowledge, no model of the space occupied by the cables has been proposed in the literature beside the pose-dependent assessment of collisions mentioned in the papers above.

In order to deal with this problem, the *cable span* is introduced in this paper. The cable span for one cable is the space occupied by this cable while the platform travels through the robot's workspace. As shown in the remainder of this paper, the cable span is a spatial geometrical object that can be described by a generalized cone. Clearly, this region must be free of obstacles to avoid a collision with the cable. Furthermore, the cable span allows to derive the deflection requirements of the guiding pulleys in the design of the robot.

The rest of the paper is organized as follows. Sect. 2 recalls the basic kinematic and workspace issues used in this paper. In Sect. 3, some data structures for workspace computation are discussed that are essential for the computation of the cable span. Using the proposed cable span, two applications of the cable span for robot analysis and design are proposed in Sect. 4. Conclusions are closing the paper.

2 Background

The kinematic background for the cable robots is briefly reviewed for the sake of completeness. The kinematic scheme of a cable-driven parallel robot with m proximal anchor points $\mathbf{a}_i$ and distal anchor points $\mathbf{b}_i$ is depicted in Fig. 1. The vector $\mathbf{l}_i$ represents the cable and it is oriented to start at the platform and point towards the robot frame. The pose of the platform is represented by the position vector $\mathbf{r}$ and the rotation matrix $\mathbf{R}$ which transforms platform coordinates from the platform frame $\mathscr{K}_P$ to world coordinates $\mathscr{K}_0$. The considerations in this paper do not assume a particular parameterization of the rotation matrix, so

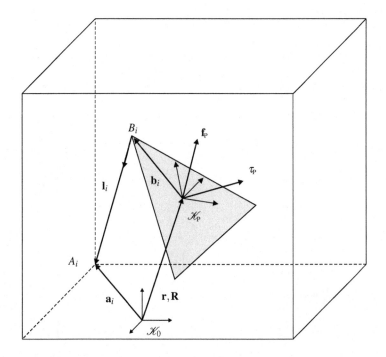

Fig. 1. Definition of the geometry and kinematics of a general cable robot

an arbitrary parameterization can be used. The applied wrench $\mathbf{w}_\mathrm{P} = [\mathbf{f}_\mathrm{P}^\mathrm{T}, \boldsymbol{\tau}_\mathrm{P}^\mathrm{T}]^\mathrm{T}$ is composed from the applied force $\mathbf{f}_\mathrm{P}$ and the applied torque $\boldsymbol{\tau}_\mathrm{P}$. Thus, the kinematic closure equation reads

$$\mathbf{l}_i = \mathbf{a}_i - \mathbf{r} - \mathbf{R}\mathbf{b}_i, \qquad i = 1,\ldots,m. \tag{1}$$

As the cable span discussed in this paper is closely related to the workspace of the cable robot, a criterium needs to be considered to decide if a given pose $(\mathbf{r}, \mathbf{R})$ belongs to the workspace. A couple of criteria are known to account for properties such as wrench-closure [2,4], wrench feasibility [2,15], generation of a given wrench set [1], or absence of cable-cable interference [9]. For the procedure discussed in this paper, the kind and number of criteria is irrelevant. In the case study, we employ a simple wrench-feasibility test [12]

$$\mathbf{f}_\mathrm{Min} \preceq \frac{1}{2}(\mathbf{f}_\mathrm{Min} + \mathbf{f}_\mathrm{Max}) - \mathbf{A}(\mathbf{r},\mathbf{R})^{+\mathrm{T}}\left(\mathbf{w}_\mathrm{P} + \mathbf{A}(\mathbf{r},\mathbf{R})^\mathrm{T}\frac{1}{2}(\mathbf{f}_\mathrm{Min} + \mathbf{f}_\mathrm{Max})\right) \preceq \mathbf{f}_\mathrm{Max}, \tag{2}$$

where $\mathbf{f}_\mathrm{Max}, \mathbf{f}_\mathrm{Max}$ are the vectors of minimum and maximum admissible cable forces and $\mathbf{A}^\mathrm{T}$ is the pose-dependent structure matrix [14]. The Moore-Penrose pseudo inverse matrix of $\mathbf{A}^\mathrm{T}$ is denoted given by $\mathbf{A}^{+\mathrm{T}} = \mathbf{A}(\mathbf{A}^\mathrm{T}\mathbf{A})^{-1}$. Based on the definitions above, a pose-depended evaluation of the workspace can be made.

3 Determination of the Cable Span

A well-known disadvantage of parallel robots and especially of cable robots is that the installation space of the robot is large compared to the workspace. One reason for this drawback is that the cables occupy a huge volume if the workspace of the robot is large. This volume that is at least temporarily taken be the cables

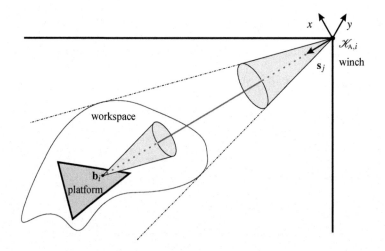

Fig. 2. Definition of the cable span based on the hull

is called *cable span* (Fig. 2). The cable span is a volumetric object for all spatial robots and a flat area like a fan for a planar robot. Based on the assumption of the standard cable model that the proximal anchor point is a pose-independent point in space, it is clear that the cable span for the i-th cable has some kind of apex at the point A_i.

3.1 Generation of Relevant Poses

In this paper, we assume the workspace that is assessed for the cable span is determined through discretization. In particular, we assume that the workspace is either computed by sampling a grid of positions for the translational workspace or by computing the boundary of the workspace with the hull algorithm [10].

For efficiency reasons, the workspace hull is used in the numerical study to compute the boundary of the workspace as the boundary is characteristic for the maximum cable deflections. As the surface of the workspace represents the extremal region of the workspace, the vertices of the hull include the relevant points. However, not every point on the workspace boundary is relevant for the cable span determination.

One can employ the constant orientation workspace or the total orientation workspace as basis for the determination of the cable span. For the sake of simplicity, we restrict the following considerations to the constant orientation workspace with a fixed orientation $\mathbf{R}_0$ of the platform and varying position $\mathbf{r}$. Thus, we assume the workspace $\mathscr{W}$ to be given as a set of positions $\mathbf{r}$

$$\mathscr{W} = \{\mathbf{r} \in \mathbb{R}^3 \,|\, g(\mathbf{r}) > 0\}\,, \tag{3}$$

where $g(\mathbf{r})$ is a function that evaluates to a positive number if the position $\mathbf{r}$ belongs to the workspace. In this paper, Eq. (2) is used but any other workspace test can be employed instead. Respectively, one can also vary the orientation to caption the orientation workspace or the total orientation workspace. As the follow consideration is purely based on efficiently sorting a set of cable vectors $\mathbf{l}_i$, it is straight-forward to extend the algorithm to a discretization of the orientation workspace.

Using Eq. (1), one receives for each cable i and for the k-th position $\mathbf{r}$ in the set $\mathscr{W}$ the respective line $\mathbf{l}_i^{(k)}$ vectors with $k = 1, \ldots, N$ where N is the number of positions in the set $\mathscr{W}$. Geometrically speaking, this set of curves consists of line segments starting at the proximal anchor point $\mathbf{a}_i$ and pointing towards the workspace. This set basically contains all required information about the space occupied by the cables. However, as one usually employs many points for sampling the workspace, an estimate for a volumetrical object is sought that can be handle more efficiently than all the lines.

In the following we omit the index of the line (i) for the sake of readability and abbreviate the set of lines $\mathbf{v}_k = -\mathbf{l}_i^{(k)}$ which shall represent the common starting point at $\mathbf{a}_i$. Thus, a set of N points $\mathscr{V} = \{\mathbf{v}_1, \ldots, \mathbf{v}_N\}$ is the input from workspace determination and the cable span shall be computed.

3.2 Sorting the Line Segments

Once the relevant lines from the proximal anchor point $\mathbf{a}_i$ to the respective world coordinates of the distal anchor points $\mathbf{b}_i(\mathbf{r}, \mathbf{R})$ are determined, the structure of this bunch of lines has to be determined.

In this paper, an algorithm called *polar sorting* is proposed to extract the relevant structure from the lines. The main steps are the following

1. Determine an estimate of the cone axis
2. Construct a coordinate system located at $\mathbf{a}_i$ with its z-axis aligned with the cone axis
3. Transform all lines $\mathbf{v}_i$ in $\mathscr{V}$ into this coordinate frame
4. Compute polar coordinates for the lines $\mathscr{V}$
5. Sort the vertices in $\mathscr{V}$ by the azimuth angle φ
6. Cluster the vertices in n_s equal classes by intervals of the azimuth angle and approximate the enclosing cone by extracting the largest deflection angle in that interval
7. Return the surface of the generalized cone consisting of the apex at $\mathbf{a}_i$ and the n_s characteristic vertices on its mantle of the cone

After this procedure, one has a simple triangulation with n_s triangles of the cable span that can be used in a number of applications.

3.3 Computing the Cable Span

From the structure of the workspace hull, we have an estimate used as projection center $\mathbf{m}$ or can compute the barycenter of the workspace $\mathscr{W}$. If the center is unknown or a grid computation was employed, the mean value of the positions

$$\mathbf{m} = \frac{1}{N} \sum_{j=1}^{N} \mathbf{r}_i \qquad (4)$$

is used instead. Now, the central line from $\mathbf{a}_i$ to $\mathbf{m}$ is employed as axis of the cone and the polar decomposition aims at sorting all the lines in the span around this central line.

Then, a coordinate frame $\mathscr{K}_{\mathrm{A},i}$ is constructed at point A_i which z-axis is aligned with the (estimated) cone axis $\mathbf{e}_z = \mathbf{m} - \mathbf{a}_i$. The x-axis represented by the vector $\mathbf{e}_x$ is perpendicular to the z-axis but has an additional degree of freedom that can be chosen arbitrarily. If a panning pulley is used for guiding the cable, it is beneficial to define the x-axis orthogonal to the first axis of the panning pulley e.g. orthogonal to the axis the pulley is panned about. The remaining y-axis is computed from the cross product $\mathbf{e}_y = \mathbf{e}_z \times \mathbf{e}_x$. The transformation matrix is then derived from the normalized vectors $(\mathbf{e}_x, \mathbf{e}_y, \mathbf{e}_z)$. This transformation is represented in terms of the transformation matrix $^{\mathrm{A},i}\mathbf{R}_0$ that maps vectors in world frame $\mathscr{K}_0$ to the local frame $\mathscr{K}_{\mathrm{A},i}$.

Based on the considerations of the workspace boundary, one can easily compute the cable span for the constant orientation workspace. The set

$\mathcal{W} = (\mathbf{r}_1, \ldots, \mathbf{r}_N)$ contains the N position of the workspace as introduced above. Then, one receives the world coordinates for all possible locations for the point B_i from

$$\mathbf{v}'_j = \mathbf{R}_0 \mathbf{b}_i + \mathbf{r}_j \quad \text{for} \quad j = 1, \ldots, N \tag{5}$$

which is simply a translation of the hull by $\mathbf{b}_i$. The cable span is approximated from connecting the point A_i with each of the vectors $\mathbf{v}'_j$. The resulting geometrical object is a generalized cone with a noncircular cross-section where most of the lines defined above are lying inside the cone. To normalize the representation and also to reduce the amount of data, a *polar decomposition* of the lines is proposed and described below.

The N lines of the span are distributed in n_s polar segments (Fig. 3) in the frame $\mathcal{K}_{A,i}$. Firstly, each line is transformed into the local frame $\mathcal{K}_{A,i}$ by

$$^{A,i}\mathbf{s}_j = {^{A,i}}\mathbf{R}(\mathbf{v}'_j - \mathbf{a}_i). \tag{6}$$

Then, the spherical coordinates $\mathbf{s}_j^{(c)}$ are computed from

$$\mathbf{s}_j^{(c)} = \begin{bmatrix} r \\ \theta \\ \varphi \end{bmatrix}_j = \begin{bmatrix} \sqrt{s_X^2 + s_Y^2 + s_Z^2} \\ \arccos \dfrac{s_Z}{\sqrt{s_X^2 + s_Y^2 + s_Z^2}} \\ \arctan 2(s_Y, s_X) \end{bmatrix}_j \quad \text{with} \quad {^{A,i}}\mathbf{s}_j = [s_X, s_Y, s_Z]^\mathrm{T}. \tag{7}$$

These spherical coordinates $\mathbf{s}_j^{(c)}$ allow for a simple extraction of the cable span. The N line vectors are sorted in ascending order of their φ-value (Fig. 3). This sorting may effortlessly be done by storing the data in an associative container

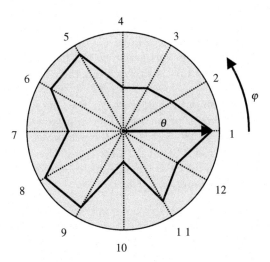

Fig. 3. Polar decomposition of the cable vector to compute the cone of the cable span for $n_\text{s} = 12$ segments

offered by most programming languages[1]. Then, n_s segments of equal size are chosen for the angle φ that represent the ranges

$$\mathscr{S}_i = \left[\frac{i\,2\pi}{n_\mathrm{s}}, \frac{(i+1)2\pi}{n_\mathrm{s}}\right]_i \quad i = 0,\ldots,(n_\mathrm{s}-1). \tag{8}$$

Finally, one loops through the sorted list $\mathbf{s}_j^{(\mathrm{c})}$ of cylinder coordinates and extracts for each range $\mathscr{S}_i$ the matching element with

$$\left.\mathbf{s}_j^{(\mathrm{c})}\right|_{\varphi \in \mathscr{S}_i} \tag{9}$$

and stores the largest angle θ for all line vectors that belong to the respective segment. After this procedure, one has a sorted list of n_s characteristic vectors of the surface of the cable span. Connecting two neighboring vectors with the apex at $\mathscr{K}_{\mathrm{A},i}$ gives a triangulation of the surface of the cable span. Exporting this triangulation to STL or VRML is straight forward and allows to use results within CAD systems. The list of the angles over the polar coordinate is basically a look-up table to check if a vector is inside the cone.

4 Application

In this section, a case study for determination of the cable span is presented. The case study is based on the IPAnema 1 robot geometry as given in Table 1. For the case studies, the translational workspace with a constant orientation of $\mathbf{R}_0 = \mathbf{I}$ is considered.

4.1 Geometric Cable Span

In Fig. 4 the cable span is visualized in polar coordinates. The points in the plot indicate the unfiltered data (258 vertices) received from the workspace evaluation. The circumferential red line is drawn from the 36 vertices received from polar sorting.

In order to assess the computational performance, the same computation was executed with a higher number of vertices. The computation time for the cable span of all eight cables for 16 386 vertices on the workspace hull was determined to be 51 ms (on Intel Core i5-5200U at 2.2 GHz) while the workspace determination consumed some 1050 ms of CPU time. Thus, the determination of the cable span is cheap in terms of computation efforts.

---

[1] The effort for this kind of sorting is $\log(N)$ for each element and it is internally done when using associative containers such as `dict` in Python or `map` in C++.

Table 1. Nominal geometric parameters in terms of base vectors $\mathbf{a}_i$ and platform vectors $\mathbf{b}_i$ for the IPAnema 1 robot.

Cable i	Base vector $\mathbf{a}_i$	Platform vector $\mathbf{b}_i$
1	$[-2.0, 1.5, 2.0]^\mathrm{T}$	$[-0.06, 0.06, 0.0]^\mathrm{T}$
2	$[2.0, 1.5, 2.0]^\mathrm{T}$	$[0.06, 0.06, 0.0]^\mathrm{T}$
3	$[2.0, -1.5, 2.0]^\mathrm{T}$	$[0.06, -0.06, 0.0]^\mathrm{T}$
4	$[-2.0, -1.5, 2.0]^\mathrm{T}$	$[-0.06, -0.06, 0.0]^\mathrm{T}$
5	$[-2.0, 1.5, 0.0]^\mathrm{T}$	$[-0.06, 0.06, 0.0]^\mathrm{T}$
6	$[2.0, 1.5, 0.0]^\mathrm{T}$	$[0.06, 0.06, 0.0]^\mathrm{T}$
7	$[2.0, -1.5, 0.0]^\mathrm{T}$	$[0.06, -0.06, 0.0]^\mathrm{T}$
8	$[-2.0, -1.5, 0.0]^\mathrm{T}$	$[-0.06, -0.06, 0.0]^\mathrm{T}$

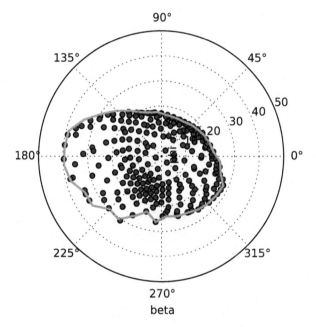

Fig. 4. Plot of the cable span in polar coordinates of frame $\mathcal{K}_{\mathrm{A},1}$ (Color figure online)

4.2 Deflection Angles for Anchor Points

Another application of the cable span lies in the dimensioning of the panning pulley unit on the machine frame of the robot. This supports the mechanical design of the cable robot when the initial position of the cable guiding system needs to be defined. Using the computation of the cable span, one maps the extremal values with the pulley kinematics function [11].

Figures 5 and 6 show the actually occurring deflection angles $\beta_{\mathrm{R},1}$ and $\gamma_{\mathrm{R},1}$ in the pulley mechanism for the IPAnema 1 robot. The sample poses are chosen from the hull of the workspace thus covering the extremal positions of the pulley.

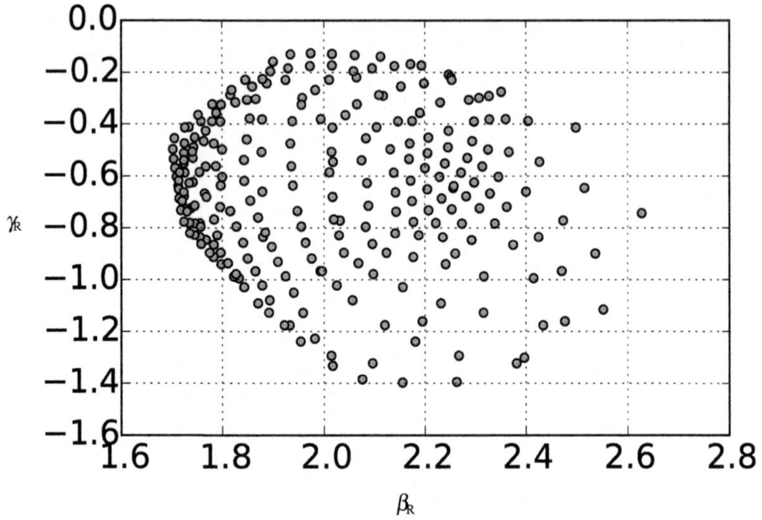

Fig. 5. Deflection angles β_R and γ_R for the first winches of the IPAnema 1 robot throughout the workspace.

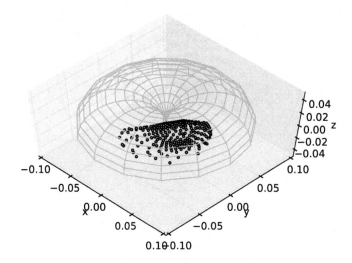

Fig. 6. Actual proximal anchor points C_1 where the cable leaves the pulley in the local frame $\mathcal{K}_{A,1}$

One can see that the panning angle γ_R of the pulley is in the range $[-\frac{\pi}{2}; 0]$ thus pointing to the inside of the machine frame. The considered winch $i = 1$ is an upper winch located at the top of the robot frame. Thus, the wrapping angle is $\beta_{R,1} \in [\frac{\Pi}{2}; \pi]$ where the cable always wraps at least a quarter of the pulley. Only a small part of the toroidal surface is actually used. The region where the point C_i may be located is notably smaller than the torus.

A similar consideration is applied to the platform. If we consider only the translational workspace, the deflection angles on the platform are essentially the same as for the proximal anchor point but with inverted sign (see Fig. 2). For the distal anchor point B_i, two mechanical constructions are widespread, the use of universal/spherical joints at the end of the cable and swivel bolts (see Fig. 7). Their main difference lies in the admissible deflection of the cable with respect to the installation orientation. For universal and spherical joints, the preferred attack angle is within a cone where an attack angle of $0°$ is optimal. This installation orientation of the joint shall thus be aligned with the central axis of the cable span for proper operation and the deflection angles shall lie inside the cable span. In contrast, swivel bolts allow for very large deflection angles which can even exceed $90°$. However, the swivel bolt has a singular configuration when the direction of the cable and the first axis of the swivel bolt is aligned. Therefore, a sufficient diagonal pull on the swivel bolt must be guaranteed (see Fig. 7). In this setting, the generalized cone computed from the cable span must be placed within the range of γ_{Min} and γ_{Max}.

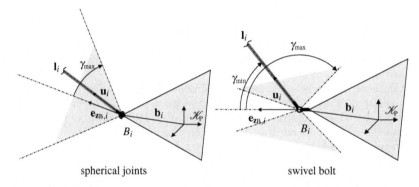

Fig. 7. Feasible deflection angles (gray area left of the platform) for the cable on the platform for spherical joints and swivel bolts

5 Conclusion

In this paper, the concept cable span is introduced and an efficient algorithm for the determination of the region occupied by the cables is presented. As the geometric structure of the cable span is a generalized cone, it can be represented as a triangulation of its shell surface. This object can be used in CAD planning to study interference with other equipment. Furthermore, the cable span is a useful concept to design the cable guiding system in order to choose feasible orientation value for the axis of the pulleys.

Acknowledgements. The authors would like to thank the German Research Foundation (DFG) for financial support of the project within the Cluster of Excellence in Simulation Technology (EXC 310/1) at the University of Stuttgart.

References

1. Bouchard, S., Moore, B., Gosselin, C.: On the ability of a cable-driven robot to generate a prescribed set of wrenches. J. Mech. Robot. **2**(1), 1–10 (2010)
2. Ebert-Uphoff, I., Voglewede, P.A.: On the connections between cable-driven parallel manipulators and grasping. In: 2004 IEEE International Conference on Robotics and Automation, New Orleans, pp. 4521–4526 (2004)
3. Ghasemi, A., Farid, M., Eghtesad, M.: Interference free workspace analysis of redundant 3D cable robots. In: 2008 World Automation Congress, pp. 1–6 (2008)
4. Gouttefarde, M., Gosselin, C.: On the properties and the determination of the wrench-closure workspace of planar parallel cable-driven mechanisms. In: Proceedings of the ASME Design Engineering Technical Conference, vol. 2, pp. 337–346 (2004)
5. Lamaury, J., Gouttefarde, M.: Control of a large redundantly actuated cable-suspended parallel robot. In: 2013 IEEE International Conference on Robotics and Automation, pp. 4659–4664 (2013)
6. Maeda, K., Tadokoro, S., Takamori, T., Hiller, M., Verhoeven, R.: On design of a redundant wire-driven parallel robot WARP manipulator. In: Proceedings of IEEE International Conference on Robotics and Automation, Detroit, MI, USA, pp. 895–900 (1999)
7. Merlet, J.P.: Analysis of the influence of wires interference on the workspace of wire robots. In: Advances in Robot Kinematics (ARK), pp. 211–218. Kluwer Academic Publishers, Sestri Levante (2004)
8. Miermeister, P., Lächele, M., Boss, R., Masone, C., Schenk, C., Tesch, J., Kerger, M., Teufel, H., Pott, A., Bülthoff, H.: The CableRobot simulator: large scale motion platform based on cable robot technology. In: IEEE/RSJ International Conference on Intelligent Robots and Systems, Daejeon, South Korea (2016)
9. Perreault, S., Cardou, P., Gosselin, C., Otis, M.J.D.: Geometric determination of the interference-free constant-orientation workspace of parallel cable-driven mechanisms. ASME J. Mech. Robot. **2**(3) (2010)
10. Pott, A.: Forward kinematics and workspace determination of a wire robot for industrial applications. In: Advances in Robot Kinematics (ARK), pp. 451–458. Springer (2008)
11. Pott, A.: Influence of pulley kinematics on cable-driven parallel robots. In: Advances in Robot Kinematics (ARK), pp. 197–204 (2012)
12. Pott, A., Bruckmann, T., Mikelsons, L.: Closed-form force distribution for parallel wire robots. In: Computational Kinematics, pp. 25–34. Springer, Heidelberg (2009)
13. Pott, A., Mütherich, H., Kraus, W., Schmidt, V., Miermeister, P., Verl, A.: IPAnema: a family of cable-driven parallel robots for industrial applications. In: Cable-Driven Parallel Robots, Mechanisms and Machine Science, pp. 119–134. Springer (2012)
14. Verhoeven, R.: Analysis of the workspace of tendon-based Stewart platforms. PhD thesis, University of Duisburg-Essen, Duisburg, Germany (2004)
15. Verhoeven, R., Hiller, M.: Estimating the controllable workspace of tendon-based Stewart platforms. In: Advances in Robot Kinematics (ARK), pp. 277–284. Springer, Portorož (2000)

Geometric Determination of the Cable-Cylinder Interference Regions in the Workspace of a Cable-Driven Parallel Robot

Antoine Martin[1], Stéphane Caro[2(✉)], and Philippe Cardou[3]

[1] École Centrale de Nantes–Laboratoire des Sciences du Numérique de Nantes,
1, Rue de la Noë, BP 92101, 44321 Nantes Cedex 3, France
antoine.martin@ls2n.fr

[2] CNRS–Laboratoire des Sciences du Numérique de Nantes,
1, Rue de la Noë, BP 92101, 44321 Nantes Cedex 3, France
stephane.caro@ls2n.fr, stephane.caro@irccyn.ec-nantes.fr

[3] Université Laval, 1065 Avenue de la M édecine Québec, Québec G1V 0A6, Canada
pcardou@gmc.ulaval.ca

Abstract. Cable-Driven Parallel Robots (CDPRs) are a type of parallel robots that have the particularity of using cables as legs. CDPRs have several advantages such as large workspaces, high acceleration and high payload capacity. However, CDPRs present also some drawbacks such as the possible collisions between their cables and environment. Therefore, this paper is about the geometric determination of the cable-cylinder interference regions in the workspace of a CDPR. The cables are considered massless and straight. Then, the boundaries of the interference regions onto the cylinder form a closed loop composed of arcs and straight line segments that can be expressed symbolically. Those geometric entities generate truncated cones and planes corresponding to the boundaries of the volume of interferences. Finally, a methodology is described to trace the cable-cylinder interference free constant orientation workspace of CDPRs.

1 Introduction

Cable-driven parallel robots (CDPRs) have received increasing attention from researchers during the last 20 years, in part because of their large workspace and low mass in motion. CDPR workspaces such as the Wrench Closure Workspace [1,2] or the Wrench Feasible Workspace [3] have been widely studied. These workspaces allow the visualisation of the volume over which the moving-platform can sustain certain external wrenches. The Twist Feasible Workspace [4] and the Dynamic Feasible Workspace [5] have also been defined in the literature in order to evaluate the capability of a CDPR to perform required twists and accelerations.

One of the main drawbacks of CDPRs is the potential risk of collisions (also called interferences) between their cables and their environment. Although such collisions may occur in many industrial operations, there have been few papers in the literature dealing with this issue. Moreover interferences usually reduce the size of the CDPR workspace [6]. For instance, for industrial operations over a large workspace and in a cluttered environment such as that shown in Fig. 1, the workspace of the CDPR is dramatically reduced because of the possible collisions between the cables and the tubes of the large truss [7]. Another relevant application is detailed in reference [8], where two CDPRs are working side by side, and collisions between cables can occur. In [9], four types of interferences are defined:

1. Collisions between two cables.
2. Collisions between a cable and the platform.
3. Collisions between a cable and the environment.
4. Collisions between the platform and the environment.

The first two types of collisions were studied in [9,10] for a moving-platform described by a set of triangles. A geometric method to compute the interference-free constant-orientation workspace of CDPRs was introduced in [11] while considering the cable-cable and cable-platform interferences. This method has been used to develop the ARACHNIS interface [12], which aims to help robot designers trace CDPR workspaces and find optimal dimensions for their good under design. In the foregoing works, cables have been considered massless and straight. A recent paper from Wang et al. [13] describes an algorithm to trace the Collisions Free Force Closure Workspace of a CDPR working in a cluttered environment. This algorithm, which uses the convex hull approach, turns out to be time consuming, mainly when the cluttered environment is known beforehand.

This paper deals with the geometric determination of cable-cylinder interference regions in the CDPR workspace. A mathematical description of the

Fig. 1. CAROCA prototype: a reconfigurable cable-driven parallel robot working in a cluttered environment (Courtesy of IRT Jules Verne and STX France)

boundaries of those regions is obtained for a constant orientation of the moving-platform. The paper is organized as follows. Section 2 deals with the determination of interference region between one cable and a cylinder. Section 3 describes the method proposed to find the interference-free constant-orientation workspace of a cable-driven parallel robot while considering the collisions between its cables and a cylinder. Finally, the determination of the interference regions between the cables of the CAROCA prototype [7] and a cylinder highlights the contributions of this paper.

2 Interference Region Between a Cable and a Cylinder

Due to their large workspaces, CDPRs can be used to perform maintenance operations on large structures such as bridges. For instance, Fig. 1 shows a CPDR working inside a tubular truss. In those cases, interferences between the CDPR cables and the cylinders forming the structure are the main limitation on the workspace.

2.1 Parametrisation

As shown in Fig. 2, the ith cable of a CDPR is attached to the moving-platform at anchor point B_i. Its length is controlled by an actuated reel whose exit point A_i is fixed to the base. The base frame is denoted $\mathscr{F}_b = (O, \mathbf{x}_b, \mathbf{y}_b, \mathbf{z}_b)$. Frame $\mathscr{F}_p = (P, \mathbf{x}_p, \mathbf{y}_p, \mathbf{z}_p)$ is attached to the moving-platform. The cylinder $\mathscr{C}$ with base point C, radius r_c, length l_c and axis $\mathscr{A}_C$ is located inside the working area of a CDPR, fixed to its base.

This section aims at determining for a given orientation of the moving-platform, the positions of P that lead to collisions between cable $\mathscr{C}_i$ and cylinder $\mathscr{C}$. Here, the cable is assumed to be straight, with a negligible cross-section. The interference point between cable $\mathscr{C}_i$ and cylinder $\mathscr{C}$ is named I. $\mathbf{p}_{c,i}$ denotes the Cartesian coordinate vector of point I expressed in the base frame $\mathscr{F}_b$.

The proposed strategy for determining and tracing the region of interference between $\mathscr{C}_i$ and $\mathscr{C}$ consists in moving point P while maintaining contact I between the line segment and the cylinder. In doing so, point P sweeps a conical surface in space, while point I makes a closed loop on the surface of $\mathscr{C}$. Interferences between $\mathscr{C}_i$ and $\mathscr{C}$ can occur in two ways: either the interference point I lies on the cylindrical surface of $\mathscr{C}$, either it lies on one of its two circular edges.

2.2 Boundaries of the Cable-Cylinder Interference Region

The type of closed-loop trajectory followed by I through this sweeping motion depends on the location of A_i with respect to $\mathscr{C}$. One finds five zones for the location of A_i, which correspond to three types of closed loop trajectories of I over the surface of $\mathscr{C}$. Figure 3 represents a section of the top half of the cylinder with the five zones in question. Its bottom half is the mirror image of the top one with respect to axis $\mathscr{A}_C$.

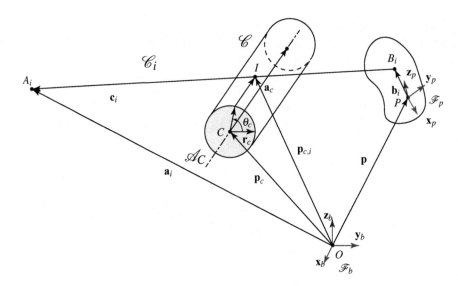

Fig. 2. ith cable $\mathscr{C}_i$, CDPR moving-platform and cylinder $\mathscr{C}$

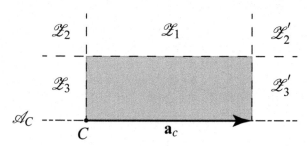

Fig. 3. Position of exit point A_i relatively to cylinder $\mathscr{C}$

It is noteworthy that the types of point I trajectories when exit point A_i belongs to zones $\mathscr{Z}_2'$ and $\mathscr{Z}_3'$ are the same as when exit point A_i belongs to zones $\mathscr{Z}_2$ and $\mathscr{Z}_3$, respectively. For zones $\mathscr{Z}_1$ and $\mathscr{Z}_2$, the interference regions between the ith cable $\mathscr{C}_i$ and the cylindrical part of $\mathscr{C}$ are considered. Then, the interference regions between $\mathscr{C}_i$ and the endcaps of $\mathscr{C}$ are obtained. For zone $\mathscr{Z}_3$, only this second part is needed.

2.2.1 Interferences Between Cable $\mathscr{C}_i$ and the Cylindrical Part of $\mathscr{C}$

The boundaries of the interference region between cable $\mathscr{C}_i$ and the cylindrical part of cylinder $\mathscr{C}$ describe the points where a straight line coming from exit point A_i is tangent to the cylinder $\mathscr{C}$. As the interference point I belongs to the cylinder $\mathscr{C}$, we can write:

$$\mathbf{p}_{c,i} = \mathbf{p}_c + \alpha_c \mathbf{a}_c + \mathbf{Q}(\mathbf{a}_c, \theta_c)\mathbf{r}_c \qquad (1)$$

where $\mathbf{p}_c$ is the Cartesian coordinate vector of point C, depicted in Fig. 2, α_c is a scalar bounded between 0 and 1 and represents the location of point I along the cylinder axis $\mathscr{A}_C$. $\mathbf{r}_c$ is a vector normal to axis $\mathscr{A}_C$. Therefore, the following equation holds:
$$\mathbf{a}_c^T \mathbf{r}_c = 0 \tag{2}$$
$\mathbf{Q}(\mathbf{a}_c, \theta_c)$ is the rotation matrix of vector $\mathbf{r}_c$ about $\mathscr{A}_C$ by an angle θ_c.

Accordingly, the tangency condition between the ith cable $\mathscr{C}_i$ and the cylindrical part of cylinder $\mathscr{C}$ is expressed as:
$$(\mathbf{p}_{c,i} - \mathbf{a}_i)^T (\mathbf{Q}(\mathbf{a}_c, \theta_c)\mathbf{r}_c) = 0 \tag{3}$$

From Eqs. (2) and (3), angle θ_c can take two values, namely,
$$\begin{cases} \mathbf{Q}(\mathbf{a}_c, \theta_c^+)\mathbf{r}_c = \eta_c \mathbf{a}_c \times (\mathbf{p}_{c,i} - \mathbf{a}_i), \\ \mathbf{Q}(\mathbf{a}_c, \theta_c^-)\mathbf{r}_c = -\eta_c \mathbf{a}_c \times (\mathbf{p}_{c,i} - \mathbf{a}_i). \end{cases} \tag{4}$$

where $\eta_c = \dfrac{r_c}{\|\mathbf{a}_c \times (\mathbf{p}_{c,i} - \mathbf{a}_i)\|_2}$. For the sake of conciseness, $\mathbf{Q}(\mathbf{a}_c, \theta_c^+)\mathbf{r}_c$ and $\mathbf{Q}(\mathbf{a}_c, \theta_c^-)\mathbf{r}_c$ are denoted as $\mathbf{r}_c^+$ and $\mathbf{r}_c^-$, respectively. Therefore θ_c^+ and θ_c^- take the form:
$$\begin{cases} \theta_c^+ = \arccos\left(\dfrac{\mathbf{r}_c^T \mathbf{r}_c^+}{\|\mathbf{r}_c\| \cdot \|\mathbf{r}_c^+\|}\right), \\ \theta_c^- = \arccos\left(\dfrac{\mathbf{r}_c^T \mathbf{r}_c^-}{\|\mathbf{r}_c\| \cdot \|\mathbf{r}_c^-\|}\right). \end{cases} \tag{5}$$

Upon introducing the position vector of point I of Eq. (1) into Eq. (4), we obtain:
$$\begin{cases} \mathbf{r}_c^+ = \eta_c \mathbf{A}_c(\mathbf{p}_{c,i} + \alpha_c \mathbf{a}_c + \mathbf{r}_c^+ - \mathbf{a}_i), \\ \mathbf{r}_c^- = -\eta_c \mathbf{A}_c(\mathbf{p}_{c,i} + \alpha_c \mathbf{a}_c + \mathbf{r}_c^- - \mathbf{a}_i). \end{cases} \tag{6}$$

$\mathbf{A}_c$ being the cross-product matrix[1] of vector $\mathbf{a}_c$. By solving Eq. (6), $\mathbf{r}_c^+$ and $\mathbf{r}_c^-$ take the form:
$$\begin{cases} \mathbf{r}_c^+ = \dfrac{\eta_c}{1 + l_c^2}(r_c \mathbf{A}_c^2 + \mathbf{A}_c)(\mathbf{p}_{c,i} - \mathbf{a}_i), \\ \mathbf{r}_c^- = \dfrac{\eta_c}{1 + l_c^2}(r_c \mathbf{A}_c^2 - \mathbf{A}_c)(\mathbf{p}_{c,i} - \mathbf{a}_i). \end{cases} \tag{7}$$

As a result, the boundaries of the interference region between cable $\mathscr{C}_i$ and the cylindrical part of $\mathscr{C}$ can be expressed as:
$$\begin{cases} \mathbf{p}_{c,i}^+ = \mathbf{p}_c + \alpha_c \mathbf{a}_c + \mathbf{r}_c^+ \\ \mathbf{p}_{c,i}^- = \mathbf{p}_c + \alpha_c \mathbf{a}_c + \mathbf{r}_c^- \end{cases} \tag{8}$$

with α_c being a scalar lying between 0 and 1. The edges of the corresponding line segments are named I_n^+ and I_n^- for $\alpha_c = 0$, and I_f^+ and I_f^- for $\alpha_c = 1$ as illustrated in Fig. 4.

---
[1] The cross-product matrix $\mathbf{Y}$ of $\mathbf{y}$ is defined as $\delta(\mathbf{y} \times \mathbf{x}) \setminus \delta\mathbf{x}$ for any $\mathbf{x}, \mathbf{y} \in \mathbb{R}^3$

2.2.2 Interferences with the Endcaps

For the second point, the interferences with the endcaps of the cylinder are studied. The question is to find how to connect I_n^+ to I_n^- and I_f^+ to I_f^-. Those connections change depending on the zone to which A_i belongs to. In $\mathscr{Z}_1$, the curve linking I_n^+ to I_n^- is an arc starting from I_n^- and going to I_n^+. The result is the same for I_f^+ and I_f^-. We obtain the equations:

$$\begin{cases} \mathbf{p}_{c,i}^n = \mathbf{p}_c + \mathbf{Q}(\mathbf{a}_c, \theta_c)\mathbf{r}_c^-, \\ \mathbf{p}_{c,i}^f = \mathbf{p}_c + \mathbf{a}_c + \mathbf{Q}(\mathbf{a}_c, \theta_c)\mathbf{r}_c^-. \end{cases} \quad (9)$$

with θ_c lying in the interval $[0, \theta_c^+ - \theta_c^-]$. The result is shown in Fig. 5(a).

For $\mathscr{Z}_2$, the behavior is different for the nearest endcap. This time, the arc to take into account starts from I_n^+ and goes to I_n^-. The equations become:

$$\begin{cases} \mathbf{p}_{c,i}^n = \mathbf{p}_c + \mathbf{Q}(\mathbf{a}_c, \theta_c)\mathbf{r}_c^+ & \theta_c \in [0, 2\pi - (\theta_c^+ - \theta_c^-)] \\ \mathbf{p}_{c,i}^f = \mathbf{p}_c + \mathbf{a}_c + \mathbf{Q}(\mathbf{a}_c, \theta_c)\mathbf{r}_c^- & \theta_c \in [0, \theta_c^+ - \theta_c^-] \end{cases} \quad (10)$$

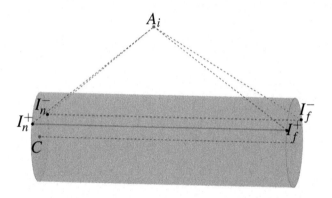

Fig. 4. Boundaries of the interference region between cable $\mathscr{C}_i$ and the cylindrical part of $\mathscr{C}$

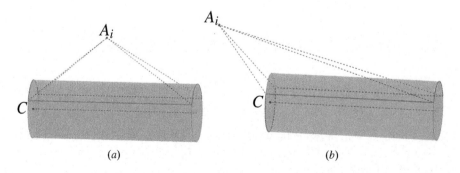

Fig. 5. Boundaries in (a) $\mathscr{Z}_1$ and (b) $\mathscr{Z}_2$

The result is shown in Fig. 5(b).

When A_i is in $\mathscr{Z}_3$, the tangency condition of Eq. (3) is no longer satisfied. In that case, the boundaries of the interference volume for the entire cylinder is the same as the one with the nearest endcap alone. The result is the circle defined by:

$$\mathbf{p}_{c,i}^n = \mathbf{p}_c + \mathbf{Q}(\mathbf{a}_c, \theta_c)\mathbf{r}_c, \quad \theta_c \in [0, 2\pi] \qquad (11)$$

In each case, the boundaries are either straight lines or arcs. The interference volume obtained from those geometric forms are detailed in the following section.

2.3 Interferences with Straight Line Segments and Arcs

As seen above, from point A_i, the closed-loop trajectory followed by I is composed of straight line and arc segments. Since a cable is considered straight without deformation, a line coming from A_i and following this shape will draw the external boundary of the interference volume.

In the case of the straight line between I_n^+ and I_f^+ (I_n^- and I_f^- respectively), the boundary is the infinite triangle originating from A_i and passing by those two points. Since the extremity B_i of the cable needs to be after the cylinder for having a possible collision, this triangle is truncated by the segment $[I_n^+ I_f^+]$.

For the arc between I_n^+ and I_n^- (I_f^+ and I_f^- respectively), the result will be an oblique cone, truncated twice:

- Once by the plane defined by points A_i, I_n^+ and I_n^-.
- Once by the plane defined by points C, I_n^+ and I_n^-.

Finally, the part of the cylinder closest to A_i and inside the lines detailed above need to be taken into account, to close the surface delimiting the interference volume. This part of the cylinder is easily obtained from the results of Sect. 2.2. Two examples are shown in Fig. 6 when the exit point of the winch is in zone $\mathscr{Z}_1$, and in zone $\mathscr{Z}_2$.

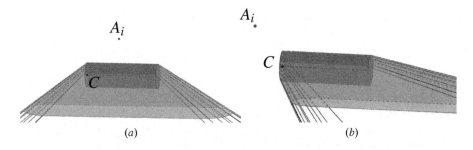

Fig. 6. Interference regions when A_i is in (a) $\mathscr{Z}_1$ and (b) $\mathscr{Z}_2$

3 Interferences Between the Cables of a CDPR and a Cylinder

In this section, the geometric model of a CDPR is detailed in Sect. 3.1. The interference volume between one cable and a cylinder obtained in Sect. 2 can then be coupled with this model to obtain in Sect. 3.2 the positions of the platform for which at least one cable is in collision with the cylinder.

3.1 Geometric Modeling

In Fig. 2, one can see the platform of a CDPR with cable $\mathscr{C}_i$ attached to it at point B_i. The exit point of the winch controlling its length is A_i. From this, by considering the cable as a straight line, a loop-closure equation gives the following geometric model:

$$\mathbf{c}_i = \mathbf{a}_i - \mathbf{p} - \mathbf{R}\mathbf{b}_i \qquad (12)$$

where $\mathbf{R}$ refers to the rotation matrix that transforms the global frame $\mathscr{F}_b$ to the frame attached to the mobile platform $\mathscr{F}_p$.

3.2 Interferences

Since the interference point between cable $\mathscr{C}_i$ and the cylinder belongs to both objects, its position can be expressed either by Eq. (1) or by:

$$\mathbf{p}_{c,i} = \mathbf{p} + \mathbf{R}\mathbf{b}_i + \gamma_i \mathbf{c}_i \qquad (13)$$

where γ_i is an scalar bounded between 0 and 1 defining the position of the interference point. 0 means that the collision occurs at point B_i, 1 in means that it happens at point A_i. The combination of Eqs. (1) and (13) leads the following equation:

$$\mathbf{p} + \mathbf{R}\mathbf{b}_i + \gamma_i \mathbf{c}_i = \mathbf{p}_c + \alpha_c \mathbf{a}_c + \mathbf{Q}(\mathbf{a}_c, \theta_c)\mathbf{r}_c \qquad (14)$$

Upon substituting the expression of $\mathbf{c}_i$ defined in Eq. (12) into Eq. (14), we obtain:

$$\mathbf{p} = \mathbf{a}_i - \mathbf{R}\mathbf{b}_i + \frac{1}{1-\gamma_i}(\mathbf{p}_c - \mathbf{a}_i) + \frac{1}{1-\gamma_i}(\alpha_c \mathbf{a}_c + \mathbf{Q}(\mathbf{a}_c, \theta_c)\mathbf{r}_c). \qquad (15)$$

This equation is valid for $\gamma_i \neq 1$. It is the case when the collision happens at the exit point of the winch, which can easily be avoided at the design phase since this point is not moving. By changing the values of α_c and θ_c according to Sect. 2, we obtain the volume in which the position $\mathbf{p}$ of the platform generates an interference with cable $\mathscr{C}_i$. This method needs to be applied for each cable-cylinder combination to obtain the interference region of the entire CDPR.

3.3 CAROCA Prototype-Cylinder Interference Region

Gagliardini *et al.* [7] introduced a method to manage the discrete reconfigurations of a CDPR, in order to use the latter for sandblasting and painting the outer part and the inner part of a large tubular structure with a minimal number of reconfigurations. This is a typical cluttered environment for which the method presented in this paper should help simplify the reconfiguration planning of the CDPR.

Fig. 7 represents a CDPR with one of the configurations obtained in [7]. The tubular structure is replaced with a single cylinder, to simplify the analysis of the

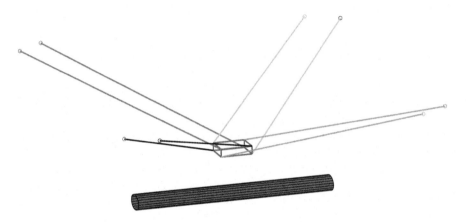

Fig. 7. One configuration of the reconfigurable CAROCA prototype and a cylinder in its working are

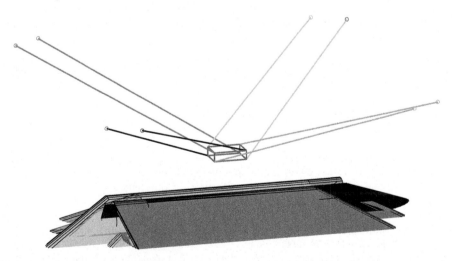

Fig. 8. Locations of the geometric center of the CAROCA mobile platform leading to some cable-cylinder interferences

interference regions. The results are shown in Fig. 8. For the cylinder included into the working area of the CDPR, eight interferences regions, namely one for each cable, are obtained. It is noteworthy that those regions are valid for a constant orientation of the mobile-platform.

4 Conclusion

The paper dealt with the geometric determination of the cable-cylinder interference regions in the workspace of a CDPR. First, a methodology was described to determine in a closed-form the boundaries of the interference region between a cable and a cylinder. The latter is first split into three parts, its two endcaps and the rest of the cylinder. By considering the cable tangent to the cylinder on the boundaries of the interference region, four points are obtained. Those points are connected by a straight segment along the cylinder, and two arcs along its endcaps depending on the position of the cable exit point with respect to the cylinder. Five zones for this position are defined, each one corresponds to a different arc segment to be taken into account to draw the boundaries of the interference region. Those line and arc segments generate truncated planes and oblique cones shapes, forming the boundaries of the interference volume with a part of the cylinder. Then, the cable-cylinder interference free constant orientation workspace of a cable-driven parallel robot can be traced while considering one or several cylinder(s) within the working area. As an illustrative example, the cable-cylinder interference free constant orientation workspace of the reconfigurable CAROCA prototype has been traced while considering a cylinder in its environment. An add-on feature to trace the cable-cylinder interference free constant orientation workspace of any cable-driven parallel robot will be implemented in ARACHNIS software [12] in future work. Finally, the contributions of this paper should ease the design of new cable-driven parallel robots working in a cluttered environment.

Acknowledgements. The financial support of the RFI ATLANSTIC 2020 CREATOR project is greatly acknowledged.

References

1. Gouttefarde, M., Gosselin, C.: On the properties and the determination of the wrench-closure workspace of planar parallel cable-driven mechanisms. In: ASME 2004 International Design Engineering Technical Conferences and Computers and Information in Engineering Conference, pp. 337–346. American Society of Mechanical Engineers (2004)
2. Stump, E., Kumar, V.: Workspaces of cable-actuated parallel manipulators. J. Mech. Des. **128**(1), 159 (2006)
3. Gouttefarde, M., Merlet, J.-P., Daney, D.: Wrench-feasible workspace of parallel cable-driven mechanisms. In: 2007 IEEE International Conference on Robotics and Automation, pp. 1492–1497. IEEE (2007)

4. Gagliardini, L., Caro, S., Gouttefarde, M.: Dimensioning of cable-driven parallel robot actuators, gearboxes and winches according to the twist feasible workspace. In: 2015 IEEE International Conference on Automation Science and Engineering (CASE), pp. 99–105. IEEE (2015)
5. Barrette, G., Gosselin, C.M.: Determination of the dynamic workspace of cable-driven planar parallel mechanisms. J. Mech. Des. **127**(2), 242 (2005)
6. Williams, R.L., Gallina, P.: Planar cable-direct-driven robots, part i: Kinematics and statics. In: Proceedings of the 2001 ASME Design Technical Conference, 27th Design Automation Conference, pp. 178–186 (2001)
7. Gagliardini, L., Caro, S., Gouttefarde, M., Girin, A.: Discrete reconfiguration planning for cable-driven parallel robots. Mech. Mach. Theory **100**, 313–337 (2016)
8. Otis, M.J.-D., Perreault, S., Nguyen-Dang, T.-L., Lambert, P., Gouttefarde, M., Laurendeau, D., Gosselin, C.: Determination and management of cable interferences between two 6-DOF foot platforms in a cable-driven locomotion interface. IEEE Trans. Syst. Man Cybern. Part A Syst. Hum. **39**(3), 528–544 (2009)
9. Nguyen, D.Q., Gouttefarde, M.: On the improvement of cable collision detection algorithms. In: Pott, A., Bruckmann, T. (eds.) Cable-Driven Parallel Robots. Mechanisms and Machine Science, vol. 32, pp. 29–40. Springer, Heidelberg (2015). doi:10.1007/978-3-319-09489-2_3
10. Merlet, J.-P.: Analysis of the influence of wires interference on the workspace of wire robots. In: On Advances in Robot Kinematics, pp. 211–218. Springer (2004)
11. Perreault, S., Cardou, P., Gosselin, C.M., Otis, M.J.-D.: Geometric determination of the interference-free constant-orientation workspace of parallel cable-driven mechanisms. J. Mech. Robot. **2**(3), 031016 (2010)
12. Ruiz, A.L.C., Caro, S., Cardou, P., Guay, F.: ARACHNIS: Analysis of robots actuated by cables with handy and neat interface software. In: Cable-Driven Parallel Robots, pp. 293–305. Springer (2015)
13. Wang, B., Zi, B., Qian, S., Zhang, D.: Collision free force closure workspace determination of reconfigurable planar cable driven parallel robot. In: 2016 Asia-Pacific Conference on Intelligent Robot Systems (ACIRS), pp. 26–30, July 2016
14. Blanchet, L.: Contribution à la modélisation de robots à câbles pour leur commande et leur conception. Ph.D. thesis, Université Nice Sophia Antipolis (2015)

Twist Feasibility Analysis of Cable-Driven Parallel Robots

Saman Lessanibahri[1], Marc Gouttefarde[2], Stéphane Caro[3](✉),
and Philippe Cardou[4]

[1] École Centrale de Nantes, Laboratoire des Sciences du Numérique de Nantes,
UMR CNRS 6004, 1, rue de la Noë, 44321 Nantes, France
Saman.Lessanibahri@irccyn.ec-nantes.fr

[2] Laboratoire d'Informatique, de Robotique et de Micro-électronique
de Montpellier (LIRMM), UM - CNRS, 161 rue Ada,
34095 Montpellier Cedex 5, France
marc.gouttefarde@lirmm.fr

[3] CNRS, Laboratoire des Sciences du Numérique de Nantes,
UMR CNRS 6004, 1, rue de la Noë, 44321, Nantes, France
stephane.caro@ls2n.fr

[4] Laboratoire de Robotique, Département de Génie Mécanique,
Université Laval, Québec, QC, Canada
pcardou@gmc.ulaval.ca

Abstract. Although several papers addressed the wrench capabilities of cable-driven parallel robots (CDPRs), few have tackled the dual question of their twist capabilities. In this paper, these twist capabilities are evaluated by means of the more specific concept of twist feasibility, which was defined by Gagliardini et al. in a previous work. A CDPR posture is called twist-feasible if all the twists (point-velocity and angular-velocity combinations), within a given set, can be produced at the CDPR mobile platform, within given actuator speed limits. Two problems are solved in this paper: (1) determining the set of required cable winding speeds at the CDPR winches being given a prescribed set of required mobile platform twists; and (2) determining the set of available twists at the CDPR mobile platform from the available cable winding speeds at its winches. The solutions to both problems can be used to determine the twist feasibility of n-degree-of-freedom (DOF) CDPRs driven by $m \geq n$ cables. An example is presented, where the twist-feasible workspace of a simple CDPR with $n = 2$ DOF and driven by $m = 3$ cables is computed to illustrate the proposed method.

1 Introduction

A cable-driven parallel robot (CDPR) consists of a base frame, a mobile platform, and a set of cables connecting in parallel the mobile platform to the base frame. The cable lengths or tensions can be adjusted by means of winches and a number of pulleys may be used to route the cables from the winches to the mobile platform. Among other advantages, CDPRs with very large workspaces, e.g.

[12,17], heavy payloads capabilities [1], or reconfiguration capabilities, e.g. [8,21] can be designed. Moreover, the moving parts of CDPRs being relatively light weight, fast motions of the mobile platform can be obtained, e.g. [15].

The cables of a CDPR can only pull and not push on the mobile platform and their tension shall not become larger than some maximum admissible value. Hence, for a given mobile platform pose, the determination of the feasible wrenches at the platform is a fundamental issue, which has been the subject of several previous works, e.g. [3,13]. A relevant issue is then to determine the set of wrench feasible poses, *i.e.*, the so-called Wrench-Feasible Workspace (WFW) [2,19], since the shape and size of the latter highly depends on the cable tension bounds and on the CDPR geometry [22]. Another issue which may strongly restrict the usable workspace of a CDPR or, divide it into several disjoint parts, are cable interferences. Therefore, software tools allowing the determination of the interference-free workspace and of the WFW have been proposed, e.g. [4,18]. Besides, recently, a study on acceleration capabilities was proposed in [5,9].

As noted in [7] and as well known, in addition to wrench feasibility, the design of the winches of a CDPR also requires the consideration of cable and mobile platform velocities since the selection of the winch characteristics (motors, gearboxes, and drums) has to deal with a trade-off between torque and speed. Twist feasibility is then the study of the relationship between the feasible mobile platform twists (linear and angular velocities) and the admissible cable coiling/uncoiling speeds. In the following, the cable coiling/uncoiling speeds are loosely referred to as cable velocities. The main purpose of this paper is to clarify the analysis of twist feasibility and of the related twist-feasible workspace proposed in [7]. Contrary to [7], the twist feasibility analysis proposed here is based on the usual CDPR differential kinematics where the Jacobian matrix maps the mobile platform twist into the cable velocities. This approach is most important for redundantly actuated CDPRs, whose Jacobian matrix is rectangular.

A number of concepts in this paper are known, notably from manipulability ellipsoids of serial robots, e.g. [23], and from studies on the velocity performance of parallel robots, e.g. [16]. A review of these works is however out of the scope of the present paper whose contribution boils down to a synthetic twist feasibility analysis of n-degrees-of-freedom (DOF) CDPRs driven by m cables, with $m \geq n$. The CDPR can be fully constrained or not, and the cable mass and elasticity are neglected.

The paper is organized as follows. The usual CDPR wrench and Jacobian matrices are defined in Sect. 2. Section 3 presents the twist feasibility analysis, which consists in solving two problems. The first one is the determination of the set of cable velocities corresponding to a given set of required mobile platform twists (Sect. 3.1). The second problem is the opposite since it is defined as the calculation of the set of mobile platform twists corresponding to a given set of cable velocities (Sect. 3.2). The twist and cable velocity sets considered in this paper are convex polytopes. In Sect. 4, a 2-DOF point-mass CDPR driven by 3 cables is considered to illustrate the twist feasibility analysis. Section 5 concludes the paper.

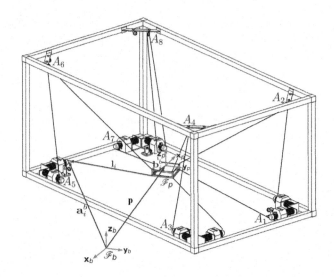

Fig. 1. Geometric description of a fully constrained CDPR

2 Wrench and Jacobian Matrices

In this section, the well-known wrench matrix and Jacobian matrix of n-DOF m-cable CDPRs are defined. The wrench matrix maps the cable tensions into the wrench applied by the cables on the CDPR mobile platform. The Jacobian matrix relates the time derivatives of the cable lengths to the twist of the mobile platform. These two matrices are essentially the same since one is minus the transpose of the other.

Some notations and definitions are first introduced. As illustrated in Fig. 1, let us consider a fixed reference frame, $\mathscr{F}_b$, of origin O_b and axes $\mathbf{x}_b$, $\mathbf{y}_b$ and $\mathbf{z}_b$. The coordinate vectors $^b\mathbf{a}_i, i = 1, \ldots, m$ define the positions of the exit points, $A_i, i = 1, \ldots, m$, with respect to frame $\mathscr{F}_b$. A_i is the point where the cable exits the base frame and extends toward the mobile platform. In this paper, the exit points A_i are assumed to be fixed, i.e., the motion of the output pulleys is neglected. A frame $\mathscr{F}_p$, of origin O_p and axes $\mathbf{x}_p$, $\mathbf{y}_p$ and $\mathbf{z}_p$, is attached to the mobile platform. The vectors $^p\mathbf{b}_i, i = 1, \ldots, m$ are the position vectors of the points B_i in $\mathscr{F}_p$. The cables are attached to the mobile platform at points B_i.

The vector $^b\mathbf{l}_i$ from B_i to A_i is given by

$$^b\mathbf{l}_i = {}^b\mathbf{a}_i - \mathbf{p} - \mathbf{R}\,^p\mathbf{b}_i, \ i = 1, \ldots, m \tag{1}$$

where $\mathbf{R}$ is the rotation matrix defining the orientation of the mobile platform, i.e., the orientation of $\mathscr{F}_p$ in $\mathscr{F}_b$, and $\mathbf{p}$ is the position vector of $\mathscr{F}_p$ in $\mathscr{F}_b$. The length of the straight line segment A_iB_i is $l_i = ||^b\mathbf{l}_i||_2$ where $||\cdot||_2$ is the Euclidean norm. Neglecting the cable mass, l_i corresponds to the length of the cable segment from point A_i to point B_i. Moreover, neglecting the cable elasticity, l_i is the "active" length of the cable that should be unwound from the winch drum. The unit vectors along the cable segment A_iB_i is given by

$$^b\mathbf{d}_i = {}^b\mathbf{l}_i/l_i, \quad i = 1,\ldots,m \qquad (2)$$

Since the cable mass is neglected in this paper, the force applied by the cable on the platform is equal to $\tau_i{}^b\mathbf{d}_i$, τ_i being the cable tension. The static equilibrium of the CDPR platform can then be written [14, 20]

$$\mathbf{W}\boldsymbol{\tau} + \mathbf{w}_e = 0 \qquad (3)$$

where $\mathbf{w}_e$ is the external wrench acting on the platform, $\boldsymbol{\tau} = [\tau_1,\ldots,\tau_m]^T$ is the vector of cable tensions, and $\mathbf{W}$ is the wrench matrix. The latter is an $n \times m$ matrix defined as

$$\mathbf{W} = \begin{bmatrix} {}^b\mathbf{d}_1 & {}^b\mathbf{d}_2 & \cdots & {}^b\mathbf{d}_m \\ \mathbf{R}^p\mathbf{b}_1 \times {}^b\mathbf{d}_1 & \mathbf{R}^p\mathbf{b}_2 \times {}^b\mathbf{d}_2 & \cdots & \mathbf{R}^p\mathbf{b}_m \times {}^b\mathbf{d}_m \end{bmatrix} \qquad (4)$$

The differential kinematics of the CDPR establishes the relationship between the twist $\mathbf{t}$ of the mobile platform and the time derivatives of the cable lengths $\dot{\mathbf{l}}$

$$\mathbf{J}\mathbf{t} = \dot{\mathbf{l}} \qquad (5)$$

where $\mathbf{J}$ is the $m \times n$ Jacobian matrix and $\dot{\mathbf{l}} = \left[\dot{l}_1,\ldots,\dot{l}_m\right]^T$. The twist $\mathbf{t} = [\dot{\mathbf{p}}, \boldsymbol{\omega}]^T$ is composed of the velocity $\dot{\mathbf{p}}$ of the origin of frame $\mathscr{F}_p$ with respect to $\mathscr{F}_b$ and of the angular velocity $\boldsymbol{\omega}$ of the mobile platform with respect to $\mathscr{F}_b$. Moreover, the well-known kineto-statics duality leads to

$$\mathbf{J} = -\mathbf{W}^T \qquad (6)$$

In the remainder of this paper, $\dot{\mathbf{l}}$ is loosely referred to as cable velocities. The wrench and Jacobian matrices depend on the geometric parameters $\mathbf{a}_i$ and $\mathbf{b}_i$ of the CDPR and on the mobile platform pose, namely on $\mathbf{R}$ and $\mathbf{p}$.

3 Twist Feasibility Analysis

This section contains the contribution of the paper, namely, a twist feasibility analysis which consists in solving the following two problems.

1. For a given pose of the mobile platform of a CDPR and being given a set $[\mathbf{t}]_r$ of required mobile platform twists, determine the corresponding set of cable velocities $\dot{\mathbf{l}}$. The set of cable velocities to be determined is called the *Required Cable Velocity Set* (RCVS) and is denoted $\left[\dot{\mathbf{l}}\right]_r$. The set $[\mathbf{t}]_r$ is called the *Required Twist Set* (RTS).

2. For a given pose of the mobile platform of a CDPR and being given a set $\left[\dot{\mathbf{l}}\right]_a$ of available (admissible) cable velocities, determine the corresponding set of mobile platform twists $\mathbf{t}$. The former set, $\left[\dot{\mathbf{l}}\right]_a$, is called the *Available Cable Velocity Set* (ACVS) while the latter is denoted $[\mathbf{t}]_a$ and called the *Available Twist Set* (ATS).

In this paper, the discussion is limited to the cases where both the RTS $[\mathbf{t}]_r$ and the ACVS $[\mathbf{\dot{l}}]_a$ are *convex polytopes*.

Solving the first problem provides the RCVS from which the maximum values of the cable velocities required to produce the given RTS $[\mathbf{t}]_r$ can be directly deduced. If the winch characteristics are to be determined, the RCVS allows to determine the required speeds of the CDPR winches. If the winch characteristics are already known, the RCVS allows to test whether or not the given RTS is feasible.

Solving the second problem provides the ATS which is the set of twists that can be produced at the mobile platform. It is thus useful either to determine the velocity capabilities of a CDPR or to check whether or not a given RTS is feasible.

Note that the feasibility of a given RTS can be tested either in the cable velocity space, by solving the first problem, or in the space of platform twists, by solving the second problem. Besides, note also that the twist feasibility analysis described above does not account for the dynamics of the CDPR.

3.1 Problem 1: Required Cable Velocity Set (RCVS)

The relationship between the mobile platform twist $\mathbf{t}$ and the cable velocities $\mathbf{\dot{l}}$ is the differential kinematics in (5). According to this equation, the RCVS $[\mathbf{\dot{l}}]_r$ is defined as the image of the convex polytope $[\mathbf{t}]_r$ under the linear map $\mathbf{J}$. Consequently, $[\mathbf{\dot{l}}]_r$ is also a convex polytope [24].

Moreover, if $[\mathbf{t}]_r$ is a box, the RCVS $[\mathbf{\dot{l}}]_r$ is a particular type of polytope called a zonotope. Such a transformation of a box into a zonotope has previously been studied in CDPR wrench feasibility analysis [3,10,11]. Indeed, a box of admissible cable tensions is mapped by the wrench matrix $\mathbf{W}$ into a zonotope in the space of platform wrenches. However, a difference lies in the dimensions of the matrices $\mathbf{J}$ and $\mathbf{W}$, $\mathbf{J}$ being of dimensions $m \times n$ while $\mathbf{W}$ is an $n \times m$ matrix, where $n \leq m$. When $n < m$, on the one hand, $\mathbf{W}$ maps the m-dimensional box of admissible cable tensions into the n-dimensional space of platform wrenches. On the other hand, $\mathbf{J}$ maps n-dimensional twists into its range space which is a linear subspace of the m-dimensional space of cable velocities $\mathbf{\dot{l}}$. Hence, when $\mathbf{J}$ is not singular, the n-dimensional box $[\mathbf{t}]_r$ is mapped into the zonotope $[\mathbf{\dot{l}}]_r$ which lies into the n-dimensional range space of $\mathbf{J}$, as illustrated in Fig. 3. When $\mathbf{J}$ is singular and has rank r, $r < n$, the n-dimensional box $[\mathbf{t}]_r$ is mapped into a zonotope of dimension r.

When an ACVS $[\mathbf{\dot{l}}]_a$ is given, a pose of the mobile platform of a CDPR is twist feasible if

$$[\mathbf{\dot{l}}]_r \subseteq [\mathbf{\dot{l}}]_a \qquad (7)$$

Since $[\dot{\mathbf{i}}]_a$ is a convex polytope, (7) is verified whenever all the vertices of $[\dot{\mathbf{i}}]_r$ are included in $[\dot{\mathbf{i}}]_a$. Moreover, it is not difficult to prove that $[\dot{\mathbf{i}}]_r$ is the convex hull of the images under $\mathbf{J}$ of the vertices of $[\mathbf{t}]_r$. Hence, a simple method to verify if a CDPR pose is twist feasible consists in verifying whether the images of the vertices of $[\mathbf{t}]_r$ are all included into $[\dot{\mathbf{i}}]_a$.

3.2 Problem 2: Available Twist Set (ATS)

The problem is to determine the ATS $[\mathbf{t}]_a$ corresponding to a given ACVS $[\dot{\mathbf{i}}]_a$.

In the most general case considered in this paper, $[\dot{\mathbf{i}}]_a$ is a convex polytope. By the Minkowski-Weyl's Theorem, a polytope can be represented as the solution set of a finite set of linear inequalities, the so-called (halfspace) *H-representation* of the polytope [6, 24], i.e.

$$[\dot{\mathbf{i}}]_a = \{\, \dot{\mathbf{i}} \mid \mathbf{C}\dot{\mathbf{i}} \leq \mathbf{d} \,\} \tag{8}$$

where matrix $\mathbf{C}$ and vector $\mathbf{d}$ are assumed to be known.

According to (5), the ATS is defined as

$$[\mathbf{t}]_a = \{\, \mathbf{t} \mid \mathbf{J}\mathbf{t} \in [\dot{\mathbf{i}}]_a \,\} \tag{9}$$

which, using (8), implies that

$$[\mathbf{t}]_a = \{\, \mathbf{t} \mid \mathbf{C}\mathbf{J}\mathbf{t} \leq \mathbf{d} \,\} \tag{10}$$

The latter equation provides an H-representation of the ATS $[\mathbf{t}]_a$.

In practice, when the characteristics of the winches of a CDPR are known, the motor maximum speeds limit the set of possible cable velocities as follows

$$\dot{l}_{i,min} \leq \dot{l}_i \leq \dot{l}_{i,max} \tag{11}$$

where $\dot{l}_{i,min}$ and $\dot{l}_{i,max}$ are the minimum and maximum cable velocities. Note that, usually, $\dot{l}_{i,min} = -\dot{l}_{i,max}$, $\dot{l}_{1,min} = \dot{l}_{2,min} = \ldots = \dot{l}_{m,min}$, and $\dot{l}_{1,max} = \dot{l}_{2,max} = \ldots = \dot{l}_{m,max}$. In other words, $\mathbf{C}$ and $\mathbf{d}$ in (8) are defined as

$$\mathbf{C} = \begin{bmatrix} \mathbf{1} \\ -\mathbf{1} \end{bmatrix} \quad \text{and} \quad \mathbf{d} = \begin{bmatrix} \dot{l}_{1,max}, \ldots, \dot{l}_{m,max}, -\dot{l}_{1,min}, \ldots, -\dot{l}_{m,min} \end{bmatrix}^T \tag{12}$$

where $\mathbf{1}$ is the $m \times m$ identity matrix. Equation (10) can then be written as follows

$$[\mathbf{t}]_a = \{\, \mathbf{t} \mid \dot{\mathbf{l}}_{min} \leq \mathbf{J}\mathbf{t} \leq \dot{\mathbf{l}}_{max} \,\} \tag{13}$$

where $\dot{\mathbf{l}}_{min} = \begin{bmatrix} \dot{l}_{1,min}, \ldots, \dot{l}_{m,min} \end{bmatrix}^T$ and $\dot{\mathbf{l}}_{max} = \begin{bmatrix} \dot{l}_{1,max}, \ldots, \dot{l}_{m,max} \end{bmatrix}^T$.

When a RTS $[\mathbf{t}]_r$ is given, a pose of the mobile platform of a CDPR is twist feasible if
$$[\mathbf{t}]_r \subseteq [\mathbf{t}]_a \qquad (14)$$
In this paper, $[\mathbf{t}]_r$ is assumed to be a convex polytope. Hence, (14) is verified whenever all the vertices of $[\mathbf{t}]_r$ are included in $[\mathbf{t}]_a$. With the H-representation of $[\mathbf{t}]_a$ in (10) (or in (13)), testing if a pose is twist feasible amounts to verifying if all the vertices of $[\mathbf{t}]_r$ satisfy the inequality system in (10) (or in (13)). Testing twist feasibility thereby becomes a simple task as soon as the vertices of $[\mathbf{t}]_r$ are known.

Finally, let the twist feasible workspace (TFW) of a CDPR be the set of twist feasible poses of its mobile platform. It is worth noting that the boundaries of the TFW are directly available in closed form from (10) or (13). If the vertices of the (convex) RTS are denoted $\mathbf{t}_j$, $j = 1, \ldots, k$, and the rows of the Jacobian matrix are $-\mathbf{w}_i^T$, according to (13), the TFW is defined by $\dot{l}_{i,min} \leq -\mathbf{w}_i^T \mathbf{t}_j$ and $-\mathbf{w}_i^T \mathbf{t}_j \leq \dot{l}_{i,max}$, for all possible combinations of i and j. Since $\mathbf{w}_i$ contains the only variables in these inequalities that depend on the mobile platform pose, and because the closed-form expression of $\mathbf{w}_i$ as a function of the pose is known, the expressions of the boundaries of the TFW are directly obtained.

4 Case Study

This section deals with the twist feasibility analysis of the two-DOF point-mass planar CDPR driven by three cables shown in Fig. 2. The robot is 3.5 m long and 2.5 m high. The three exit points of the robot are named A_1, A_2 are A_3, respectively. The point-mass is denoted P. ${}^b\mathbf{d}_1$, ${}^b\mathbf{d}_2$ and ${}^b\mathbf{d}_3$ are the unit vectors, expressed in frame $\mathscr{F}_b$, of the vectors pointing from point-mass P to cable exit points A_1, A_2 are A_3, respectively. The 3×2 Jacobian matrix $\mathbf{J}$ of this planar CDPR takes the form:
$$\mathbf{J} = - \begin{bmatrix} {}^b\mathbf{d}_1 & {}^b\mathbf{d}_2 & {}^b\mathbf{d}_3 \end{bmatrix}^T \qquad (15)$$

Figure 3 is obtained by solving the Problem 1 formulated in Sect. 3. For the robot configuration depicted in Fig. 3a and the given RTS of the point-mass P represented in Fig. 3b, the RCVS for the three cables of the planar CDPR are illustrated in Figs. 3c and d. Note that the RTS is defined as:
$$-1 \text{ m.s}^{-1} \leq \dot{x}_P \leq 1 \text{ m.s}^{-1} \qquad (16)$$
$$-1 \text{ m.s}^{-1} \leq \dot{y}_P \leq 1 \text{ m.s}^{-1} \qquad (17)$$
where $[\dot{x}_P, \dot{y}_P]^T$ is the velocity of P in the fixed reference frame $\mathscr{F}_b$.

Figure 4 depicts the isocontours of the Maximum Required Cable Velocity (MRCV) over the Cartesian space (a) for cable 1 and (b) for all cables combined, for the required twist set shown in Fig. 3b. Those results are obtained by solving Problem 1 for all positions of point P. It is apparent that P RTS is satisfied through the Cartesian space as long as the maximum velocity of each

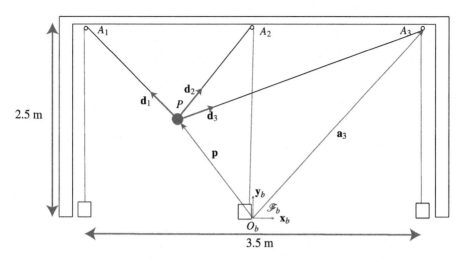

Fig. 2. A two-DOF point-mass planar cable-driven parallel robot driven by three cables

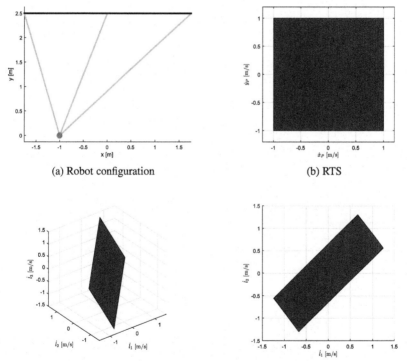

(a) Robot configuration

(b) RTS

(c) Image of the RTS into the three-dimensional cable velocity space

(d) Image of the RTS into the ($\dot{l}_1$, $\dot{l}_2$)-space

Fig. 3. Required Twist Set (RTS) of the point-mass P and corresponding required cable velocity set for the three cables of the CDPR in a given robot configuration

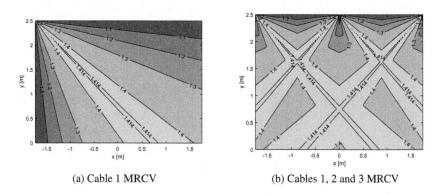

(a) Cable 1 MRCV

(b) Cables 1, 2 and 3 MRCV

Fig. 4. Maximum Required Cable Velocity (MRCV) (*a*) of cable 1 alone and (*b*) of cables 1, 2 and 3 combined over the cartesian space for the required twist set of Fig. 3b

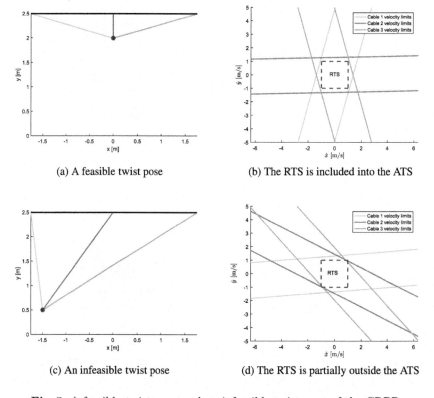

(a) A feasible twist pose

(b) The RTS is included into the ATS

(c) An infeasible twist pose

(d) The RTS is partially outside the ATS

Fig. 5. A feasible twist pose and an infeasible twist pose of the CDPR

cable is higher than $\sqrt{2}$ m.s^{-1}, namely, $\dot{l}_{1,max} = \dot{l}_{2,max} = \dot{l}_{3,max} = \sqrt{2}$ m.s^{-1} with $\dot{l}_{i,min} = -\dot{l}_{i,max}$, $i = 1, 2, 3$.

For the Available Cable Velocity Set (ACVS) defined by inequalities (11) with
$$\dot{l}_{i,max} = 1.3 \text{ m.s}^{-1}, i = 1, 2, 3 \qquad (18)$$

Figure 5 is obtained by solving the Problem 2 formulated in Sect. 3.

For the two robot configurations illustrated in Fig. 5a and c, the Available Twist Set (ATS) associated to the foregoing ACVS is determined from Eq. (13). It is noteworthy that the ATS in each configuration in delimited by three pairs of lines normal to three cables, respectively. It turns out that the first robot configuration is twist feasible for the RTS defined by Eqs. (16) and (17) because the latter is included into the ATS as shown Fig. 5b. Conversely, the second robot configuration is not twist feasible as the RTS is partially outside the ATS as shown Fig. 5d.

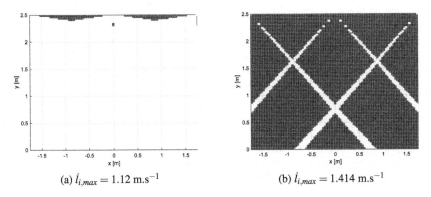

Fig. 6. TFW of the planar CDPR for two maximum cable velocities and for the RTS shown in Fig. 3b

Finally, Fig. 6 shows the TFW of the planar CDPR for two maximum cable velocities and for the RTS shown in Fig. 3b. All robot poses turned out to be twist feasible when the maximum cable velocities were set to values higher than $\sqrt{2}$ m.s^{-1} for the three cables.

5 Conclusion

In summary, this paper presents two methods of determining the twist-feasibility of a CDPR. The first method uses a set of required mobile platform twists to compute the corresponding required cable velocities, the latter corresponding to cable winding speeds at the winches. The second method takes the opposite route, i.e., it uses the available cable velocities to compute the corresponding set of available mobile platform twists. The second method can be applied to compute the twist-feasible workspace, i.e., to determine the set of mobile platform poses where a prescribed polyhedral required twist set is contained within the available twist set. This method can thus be used to analyze the CDPR speed capabilities over its workspace, which should prove useful in high-speed CDPR applications.

The proposed method can be seen as a dual to the one used to compute the wrench-feasible workspace of a CDPR, just as the velocity equations may be seen as dual to static equations. From a mathematical standpoint, however, the problem is much simpler in the case of the twist-feasible workspace, as the feasibility conditions can be obtained explicitly. Nevertheless, the authors believe that the present paper complements nicely the previous works on wrench feasibility.

Finally, we should point out that the proposed method does not deal with the issue of guaranteeing the magnitudes of the mobile platform point-velocity or angular velocity. In such a case, the required twist set becomes a ball or an ellipsoid, and thus is no longer polyhedral. This ellipsoid could be approximated by a polytope in order to apply the method proposed in this paper. However, since the accuracy of the approximation would come at the expense of the number of conditions to be numerically verified, part of our future work will be dedicated to the problem of determining the twist-feasibility of CDPRs for ellipsoidal required twist sets.

Acknowledgements. The financial support of the ANR under grant ANR-15-CE10-0006-01 (DexterWide project) is greatly acknowledged. This research work was also part of the CAROCA project managed by IRT Jules Verne (French Institute in Research and Technology in Advanced Manufacturing Technologies for Composite, Metallic and Hybrid Structures) and of the RFI ATLANSTIC 2020 CREATOR project.

References

1. Albus, J., Bostelman, R., Dagalakis, N.: The NIST robocrane. J. Robot. Syst. **10**(2), 709–724 (1993)
2. Bosscher, P., Riechel, A.T., Ebert-Uphoff, I.: Wrench-feasible workspace generation for cable-driven robots. IEEE Trans. Robot. **22**(5), 890–902 (2006)
3. Bouchard, S., Gosselin, C.M., Moore, B.: On the ability of a cable-driven robot to generate a prescribed set of wrenches. ASME J. Mech. Robot. **2**(1), 1–10 (2010)
4. Ruiz, A.L.C., Caro, S., Cardou, P., Guay, F.: ARACHNIS: Analysis of robots actuated by cables with handy and neat interface software. In: Mechanisms and Machine Science, vol. 32, pp. 293–305 (2015)
5. Eden, J., Lau, D., Tan, Y., Oetomo, D.: Available acceleration set for the study of motion capabilities for cable-driven robots. Mech. Mach. Theory **105**, 320–336 (2016)
6. Fukuda, K.: Frequently asked questions in polyhedral computation. Technical report. http://www.ifor.math.ethz.ch/~fukuda/polyfaq/polyfaq.html
7. Gagliardini, L., Caro, S., Gouttefarde, M., Girin, A.: Dimensioning of cable-driven parallel robot actuators, gearboxes and winches according to the twist feasible workspace. In: Proceedings of the IEEE International Conference on Automation Science and Engineering, Gothenburg, Sweden (2015)

8. Gagliardini, L., Caro, S., Gouttefarde, M., Girin, A.: Discrete reconfiguration planning for cable-driven parallel robots. Mech. Mach. Theory **100**, 313–337 (2016)
9. Gagliardini, L., Gouttefarde, M., Caro, S.: Determination of a dynamic feasible workspace for cable-driven parallel robots. In: The 15th International Symposium on Advances in Robot Kinematics, Grasse, France (2016)
10. Gallina, P., Rosati, G., Rossi, A.: 3-DOF wire driven planar haptic interface. J. Intell. Robot. Syst. **32**(1), 23–36 (2001)
11. Gouttefarde, M., Krut, S.: Characterization of parallel manipulator available wrench set facets. In: Advanced in Robot Kinematics, Portoroz, Slovenia (2010)
12. Gouttefarde, M., Lamaury, J., Reichert, C., Bruckmann, T.: A versatile tension distribution algorithm for n-DOF parallel robots driven by $n+2$ cables. IEEE Trans. Robot. **31**(6), 1444–1457 (2015)
13. Hassan, M., Khajepour, A.: Analysis of bounded cable tensions in cable-actuated parallel manipulators. IEEE Trans. Robot. **27**(5), 891–900 (2011)
14. Hiller, M., Fang, S., Mielczarek, S., Verhoeven, R., Franitza, D.: Design, analysis and realization of tendon-based parallel manipulators. Mech. Mach. Theory **40**(4), 429–445 (2005)
15. Kawamura, S., Kino, H., Won, C.: High-speed manipulation by using parallel wire-driven robots. Robotica **18**, 13–21 (2000)
16. Krut, S., Company, O., Pierrot, F.: Velocity performance indices for parallel mechanisms with actuation redundancy. Robotica **22**, 129–139 (2004)
17. Lambert, C., Nahon, M., Chalmers, D.: Implementation of an aerostat positioning system with cable control. IEEE/ASME Trans. Mechatron. **12**(1), 32–40 (2007)
18. Perreault, S., Cardou, P., Gosselin, C., Otis, M.: Geometric determination of the interference-free constant-orientation workspace of parallel cable-driven mechanisms. ASME J. Mech. Robot. **2**(3), 031016 (2010)
19. Riechel, A.T., Ebert-Uphoff, I.: Force-feasible workspace analysis for underconstrained point-mass cable robots. In: Proceedings of the IEEE International Conference on Robotics and Automation (ICRA), New Orleans, pp. 4956–4962 (2004)
20. Roberts, R.G., Graham, T., Lippitt, T.: On the inverse kinematics, statics, and fault tolerance of cable-suspended robots. J. Robot. Syst. **15**(10), 581–597 (1998)
21. Rosati, G., Zanotto, D., Agrawal, S.K.: On the design of adaptive cable-driven systems. J. Mech. Robot. **3**(2), 021004 (2011)
22. Verhoeven, R.: Analysis of the Workspace of Tendon-Based Stewart-Platforms. Ph.D. thesis, University of Duisburg-Essen, Germany (2004)
23. Yoshikawa, T.: Foundations of Robotics, 2nd edn. MIT Press, Boston (1990)
24. Ziegler, G.M.: Lectures on Polytopes. Graduate Texts in Mathematics, vol. 152. Springer, New York (1994)

Initial Length and Pose Calibration for Cable-Driven Parallel Robots with Relative Length Feedback

Darwin Lau[✉]

The Chinese University of Hong Kong, Shatin, Hong Kong
darwinlau@cuhk.edu.hk

Abstract. Feedback of cable lengths is commonly used in the determination of the robot pose for cable-driven parallel robots (*CDPRs*). As such, accurate information on the absolute cable length is important. However, for most CDPRs equipped with relative encoders, the absolute cable lengths depend on the system's initial lengths. The initial cable length, and hence the robot's initial pose, is typically unknown. In this paper, a forward kinematics based method to determine (*calibrate*) for the initial cable lengths and robot pose is proposed. The calibration problem is solved as a non-linear least squares optimisation problem, where only the relative lengths of cables over any random trajectory are required and measured. The proposed method is generic in the sense that it can be applied to any type of CDPR. The simulation and experimental results for various robots show that the method can effectively and efficiently determine the initial cable lengths and pose of the cable robot. This is useful in order to obtain more accurate cable length data to be used for forward kinematics to determine the robot's pose.

1 Introduction

Cable-driven Parallel Robots (*CDPRs*) are a class of parallel manipulators where the rigid links are replaced by cables. The advantages of CDPRs include: high payload to weight ratio, large operational distances, ease of reconfigurability, ease of transportability and naturally bio-inspired. An important characteristic of CDPRs is that the cables can only apply forces in tension (*positive cable force*). This constraint results in the need of actuation redundancy for a CDPR to be fully constrained, creating challenges in the modelling and analysis of CDPRs.

For CDPRs, the forward kinematics (*FK*) problem refers to the determination of the robot pose when provided with the cable lengths and is fundamentally important in the study of CDPRs. As a parallel manipulator, the FK problem is challenging as there is no closed form analytical solution in general. Furthermore, even when solving the problem numerically there may be either no valid solutions or the existence of multiple solutions. For an n degrees-of-freedom (*DoFs*) CDPR actuated by m cables, the FK problem requires the determination of n unknowns (the number of DoFs) from the m equations that relate the length of cables and the system pose. As such, for fully constrained CDPRs ($m \geq n+1$) there will be more equations than unknowns for the FK problem.

The two primary types of approaches to solve the FK of CDPRs are analytical or numerical techniques. Analytical techniques are difficult to apply due to the nonlinearity and complexity of the kinematic equations, and have only been used for simpler CDPR systems [1,2]. Numerical methods consider the FK relationship generically and can be used for any type of CDPR. The most common numerical approach is to solve the FK as a non-linear least squares problem [3,4]. Other numerical techniques include using neural networks [5] and interval analysis [6].

In addition to using the forward kinematics to determine the robot pose, FK has also been used in the calibration of kinematic parameters of CDPRs. *Calibration* is used in CDPRs to correct for any kinematic or dynamic modelling uncertainties or errors. In [7–11], the attachment locations of the cables were calibrated using the cable length feedback and FK. Some studies also considered dynamic parameters such as mass and cable stiffness [12,13].

Previous studies have focused on the calibration of static system parameters, such as the cable attachment locations and cable elasticity, which do not change significantly over time. Such parameters are slow changing and hence only need to be calibrated infrequently. However, some parameters, such as the initial robot pose and cable lengths, may be different each time the system is turned on. A majority of CDPRs use motors that equipped with encoders to obtain feedback of the cable lengths at each instance in time. While some use multi-turn absolute encoders [3], most possess only relative encoders. As a result, the initial cable lengths and robot pose are typically not known and is different each time.

One simple approach that has been used to know the initial cable lengths is to place the robot in a known pose, referred to as the *initial pose*, before enabling the robot. However, in some applications it may be difficult to set up the robot consistently and accurately in this way. Another approach is to employ external sensors such as camera tracking systems (*external calibration*). However, it is normally preferred to perform *internal calibration* using the CDPR's internal sensors. In [2], the initial cable lengths calibration for a 2-DoF point mass planar CDPR actuated by 4 cables is performed. The method is based on a "jitter" approach where the lengths of two cables are perturbed and the measured lengths of the remaining two cables provide information to solve the initial pose of the system. Although effective, the approach requires the closed-form analytical solution to the forward kinematics. As such, the method would only work for simple systems, such as the 2-DoF CDPR [2].

Accurate knowledge of the initial lengths is important for two purposes. First, the initial length can provide knowledge of the initial position of the robot end-effector. Second, for CDPRs equipped with relative encoders, the initial length must be used to compute the absolute cable lengths. This absolute cable length is then used to determine the pose of the robot through FK. As such, inaccuracies in the initial length would result in error in the forward kinematics, and hence robot pose, which cannot be eliminated. In summary, the determination of the initial cable lengths in a generic manner for any type of CDPR without requiring the system to begin at a known position has not been studied thus far.

In this paper, a generic method to calibrate for the initial cable lengths for a CDPR with relative encoders using a forward kinematics approach is proposed. Without assuming any initial robot pose, the CDPR is commanded to perform any random motion in a way that excites all of the system's degrees-of-freedom. The resulting relative changes in cable lengths are captured and then used to calibrate for the initial cable lengths and initial pose. This method assumes that the attachment locations of the cables are known beforehand. The proposed approach is validated both in simulation and hardware experiments for different CDPRs to show its effectiveness and ability to be generically used on different systems. Furthermore, the proposed algorithm is implemented in the open-source cable-robot software CASPR [14] and the source-code is publicly available[1].

2 Numerical Forward Kinematics Formulations

It was shown in [15] that the kinematics of any generalised CDPR models (n DoF actuated by m cables), as shown in Fig. 1, could be expressed as

$$\mathbf{l} = \mathbf{f}(\mathbf{q}) \tag{1}$$

where $\mathbf{l} = [l_1 \cdots l_m]^T \in \mathbb{R}^m$ and $\mathbf{q} = [q_1 \cdots q_n]^T \in \mathbb{R}^n$ are the vector of cable lengths and pose of the system, respectively.

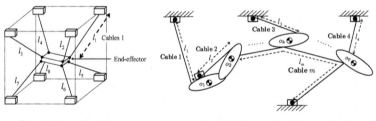

(a) Single link cable-driven robot (b) Multilink cable-driven robot

Fig. 1. CDPR models and the cable lengths

Taking the time derivative of (1) results in the well established relationship

$$\dot{\mathbf{l}} = L(\mathbf{q})\dot{\mathbf{q}} \tag{2}$$

where $L \in \mathbb{R}^{m \times n}$ is the Jacobian matrix relating the pose and cable length derivatives.

---

[1] CASPR and the presented work can be accessed at https://www.github.com/darwinlau/CASPR.

The inverse kinematics problem, the determination of cable lengths l for a given pose $\mathbf{q}$, is a trivial problem using (1). However, the forward kinematics (*FK*) problem is much more challenging as the inverse of the kinematic relationship $\mathbf{q} = \mathbf{f}^{-1}(\mathbf{l})$ does not have an analytical closed-form solution in general. One common way to solve the FK problem is to formulate the optimisation problem:

$$\mathbf{q}^* = \arg\min_{\mathbf{q}} \|\mathbf{l} - \mathbf{f}(\mathbf{q})\|^2. \qquad (3)$$

In general, the problem in (3) is a non-linear least squares problem that can be solved using techniques such as the Levenberg-Marquardt Algorithm. The poses which result in a zero objective function value are solutions to the FK problem since $\|\mathbf{l} - \mathbf{f}(\mathbf{q})\|^2 = 0 \Leftrightarrow \mathbf{l} = \mathbf{f}(\mathbf{q})$. However, it is difficult to achieve a zero objective function value in real systems due to the presence of sensor noise. As such, in practice the solution (minimum) of (3) with a small objective function value is taken as the solution to the FK problem since $\|\mathbf{l} - \mathbf{f}(\mathbf{q})\|^2 \approx 0 \Leftrightarrow \mathbf{l} \approx \mathbf{f}(\mathbf{q})$. Note again that the FK problem, from (1), has m equations and n unknown variables. For fully constrained systems $m \geq n + 1$, resulting in an overdetermined problem.

3 Least Squares Problems for Initial Lengths

In the FK problem presented in (3), it is assumed that the *absolute length* of the cables $\mathbf{l}(t)$ is known at all times. However, for CDPRs with relative encoders, only the *relative length* of the cables $\mathbf{l}_r(t)$ since $t = 0$ is known. The relationship between the absolute and relative cable lengths at any instance in time t can be described as

$$\mathbf{l}(t) = \mathbf{l}_0 + \mathbf{l}_r(t) = \mathbf{f}(\mathbf{q}(t)) \qquad (4)$$

where $\mathbf{l}_0 = \mathbf{l}(0)$ is the vector of initial cable lengths of the CDPR since $\mathbf{l}_r(0) = \mathbf{0}$. It can be observed that (4) contains $n+m$ unknowns ($\mathbf{q}$ and $\mathbf{l}_0$) but only with m equations. As such, there are not enough equations to uniquely determine both the initial cable length and pose. One important property that can be taken advantage of is that after the system is turned on, $\mathbf{l}_0$ is time invariant until the system is restarted.

As such, initial length calibration and FK problem can be simultaneously solved by considering the problem in (4) over a set of different time instances, or samples, for a trajectory. Assuming that p different instances in time $t \in \{t_1, t_2, \cdots, t_p\}$ are selected to solve for the initial length and FK problem, the following non-linear system of equations can be expressed as

$$\mathbf{l}_0 + \mathbf{l}_r(t_i) = \mathbf{f}(\mathbf{q}(t_i)), \quad i = 1, \cdots, p \qquad (5)$$

The equations in (5) possess a total of $m + n \times p$ unknowns (the initial length $\mathbf{l}_0$ and the poses $\mathbf{q}(t)$ at each time instance $t \in \{t_1, t_2, \cdots, t_p\}$) and

$m \times p$ equations. The vector of unknown variables for (5) can be denoted as $\mathbf{x} = [\mathbf{l}_0^T \; \mathbf{q}^T(t_1) \; \mathbf{q}^T(t_2) \; \cdots \; \mathbf{q}^T(t_p)]^T$.

In a similar manner to (3), the initial length calibration problem can be solved through the non-linear least squares (*NLLS*) optimisation problem

$$\mathbf{x}^* = \{\mathbf{l}_0^*, \mathbf{q}_1^*, \cdots, \mathbf{q}_p^*\} = \underset{\{\mathbf{l}_0, \mathbf{q}_1, \cdots, \mathbf{q}_p\}}{\arg\min} \sum_{i=1}^{p} \|\mathbf{l}_r(t_i) + \mathbf{l}_0 - \mathbf{f}(\mathbf{q}_i)\|^2 \qquad (6)$$

where $\mathbf{q}_i := \mathbf{q}(t_i)$. Since the problem (6) is an NLLS problem, numerical methods such as the Levenberg-Marquardt algorithm can be employed. For such problems and approaches, the initial guess of the solution and the Jacobian value of the NLLS objective function would significantly increase the computational time and accuracy of the non-linear optimisation problem. Expressing (6) in the standard form

$$\mathbf{x}^* = \underset{\mathbf{x}}{\arg\min} \; \|\mathbf{g}(\mathbf{x})\|^2 \qquad (7)$$

the non-linear vector function can be equivalently expressed from (6) as

$$\mathbf{g}(\mathbf{x}) = \begin{bmatrix} \mathbf{l}_0 + \mathbf{l}_r(t_1) - \mathbf{f}(\mathbf{q}_1) \\ \vdots \\ \mathbf{l}_0 + \mathbf{l}_r(t_p) - \mathbf{f}(\mathbf{q}_p) \end{bmatrix}. \qquad (8)$$

From (8), the problem Jacobian $\frac{\partial \mathbf{g}}{\partial \mathbf{x}}$ can be expressed analytically as

$$\frac{\partial \mathbf{g}}{\partial \mathbf{x}} = \begin{bmatrix} I_{m \times m} & -L(\mathbf{q}_1) & 0_{m \times n} & \cdots & 0_{m \times n} \\ I_{m \times m} & 0_{m \times n} & -L(\mathbf{q}_2) & \cdots & 0_{m \times n} \\ \vdots & \vdots & \vdots & \ddots & \vdots \\ I_{m \times m} & 0_{m \times n} & 0_{m \times n} & \cdots & -L(\mathbf{q}_p) \end{bmatrix}. \qquad (9)$$

It will be shown in the results of Sect. 4 that the use of the Jacobian matrix (9) significantly reduces the required optimisation time to perform the initial length calibration and also is more robust to inaccurate initial guesses of $\mathbf{x}$.

In order to solve (6), a sufficient number of points p in the random motion should be sampled such that the number of equations mp is equal to or greater than the number of unknowns $m + np$, that is, $mp \geq m + np$. As such, the minimum number of trajectory points $p \in \mathbb{Z}$ that should be selected should be $p \geq m/(m-n)$. As will be shown in Sect. 4, the selection of the number of sample trajectory points p for the initial length calibration has a significant effect on the computational speed and effectiveness. Moreover, it is important to note that the sampled motion points of the random motion must excite the different degrees-of-freedom such that the NLLS optimisation of (6) have sufficient measurements to recover $\mathbf{x}$.

In summary, the proposed method uses the fact that the lengths of different cables for a trajectory must be related (kinematically consistent) due to the CDPR cable attachments. As such, the calibration problem is equivalent to determining the set of initial lengths that produce kinematically consistent cable lengths.

4 Simulation Results

The two simulation examples, performed through CASPR [14], aim to show: (1) the ability to calibrate for uncertain initial poses; (2) the effectiveness and efficiency for different number of samples; and (3) the short calibration time required.

4.1 Planar Robot Example

Figure 2(a) shows the 3-DoF planar robot model actuated by 4 cables, where $\mathbf{q} = [x \ y \ \theta]^T$ represent translations (x and y) and orientation θ of the robot, respectively.

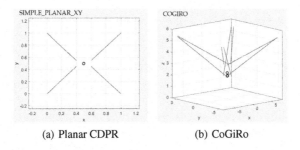

(a) Planar CDPR   (b) CoGiRo

Fig. 2. CDPR models used in the simulation examples

To demonstrate the initial length calibration, the reference trajectory as shown in Fig. 3(a), with initial pose $\mathbf{q}(t = 0) = [0.3 \ 0.6 \ 0.1]^T$ and final pose $\mathbf{q}(t = 4) = [0.2 \ 0.3 \ 0.2]^T$, will be used. At pose the initial pose $\mathbf{q}(t = 0)$, the cable lengths can be determined using inverse kinematics as $\mathbf{l}_0 = [0.578 \ 0.8176 \ 0.6946 \ 0.4099]^T$. Using $\mathbf{l}_0$ and the cable length trajectory $\mathbf{l}(t)$ obtained from computing the inverse kinematics $\mathbf{q}(t)$ (Fig. 3(a)), the relative cable lengths $\mathbf{l}_r(t)$ (emulating the cable length feedback from relative encoders) can be determined using (4).

Since the initial length is unknown, without loss of generality it will be assumed for this example that $\bar{\mathbf{l}}_0 = [0.5 \ 0.5 \ 0.5 \ 0.5]^T$. Using this, the absolute cable lengths trajectory with the erroneous initial length $\bar{\mathbf{l}}(t)$ can be determined as $\bar{\mathbf{l}}(t) = \bar{\mathbf{l}}_0 + \mathbf{l}_r(t)$. Using $\bar{\mathbf{l}}(t)$, the resulting joint space trajectory $\bar{\mathbf{q}}(t)$, shown in Fig. 3(b), was determined using the NLLS FK method from (3). As expected, it can be clearly observed that the resulting trajectory $\bar{\mathbf{q}}(t)$ is significantly different to that of $\mathbf{q}(t)$, demonstrating the impact in FK feedback when the initial lengths

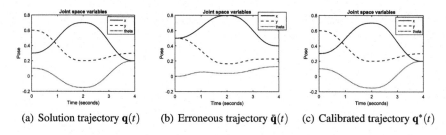

Fig. 3. Joint space trajectory results $q(t)$ for planar CDPR simulation

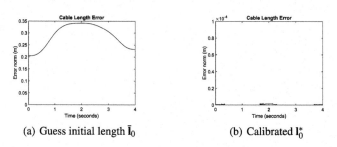

Fig. 4. Error profile in FK using different initial cable lengths for planar robot

and pose are unknown. Figure 4(a) shows the error value of the FK optimisation result $\bar{\mathbf{e}}(t) = \|\bar{\mathbf{l}}(t) - \mathbf{f}(\bar{\mathbf{q}}(t))\|$, confirming the error observed in the trajectory of $\bar{\mathbf{q}}(t)$.

Using the calibration method presented in Sect. 3, the initial cable lengths $\mathbf{l}_0^*$ is determined using only the relative cable length $\mathbf{l}_r(t)$ as the input to the NLLS problem (6). The relative length profile $\mathbf{l}_r(t)$ consists of a total of $p = 401$ time samples (from $t = 0$ s to $t = 4$ s at $\Delta t = 0.01$ s). With this sample size, the problem (6) will have 1207 unknowns. To solution $\mathbf{x}^*$ to calibration problem resulted in the initial cable lengths solution of $\mathbf{l}_0^* = [0.578 \ 0.8176 \ 0.6946 \ 0.4099]^T$, with an error norm compared with the nominal solution $\mathbf{l}_0$ of $\|\mathbf{l}_0 - \mathbf{l}_0^*\| = 4.7868 \times 10^{-9}$. Using this solution and FK, the absolute cable length $\mathbf{l}^*(t) = \mathbf{l}_0^* + \mathbf{l}_r(t)$ produced the joint space trajectory of the calibration motion $\mathbf{q}^*(t)$ and the resulting FK error norm $\mathbf{e}^*(t) = \|\mathbf{l}^*(t) - \mathbf{f}(\mathbf{q}^*(t))\|$ shown in Figs. 3(c) and 4(b), respectively. These results show that the proposed method is able to determine the initial cable lengths, initial robot pose and also the calibration trajectory motion without prior knowledge of the initial state or the calibration motion to be performed.

In the results of Figs. 3 and 4, every point on the calibration motion $p = 401$ was used in the calibration optimisation. However, it is also possible to take a subset sample of $\mathbf{l}_r(t)$ to use within the calibration. Table 1 shows the properties of the calibration method for different sample frequency (every N of the trajectory points are taken). The comparison results show that if not enough samples are taken, such as $p = 5$ and 9 for $N = 100$ and 50, respectively, the initial cable lengths cannot be correctly determined. It can also be observed that for larger samples the computational time is higher, although no significant increase in accuracy is achieved. However, it is worth noting that the calibration time of $p = 401$ over a 4 s motion is only 5.28 s, making this method very practical in use within real applications.

Table 1. Initial length calibration for different sample frequencies

	Sample frequency N					
	1	5	20	30	50	100
No. of samples (p)	401	81	21	11	9	5
Dimension of $\mathbf{x}$	1207	247	67	37	31	19
Calibration time (s)	5.28	1.38	0.60	0.20	0.40	0.10
Initial length error	4.8×10^{-9}	2.58×10^{-9}	3.2×10^{-9}	2.84×10^{-8}	0.18	0.20
FK error $\sum_t \mathbf{e}^*(t)$	1.04×10^{-5}	1.04×10^{-5}	1.04×10^{-5}	1.04×10^{-5}	0.01	0.01

As discussed above, the effectiveness and efficiency of the NLLS optimisation problem can be improved by providing an initial guess $\mathbf{x}_{guess}$ and the Jacobian matrix (9). In the above simulations, the initial cable lengths were simply always set as a constant value of $\bar{\mathbf{l}}_0 = [0.5 \ 0.5 \ 0.5 \ 0.5]^T$ regardless of the calibration motion. For the initial guess of the trajectory, the erroneous trajectory $\bar{\mathbf{q}}(t)$ determined from $\bar{\mathbf{l}}_0$ and FK (Fig. 3(b)). This shows that no trajectory specific information about the initial cable lengths or pose are required for the calibration method.

4.2 Spatial Robot Example

As no assumptions on the robot type are required, the proposed approach can be used for any CDPR. In this example, the initial length calibration of the 6-DoF spatial cable robot CoGiRo [16] actuated by 8 cables, as shown in Fig. 2(b), is demonstrated. The generalised coordinates of the robot can be described by $\mathbf{q} = [x \ y \ z \ \alpha \ \beta \ \gamma]^T$, where x, y, z are the translational DoFs and α, β, γ are the xyz-Euler angles that represent the system orientation. The initial length calibration for the CoGiRo robot is demonstrated for three different trajectories beginning at different poses $\mathbf{q}_0$ with different initial cable lengths $\mathbf{l}_0$, as observed in Fig. 5.

In a similar manner as Sect. 4.1, the calibration using the relative lengths of the trajectories in Fig. 5 was performed. For the calibration of the CoGiRo, a sample frequency of $N = 10$ was used, such that each calibration trajectory motion has $p = 41$ sample points. The results for all three trajectories are summarised in Table 2 and Fig. 6, and it can be observed that the calibration method successfully determined the initial cable lengths $\mathbf{l}_0$ for different robots and calibration trajectories.

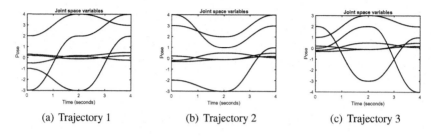

(a) Trajectory 1   (b) Trajectory 2   (c) Trajectory 3

Fig. 5. Different joint space trajectories $\mathbf{q}(t)$ for CoGiRo simulations

Table 2. Result of calibration for CoGiRo robot on different trajectories

	Comp. time (s)	l_0 error	FK error $\sum_t \mathbf{e}^*(t)$
Trajectory 1 (Fig. 5(a))	2.2	1.19×10^{-8}	6.33×10^{-6}
Trajectory 2 (Fig. 5(b))	1.27	3.65×10^{-10}	3.19×10^{-6}
Trajectory 3 (Fig. 5(c))	1.16	5.29×10^{-10}	2.80×10^{-6}

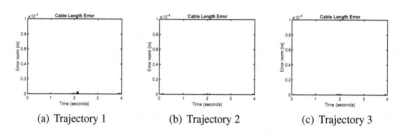

(a) Trajectory 1   (b) Trajectory 2   (c) Trajectory 3

Fig. 6. Error profile in FK using different initial cable lengths for CoGiRo simulation

5 Experimental Results

This section illustrates the proposed approach on a real CDPR, the 2-link 4-DoF 6-cable BioMuscular Arm (*BM-Arm*), as shown in Fig. 7(a), actuated by the MYO-muscle actuators [17]. The robot consists of two links, connected by a spherical joint and a revolute joint. Hence the generalised coordinates of the BM-Arm can be expressed as $\mathbf{q} = [\alpha \ \ \beta \ \ \gamma \ \ \theta]^T$, representing the xyz-Euler angles of the spherical joint and the angle of the revolute joint, respectively.

For the BM-Arm experiment, the robot was set into force control mode with a low constant force value in each cable in order to maintain the robot in equilibrium or a slow moving state. The position of the robot, and hence cable lengths, was unknown initially. The BM-Arm was then manipulated physically in a random motion, while the relative cable lengths $\mathbf{l}_r(t)$ were measured using a relative encoder by the MYO-muscles. This procedure was performed for different robot poses and different random trajectories (of approximately 30 s each). This calibration procedure would be very similar to how the proposed method is envisioned to be used to quickly calibrate a real CDPR every time it is turned on.

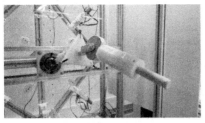
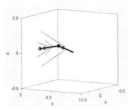

(a) BM-Arm robot with motion capture markers (b) Model in CASPR

Fig. 7. BM-Arm robot

To validate the calibration approach, an OptiTrack motion capture system with four cameras was installed onto the BM-Arm to capture the orientation ($\mathbf{q} = [\alpha\ \beta\ \gamma]^T$, the xyz-Euler angles) of the link 1 (markers are shown in Fig. 7(a)) during the manual motion (as the calibration trajectory). Table 3 shows the comparison results of four different trajectories where the CDPR begins in various poses.

Table 3. Comparison results between motion capture and calibration algorithm for BM-Arm experiments (link 1 only). Value d is the norm of the differences between $\mathbf{l}_0$ from the motion capture system ($\mathbf{l}_0^c$) and the calibration algorithm ($\mathbf{l}_0^*$)

Trajectory	Motion capture	Calibration	$d = \left\| \mathbf{l}_0^c - \mathbf{l}_0^* \right\|$
1	$\mathbf{q}_0 = [0.04\ -0.02\ -0.001]^T$	$\mathbf{q}_0 = [0.02\ -0.01\ -0.07]^T$	0.014
	$\mathbf{l}_0 = [0.224\ 0.232\ 0.224\ 0.233]^T$	$\mathbf{l}_0 = [0.229\ 0.241\ 0.216\ 0.228]^T$	
2	$\mathbf{q}_0 = [-0.18\ 0.07\ -0.40]^T$	$\mathbf{q}_0 = [-0.18\ 0.04\ -0.48]^T$	0.014
	$\mathbf{l}_0 = [0.240\ 0.288\ 0.162\ 0.220]^T$	$\mathbf{l}_0 = [0.247\ 0.294\ 0.157\ 0.212]^T$	
3	$\mathbf{q}_0 = [-0.19\ -0.01\ 0.36]^T$	$\mathbf{q}_0 = [-0.16\ -0.03\ 0.33]^T$	0.008
	$\mathbf{l}_0 = [0.167\ 0.222\ 0.237\ 0.285]^T$	$\mathbf{l}_0 = [0.174\ 0.224\ 0.238\ 0.282]^T$	

From the results, it can be observed that the initial length calibration method is able to resolve for $\mathbf{l}_0$ even when starting pose is unknown. Although the method is able to clearly able to solve for the initial cable lengths and pose, it should be noted that some errors still existed. Such errors may exist due to various reasons, including: (1) errors in the cable length feedback measurement $\mathbf{l}_r(t)$; (2) errors in the calibration of the motion capture system and noise due to reflections and disturbances; and (3) wrapping and slack in the cables. Moreover, it is important to note that adequate calibration motion in all of the CDPR's DoFs, in order to obtain sufficient data for the NLLS optimisation, must be performed.

The joint space trajectories for trajectory 2 in Table 3 for both the motion capture system and FK using the calibrated initial cable lengths are shown in Fig. 8 and shows that indeed the calibration method is capable of determining

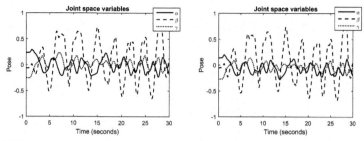

(a) Trajectory from motion capture system (b) Using calibration algorithm and FK

Fig. 8. Joint space trajectory $\mathbf{q}(t)$ (link 1 only) for BM-Arm hardware experiment

correct initial lengths by only using internal sensors (motor encoders) of the CDPR system.

6 Conclusion

In this paper, a novel FK-based calibration method for the initial cable lengths of arbitrary CDPRs is proposed. For systems with relative feedback, the initial cable lengths is required to determine the absolute lengths to be used in the FK analysis. The simulation and experiment examples show that the proposed method is effective in determining the initial cable lengths and requiring only a short random calibration motion. Furthermore, the computational time required for the calibration is short (less than 5 s), making it practical in real-life use. Future work will focus on the analysis on the requirements of suitable calibration trajectories.

Acknowledgements. The work was supported by the grants from the Early Career Scheme sponsored by the Research Grants Council (Reference No. 24200516) and the Germany/Hong Kong Joint Research Scheme sponsored by the Research Grants Council of Hong Kong and the German Academic Exchange Service of Germany (Reference No. G-CUHK410/16). Acknowledgements to the CUHK T-Stone Robotics Institute for supporting this work, and Chen Song for assisting in the collection of results of the BM-Arm.

References

1. Bosscher, P., Williams II, R.L., Bryson, L.S., Castro-Lacouture, D.: Cable-suspended robotic contour crafting system. Autom. Constr. **17**(1), 45–55 (2007)
2. Borgstrom, P.H., Jordan, B.L., Borgstrom, B.J., Stealey, M.J., Sukhatme, G.S., Batalin, M.A., Kaiser, W.J.: NIMS-PL: a cable-driven robot with self-calibration capabilities. IEEE Trans. Robot. **25**(5), 1005–1015 (2009)
3. Pott, A.: An algorithm for real-time forward kinematics of cable-driven parallel robots. In: Lenarcic, J., Stanisic, M.M. (eds.) Advances in Robot Kinematics: Motion in Man and Machine, pp. 529–538. Springer, Dordrecht (2010)

4. Pott, A., Schmidt, V.: On the forward kinematics of cable-driven parallel robots. In: Proceedings of the IEEE/RSJ International Conference on Intelligent Robots and Systems, pp. 3182–3187 (2015)
5. Schmidt, V., Müller, B., Pott, A.: Solving the forward kinematics of cable-driven parallel robots with neural networks and interval arithmetic. In: Thomas, F., Gracia, A.P. (eds.) Computational Kinematics. Mechanisms and Machine Science, vol. 15, pp. 103–110. Springer, Dordrecht (2014)
6. Merlet, J.-P., dit Sandretto, J.A.: The forward kinematics of cable-driven parallel robots with sagging cables. In: Pott, A., Bruckmann, T. (eds.) Cable-Driven Parallel Robots. Mechanisms and Machine Science, vol. 32, pp. 3–15. Springer International Publishing, Cham (2015)
7. Mustafa, S.K., Yang, G., Yeo, S.H., Lin, W., Chen, I.-M.: Self-calibration of a biologically inspired 7 DOF cable-driven robotic arm. IEEE/ASME Trans. Mechatronics **13**(1), 66–75 (2008)
8. Miermeister, P., Pott, A., Verl, A.: Auto-calibration method for overconstrained cable-driven parallel robots. In: Proceedings of the 7th German Conference on ROBOTIK, pp. 301–306 (2012)
9. Chen, Q., Chen, W., Yang, G., Liu, R.: An integrated two-level self-calibration method for a cable-driven humanoid arm. IEEE Trans. Autom. Sci. Eng. **10**(2), 380–391 (2013)
10. dit Sandretto, J.A., Daney, D., Gouttefarde, M.: Calibration of a fully-constrained parallel cable-driven robot. In: Padois, V., Bidaud, P., Khatib, O. (eds.) Romansy 19 Robot Design, Dynamics and Control. CISM International Centre for Mechanical Sciences, vol. 544, pp. 77–84. Springer, Heidelberg (2013)
11. Surdilovic, D., Radojicic, J., Bremer, N.: Efficient calibration of cable-driven parallel robots with variable structure. In: Pott, A., Bruckmann, T. (eds.) Cable-Driven Parallel Robots. Mechanisms and Machine Science, vol. 32, pp. 113–128. Springer International Publishing, Cham (2015)
12. Miermeister, P., Pott, A.: Auto calibration method for cable-driven parallel robots using force sensors. In: Lenarcic, J., Husty, M. (eds.) Latest Advances in Robot Kinematics, pp. 269–276. Springer, Dordrecht (2012)
13. Kraus, W., Schmidt, V., Rajendra, P., Pott, A.: Load identification and compensation for a cable-driven parallel robot. In: Proceedings of the IEEE International Conference on Robotics and Automation, pp. 2485–2490 (2013)
14. Lau, D., Eden, J., Tan, Y., Oetomo, D.: CASPR: a comprehensive cable-robot analysis and simulation platform for the research of cable-driven parallel robots. In: Proceedings of the IEEE/RSJ International Conference on Intelligent Robots and Systems, pp. 3004–3011 (2016)
15. Lau, D., Oetomo, D., Halgamuge, S.K.: Generalized modeling of multilink cable-driven manipulators with arbitrary routing using the cable-routing matrix. IEEE Trans. Robot. **29**(5), 1102–1113 (2013)
16. Lamaury, J., Gouttefarde, M.: Control of a large redundantly actuated cable-suspended parallel robot. In: Proceedings of the IEEE International Conference on Robotics and Automation, pp. 4659–4664 (2013)
17. Marques, H.G., Maufroy, C., Lenz, A., Dalamagkidis, K., Culha, U., Siee, M., Bremner, P., The MYOROBOTICS Project Team: MYOROBOTICS: a modular toolkit for legged locomotion research using musculoskeletal designs. In: Proceedings of the 6th International Symposium on Adaptive Motion of Animals and Machines (AMAM 2013) (2013)

Static Analysis and Dimensional Optimization of a Cable-Driven Parallel Robot

Matthew Newman[✉], Arthur Zygielbaum, and Benjamin Terry

Department of Mechanical and Materials Engineering,
University of Nebraska – Lincoln, Lincoln, NE, USA
mbnewman91@gmail.com

Abstract. A cable-driven parallel manipulator has been chosen to suspend and navigate instruments over a phenotyping research facility at the University of Nebraska. This paper addresses the static analysis and dimensional optimization of this system. Analysis of the system was performed with catenary simplification to create force equilibrium equations and define a mathematical model. The model incorporates flexibility due to catenary sag of the cables. Cable axial stiffness was not included because stiffness is dominated by catenary flexibility for the expected cable tensions. The model was used to optimize system dimensions, and a twelfth-scale system was constructed to verify the model as well as enable dynamic and control system experimentation during full-scale system construction. Miniature end-effectors were used to obtain end-effector orientation and cable tension measurements which were comparable to model predictions. The mathematical model was thereby shown to be accurate for the purpose of system static analysis.

Keywords: Parallel machines · Robot kinematics · Modeling · Manipulator motion-planning

1 Introduction

1.1 Motivation

Agricultural productivity is dependent on the development of crops which can meet certain requirements such as resilience in the face of environmental or pest stressors, or a level of productivity (yield) despite restrictions in nutrients or water. Breeding such crops is an iterative process where the result of crossing the genes of sets of plants causes measureable changes in successive generations. These changes are determined by measuring the plants phenotypes – observable characteristics. Phenotyping in a greenhouse can now be done rapidly using automated equipment. Greenhouse plants, however, are different from plants grown in a field environment. Light conditions are different. Soils are less uniform. And wind does not encourage the growth of support structure within the plants. Assuring that measurements in a greenhouse are trustworthy predictions of field performance is the holy grail of phenotyping. To this end, a field rapid phenotyping system is being developed at the University of Nebraska-Lincoln's Agricultural Research and Development Center. The system described in this paper is designed to position instruments precisely over research plots in order to rapidly and repeatedly make phenotypic measurements of sets of plant varieties and experimental treatments.

1.2 Cable-Driven Parallel Robots

A cable-driven parallel robot (CDPR) is a robotic manipulator designed to control the position and/or orientation of its end-effector within the system's workspace by use of actuated cables. CDPRs provide several benefits over traditional rigid-leg serial and rigid-leg parallel manipulators in the study of crop phenotyping. CDPRs offer minimal interference with the crops compared to rigid-support systems. Traditional serial or parallel manipulators interfere with plant growth because they are composed of large supports and machinery which reflect and obstruct light and air flow. In addition, CDPRs are generally lighter and therefore capable of greater accelerations while maintaining high energy efficiency compared to rigid-linkage robots [1]. However, CDPRs have several design challenges. Cables can only perform while in tension, which puts limitations on end-effector position and greatly influences positional accuracy and system vibrations [2, 3].

CDPRs can be broken into three basic categories based on the number of cables and the mobility of the system: fully constrained, under constrained, and over constrained. A fully constrained parallel robot requires at least one more cable than the degrees of freedom of the end effector. In the case of three-dimensional translational motion, as is the focus of this paper, a fully constrained system requires four cables for full control of position. The number of cables can be reduced if a constant external force, such as gravity, is applied to the end-effector. This force acts as an additional cable on the end-effector, reducing the number of physical cables needed to fully constrain the system [1].

This paper focuses on the suspended four-cable parallel robot. In these systems, the end-effector is supported by four cables with gravity delivering a downward force on the end-effector, behaving as a fifth cable. The four-cable configuration is beneficial over three-cable systems as the same system footprint has an expanded available workspace and the cable load is reduced by distributing the load to an additional cable. However, using four cables creates a redundancy in the support system and complicates the system modeling and control as no unique cable configuration exists for an arbitrary location in the workspace [1].

Further modeling and design considerations come from the scale of the CDPR. In many CDPRs, cables can be assumed to have negligible mass, greatly simplifying system modeling and control. However, in the case of large-scale systems, cable weight can induce catenary sag in the cables which strongly influences positional accuracy as well as system dynamics and vibration.

Significant work has been accomplished in the area of CDPRs, including kinematic design [1, 2, 15, 16] and dynamic analysis [3–7]. Additionally, a large amount of research has been conducted in the area of cable mechanics [6, 8–10]. However, limited research exists in the field of large-scale suspended CPDRs where cable sag can play a major role in system dynamics and control. One of the few examples of research into the area of cable sag in cable-driven manipulators is the FAST telescope, a newly constructed five hundred meter CDPR in China [11].

1.3 Objective

Substantial research has been performed by the FAST project on vibrations and stabilization of large scale CDPRs. However, the high speed requirements of the phenotyping system and the proportionally lower weight end-effector and cables results in significantly different system requirements and dynamics for a phenotyping system with four cables. The objective of this research is to develop a CDPR design and control scheme that can autonomously and rapidly move between crop plots. This system must be functional during harsh weather conditions, pass through the crop canopy with minimal crop interference, and provide stability for the phenotyping sensors mounted on the end-effector. The purpose of this paper is to present a static model of the system as a first step to aid future system design optimization and dynamic modeling of a CDPR for crop phenotyping. In addition, a scaled-down system is built to gather experimental results and confirm the validity of the theoretical models developed.

This paper focuses on computing the inverse kinematics and verifying these results experimentally. The solution begins with an analysis of a single cable to obtain the cable profile and tension. This solution then determines the force equilibrium equations for the four-cable system supporting a point-mass end-effector. The resulting force vectors are then applied to the end-effector model using the moment equilibrium equations to determine the orientation of the end-effector. In order to simplify calculations, cables are assumed to be inextensible due to low tension values predicted in the cables compared to their elastic modulus and the predicted dominance of cable sag on cable flexibility [9].

Until construction of the full-scale system is complete, drive and control systems tests are performed using a scaled system. Vibrations and stability of the scaled system are not thoroughly investigated due to scaling incompatibilities between the test platform and the full-scale system. Because of the difficulties associated with scaling cable properties, the dynamic experimentation is assumed to not scale to the full-scale system. As such, controls tests and system properties including system stiffness and vibration predictions are not discussed in this paper.

2 Simulation

2.1 Geometric Analysis

In flexible cables with significant mass, the weight of the cable provides varying vertical load along the length of the cable which generates a curve as defined by (1) and is illustrated by Fig. 1 [12].

$$y = A * \cosh\left(\frac{x}{A}\right) \tag{1}$$

Where A is the relationship between the constant horizontal tension seen in the cable (Th) and the linear weight of the cable (w) (Fig. 2).

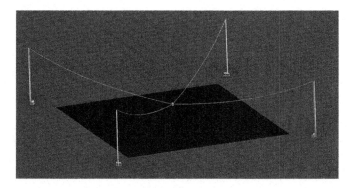

Fig. 1. Conceptual model of phenotyping system.

$$A = \frac{T_h}{w} \qquad (2)$$

Cable length (S) can then be calculated based on the arc length formula, integrating from cable end points, (x_1, y_1) and (x_2, y_2).

$$S = \int_{x_1}^{x_2} \sqrt{1 + \left(\frac{dy}{dx}\right)^2} \, dx = A * \sinh\left(\frac{x_2}{A}\right) - A * \sinh\left(\frac{x_1}{A}\right) \qquad (3)$$

The cable angle at any point along the cable (Ψ) can also be solved geometrically as,

$$\tan(\Psi) = \frac{dy}{dx} = \sinh\left(\frac{x}{A}\right) \qquad (4)$$

Provided that a cable can only experience axial load, at any point along the cable, tension (T) must be tangent to the cable curvature. Furthermore, the only horizontal forces acting on the cable are located at the end points of the cable. Therefore, T_h is constant along the length of the cable. Cable tension can then be determined for any point along the cable,

$$T = T_h \sec(\Psi) \qquad (5)$$

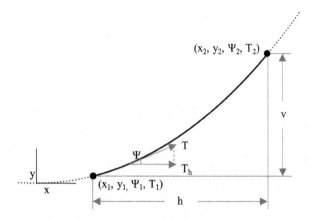

Fig. 2. Catenary curve profile

Solving (4) for Ψ, and substituting into (5),

$$T = T_h * \cosh\left(\frac{x}{A}\right) = A * w * \cosh\left(\frac{x}{A}\right) \tag{6}$$

For any given point in the field, the horizontal and vertical distances between the end-effector and the cable anchor point, h and v respectively, are known.

$$h = x_2 - x_1 \tag{7}$$

$$v = y_2 - y_1 = A * \cosh\left(\frac{x_1 + h}{A}\right) - A * \cosh\left(\frac{x_1}{A}\right) \tag{8}$$

Reducing the system of equations produces three equations with four unknowns, A, S, T1, and x1.

$$v = A * \cosh\left(\frac{x_1 + h}{A}\right) - A * \cosh\left(\frac{x_1}{A}\right) \tag{9}$$

$$S = A * \sinh\left(\frac{x_1 + h}{A}\right) - A * \sinh\left(\frac{x_1}{A}\right) \tag{10}$$

$$T_1 = A * w * \cosh\left(\frac{x_1}{A}\right) \tag{11}$$

2.2 Inverse Kinematics

Solving the inverse kinematics for CDPRs involves solving static equilibrium equations of the system. In the four-cable CDPR with a point-mass end-effector, there are three translational degrees of freedom. The system is therefore defined by the equations for static equilibrium,

$$\sum \underline{F} = 0 = \sum_{i=1}^{4} (T_i * \underline{R_i}) - \underline{W} \tag{12}$$

Where T_i is the tension value of the *ith* cable, R_i is the unit vector in the direction of force $\underline{T_i}$, and W is the weight vector of the end-effector.

As indicated in the previous section, each cable is defined by a system of three equations that, given the current known geometric variables, depend on four unknowns. In the three-cable CDPR, adding the equations for three cables to the three static equilibrium equations produces a balanced system of equations that can be solved. Except in special circumstances, numerical methods must be used to solve the system as no explicit solution exists for this system of equations.

In the four-cable CDPR, there is one more unknown value than equilibrium equations available. The use of four cables in a three degree of freedom CDPR results in a redundant cable which generally suggests no unique solution exists for any given point in the system workspace. To solve this system of equations, a constrained optimization condition must be included with the problem. In this study, it was chosen to optimize the distribution of load on the cables by increasing the load on the lowest tension cable until the ratio between the highest and lowest tension is minimized. To achieve this, the model initially selects the position in the workspace to be considered. The length of the cable anchored the furthest away from the end-effector is then set to a predefined value greater than the straight-line distance between the anchor point and the end-effector. With one cable fully defined, the system of equations and unknowns are balanced, and can be solved iteratively. By increasing the tension on the prescribed cable, its tension gradually approaches that of the next lowest cable tension, more evenly distributing load between the cables until the system is considered optimized, and the resulting tensions, cable lengths, and cable profile are recorded.

2.3 Orientation Prediction

Thus far, the system end-effector has been assumed to be a point-mass. However, a potentially important parameter of CDPR design is the predicted orientation of the end-effector in different regions of the workspace. In the phenotyping system, end-effector orientation impacts the use of sensors intended to be downward facing as well as the range of motion of the end-effector gimbal.

Orientation is predicted by utilizing the force equilibrium results, applying them to a rigid body end-effector, and solving moment equilibrium equations,

$$\sum \underline{M} = 0 = \sum_{i=1}^{4} \underline{R_i} \times \underline{F_i} \tag{13}$$

where F_i is the force vector generated by the tension in the *ith* cable and R_i is the position vector from the center-of-mass of the end-effector to the attachment point of the *ith* cable. R_i is obtained by taking the position vector of the cable attachment point according to the end-effector frame of reference, R_i^*, and passing it through three rotation matrixes representing the rotation about the system x, y, and z axis.

$$[R]_x = \begin{bmatrix} 1 & 0 & 0 \\ 0 & \cos(\alpha) & -\sin(\alpha) \\ 0 & \sin(\alpha) & \cos(\alpha) \end{bmatrix} \quad (14)$$

$$[R]_y = \begin{bmatrix} \cos(\beta) & 0 & \sin(\beta) \\ 0 & 1 & 0 \\ -\sin(\beta) & 0 & \cos(\beta) \end{bmatrix} \quad (15)$$

$$[R]_z = \begin{bmatrix} \cos(\gamma) & -\sin(\gamma) & 0 \\ 0 & \sin(\gamma) & \cos(\gamma) \\ 0 & 0 & 1 \end{bmatrix} \quad (16)$$

$$R_i = [R]_{z''} * [R]_{y'} * [R]_x * R_i^* \quad (17)$$

The three moment equilibrium equations can be solved numerically for the three angles. With an orientation of the end-effector predicted, the force equilibrium[1] and moment equilibrium equations can be iteratively solved until the orientation prediction converges.

3 Theoretical Results

3.1 Simulator Outputs

The outputs of this model can be used to predict tension along the cables, cable lengths, cable profiles, and end-effector orientation. To accelerate simulation, it is assumed that system behavior is symmetrical across the geometric symmetry planes of the system. Thus, the same tension values are predicted in each quadrant of the field, but are associated with the mirrored cables.

Based on this assumption, cable tensions are solved across one quadrant of the workspace, and the behavior of the system in each other quadrant are then extrapolated. Figure 3a displays tension for a single cable as a function of end-effector position in the field at a fixed height.[2] Figure 3b illustrates the amount that the end-effector is predicted to tilt as a function of end-effector position in the field at a fixed height.

---

[1] After the first iteration of solving the force and moment equilibrium equations is performed, the end-effector is changed from a point-mass to a rigid body, oriented based on the prediction created by the results of the first iteration of moment equations.

[2] Data given for 68 kg end-effector, 3 m above ground.

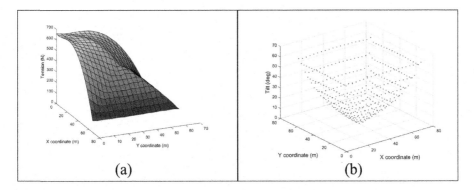

Fig. 3. (a) Theoretical cable tension. (b) Theoretical end-effector tilt

3.2 Dimensional Optimization

Modeling CDPRs requires knowledge of seven system parameters (Fig. 4):

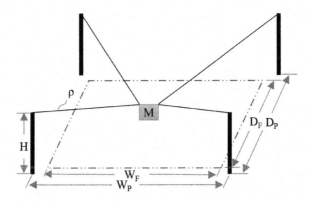

Fig. 4. System parameters of a four-cable CDPR system

- Field width, W_F
- Field depth, D_F
- End-effector mass, M
- Cable density, ρ
- Width between cable feed points, W_P
- Depth between cable feed points, D_P
- Height of cable feed points, H

Field dimensions and end-effector operational height were predetermined by the design of the phenotyping facility and are presented in Table 1. During system design, it was chosen to use a custom Kevlar cable with a fiber optic core for sensor data transmission. Use of the selected cable defines the cable density and adds an additional constraint by limiting tension in the cables.

Table 1. System parameters

Defined parameters		Variable parameters	
Field width	67 m	End-effector mass	45–90 kg
Field depth	60 m	Pole width	75–100 m
Maximum end-effector height	10 m	Pole height	15–26 m
Cable density	10 g/m		
Pole aspect ratio	10:9		
Maximum tension	1500 N		

The primary objective of this analysis is to determine the most appropriate location for the poles supporting the cable system and to determine the maximum required height for the cable-feed pulleys. The end-effector design is currently incomplete; therefore, studies investigating multiple end-effector weights are analyzed alongside of pole layout and height.

To optimize pole location and height as well as end-effector weight, three measurements must be analyzed:

- Maximum cable tension in consideration of cable strength
- Tension distribution in consideration of system stabilization
- End-effector orientation in consideration of end-effector reorientation capabilities

Many simulations were generated with different permutations of pole height, pole distancing, and end-effector mass. Selected results from these simulations are presented in Figs. 5 and 6. Figure 5 shows the influence of all three variables on the predicted maximum tensions for the system within the operational workspace.

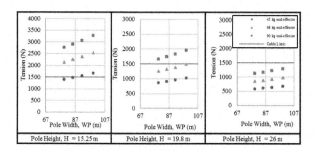

Fig. 5. Theoretical maximum tension in field

The even distribution of load between cables has a substantial impact on cable control and system vibrations [2]. The distribution of load between the cables can be parameterized by the variable η as follows:

$$\eta_{xyz} = \frac{T_{max}(x, y, z)}{T_{min}(x, y, z)} \tag{18}$$

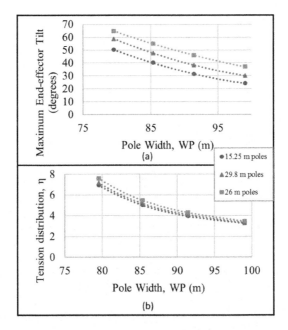

Fig. 6. (a) Theoretical end-effector tilt (b) Theoretical tension distribution

Where T_{max} and T_{min} are the highest and lowest cable tensions, respectively, for the given orientation. η_{max} is then the highest predicted η_{xyz} in the workspace for the given system configuration. Load distribution, and therefore cable performance, is expected to improve as η_{max} approaches one. Figure 6b shows the impact of pole location and height on η_{max}.[3]

As the end-effector moves radially from the center of the workspace, the uneven distribution of load on the cables causes the vertical axis of the end-effector to tilt towards the center of the field, away from the vertical axis of the workspace (Fig. 3b). This behavior can be parameterized by measuring the angle between the vertical axis of the end-effector and the vertical axis of the workspace. For a gimbaled end-effector, which is what is being used in this project, the maximum predicted angle is required to determine the required range of motion of the gimbal. In an end-effector without a gimbal, extreme angles can limit the use of sensors and equipment that are required to maintain a certain orientation. Figure 6a shows the impact of pole location and height on the end-effector inclination angle.[4]

According to preliminary designs, the end-effector with the maximum weighted sensor package will be between 45 and 68 kg. Based on the data presented in Figs. 5 and 6, the minimal system configuration that will safely support a 68 kg end-effector utilizes 19.8 m (65 ft) poles. A pole shorter than this would require placement too close to the workspace and cable performance would likely cause the system to be

---

[3] End-effector weight was found to have no impact on η_{max}.
[4] End-effector weight was found to have no impact on end-effector inclination angle.

uncontrollable. Taller poles reduce the load on the cables, which allow the poles to be placed further from the workspace, improving cable performance and reducing end-effector tilt. However, this introduces further design challenges. Moving the poles outwards expands the space requirements of the system by adding a large perimeter of empty space between the workspace and poles. Also, taller poles are more expensive and require larger footings for support.

With 19.8 m poles selected, the maximum allowable width between poles for the specified end-effector weight and cable strength is 99 m (325 ft). Positioning the poles this far from the workspace increases system footprint by 53% and generates an 18% increase in maximum tension compared to a system with similar poles placed 80 m apart. However, it also reduces η and end-effector inclination by 54% and 49% respectively, enhancing system performance. Positioning the poles any further out, however, increases cable tension, reducing the safety factor for the cables. The final recommended configuration for this system is outlined in Table 2.

Table 2. Optimized system dimensions

Parameter	Optimized dimension
Pole distance	99 × 89 m (325 × 293 ft)
Pole height	19.8 m (65 ft)
End-effector mass limit	68 kg (150 lb)

4 Experimental System

4.1 Design

A twelfth-scale model of the field phenotyping system was designed to confirm the simulator results and to test control system design, system dynamics, and end-effector stabilization hardware and controls. Scaling factors are calculated using the Buckingham Pi theory following the procedures used by Yao, et al. [2]. Dimensional Parameters are listed in Table 3 (Fig. 7).

Table 3. Scaled system parameters

Parameter	Similarity scale	Full size dimension	Model dimension
Field width	1:12	67 m	5.60 m
Field depth	1:12	60.35 m	5.03 m
Pole height	1:12	25.91 m	2.16 m
Cable density	1:55*	10.8 g/m	0.197 g/m
End-effector weight	1:144	77 kg	0.535 kg

An appropriate cable was not utilized in the twelfth-scale system due to the challenges of scaling cable properties of density, construction, and stiffness. Dyneema fishing line with a diameter of 1 mm was instead used. Due to this change, cable sag and stiffness are not similar between the twelfth-scale and full-scale systems. Thus,

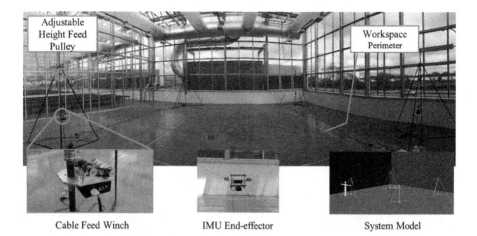

Fig. 7. Twelfth-scale system

full-scale system dynamics cannot be predicated on twelfth-scale experimentation. As a result, the twelfth-scale system is used in studying *general* CDPR behavior in the testing of stabilization and control systems. While these tests may be briefly mentioned, their results are not discussed in this paper.

The twelfth-scale system was designed to test, not only the determined optimal configuration, but an array of system configurations. As such, poles used to support the cable system were designed as collapsible tripods to allow for easy alteration to pole layouts and system scales. Cable-feed pulleys with adjustable height were mounted on the poles to experiment with multiple cable systems heights. Attached to the poles were custom winches to actuate cable feed. Each winch wirelessly communicated with the system navigational controller to drive the system with motor-mounted-encoder feedback to track cable length and approximate end-effector position.

An end-effector mounted with an inertial measurement unit (IMU) was created to measure end-effector orientation when navigated through the workspace. It was also used to observe the response to impulse disturbances on the end-effector as well as the impact of end-effector acceleration during travel on system vibration. Additionally, a gimballed end-effector equipped with load cells at the cable connection points was used to perform experiments to measure cable tensions during travel as well as to confirm tension predictions from the simulator.

4.2 Experimental Static Results

One task of the twelfth-scale system was to determine the accuracy of the mathematical model. Two primary criteria for confirming the validity of the simulator results were cable tension and end-effector orientation. Two tests were performed to determine the accuracy of the theoretical predictions. One test involved navigating the load-cell

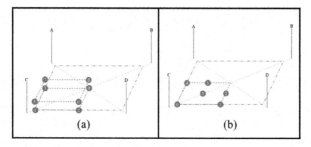

Fig. 8. (a) Tension experiment tested locations (b) Orientation experiment tested locations

end-effector through a series of points (Fig. 8a).[5] At each point, load cell readings were taken and were compared to theoretical values predicted by the simulator, as displayed in Fig. 9.[6] The second test involved navigating the IMU end-effector through a series of points (Fig. 8b) to measure end-effector orientation, which in turn was compared to simulator results, as displayed in Fig. 10. Due to the symmetry of the system, all tests are performed in one quadrant of the workspace, and the results are assumed to mirror across the symmetry planes.

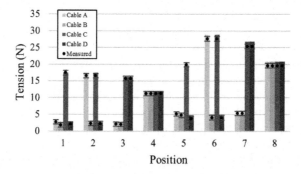

Fig. 9. Theoretical vs. experimental values of the cable tension. The bars indicate the theoretical values and the circles are the means of the measured values from the 12th-scale model. Error bars are one standard deviation of the mean.

Results from the first test show that the simulator predicted cable tensions to within an error of 0.7 N with a standard deviation of 0.5 N for an end-effector of weight 18.35 N. Results from the second test were then shown to predicted end-effector tilt to within 2.0° with a standard deviation of 1.3°. These results indicate that the designed simulator accurately predicts cable performance for the purpose of static analysis.

---

[5] For tension testing, points are located at heights of 0.25 m (lowest feasible elevation for given end-effector) and 1.14 m (maximum safe operating height for given weight).

[6] Rather than using a 0.535 kg end-effector for the tension tests, a 1.9 kg end-effector was used. This was done to increase cable tensions to a level more appropriate for the utilized load cells.

Based on these experiments, agreement between the simulator and physical model is adequate to justify the use of the simulator results in predicting the static behavior of the full-scale phenotyping system.

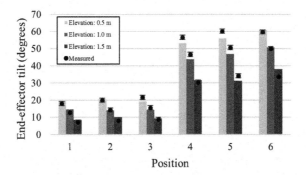

Fig. 10. Theoretical vs. experimental values of the end-effector tilt angle. The bars indicate the theoretical values and the circles are the means of the measured values from the 12th-scale model. Error bars are one standard deviation of the mean.

5 Conclusions

This paper addressed the rationale for use of a cable-driven parallel robot (CDPR) system for control of an outdoor phenotyping site. It addressed the derivation and solution of the inverse kinematics and used this model to optimize system dimensions. These simulations were compared against experimental results of a twelfth-scale system to determine the accuracy of the calculations. This research can be extended to aid in modeling of the dynamic system to predict system vibrations and to determine stabilization requirements during system control. This research can readily be adapted for other four-cable CDPR systems to predict static properties.

References

1. Hiller, M., Fang, S., Mielczarek, S., Verhoeven, R., Franitza, D.: Design, analysis and realization of tendon-based parallel manipulators. Mech. Mach. Theory **40**(4), 429–445 (2005)
2. Yao, R., Tang, X., Wang, J., Huang, P.: Dimensional optimization design of the four-cable-driven parallel manipulator in FAST. IEEE ASME Trans. Mechatron. **15**(6), 932–941 (2010)
3. Kawamura, S., Choe, W., Tanaka, S., Pandian, S.R.: Development of an ultrahigh speed robot FALCON using wire drive system. In: 1995 Proceedings of the IEEE International Conference on Robotics and Automation, vol. 1, pp. 215–220 (1995)
4. Zi, B., Duan, B.Y., Du, J.L., Bao, H.: Dynamic modeling and active control of a cable-suspended parallel robot. Mechatronics **18**(1), 1–12 (2008)

5. Fang, S., Franitza, D., Torlo, M., Bekes, F., Hiller, M.: Motion control of a tendon-based parallel manipulator using optimal tension distribution. IEEE ASME Trans. Mechatron. **9**(3), 561–568 (2004)
6. Suilu, Y., Zhao, W., Qi, L., Yixin, C.: Stiffness analysis of a wire-driven parallel manipulator. In: 2012 IEEE International Conference on Computer Science and Automation Engineering (CSAE), vol. 3, pp. 31–34 (2012)
7. Yamamoto, M., Yanai, N., Mohri, A.: Trajectory control of incompletely restrained parallel-wire-suspended mechanism based on inverse dynamics. IEEE Trans. Robot. **20**(5), 840–850 (2004)
8. Bin, L., Yinghui, L., Xuegang, Y.: Dynamic modeling and simulation of flexible cable with large sag. Appl. Math. Mech. **21**(6), 707–714 (2000)
9. Kozak, K., Zhou, Q., Wang, J.: Static analysis of cable-driven manipulators with non-negligible cable mass. IEEE Trans. Robot. **22**(3), 425–433 (2006)
10. Russell, J.C., Lardner, T.J.: Statics experiments on an elastic catenary. J. Eng. Mech. **123**(12), 1322–1324 (1997)
11. Nan, R., Li, D., Jin, C., Wang, Q., Zhu, L., Zhu, W., Zhang, H., Yue, Y., Qian, L.: The five-hundred-meter aperture spherical radio telescope (FAST) project. Int. J. Mod. Phys. D **20**(6), 989–1024 (2011)
12. Costello, E.: Length of a hanging cable. Undergrad. J. Math. Model One Two 4(1) (2011)
13. Yingjie, L., Wenbai, Z., Gexue, R.: Feedback control of a cable-driven gough-stewart platform. IEEE Trans. Robot. **22**(1), 198–202 (2006)
14. Yang, G., Pham, C.B., Yeo, S.H.: Workspace performance optimization of fully restrained cable-driven parallel manipulators. In: 2006 IEEE/RSJ International Conference on Intelligent Robots and Systems, pp. 85–90 (2006)
15. Maeda, K., Tadokoro, S., Takamori, T., Hiller, M., Verhoeven, R.: On design of a redundant wire-driven parallel robot WARP manipulator. In: 1999 Proceedings of the IEEE International Conference on Robotics and Automation, vol. 2, pp. 895–900 (1999)
16. Ebert-Uphoff, I., Voglewede, P.A.: On the connections between cable-driven robots, parallel manipulators and grasping. In: 2004 Proceedings of the IEEE International Conference on Robotics and Automation, ICRA 2004, vol. 5, pp. 4521–4526 (2004)
17. Shen, Y., Osumi, H., Arai, T.: Manipulability measures for multi-wire driven parallel mechanisms In: Proceedings of the IEEE International Conference on Industrial Technology, pp. 550–554 (1994)
18. Voglewede, P.A., Ebert-Uphoff, I.: Measuring 'closeness' to singularities for parallel manipulators. In: 2004 Proceedings of the IEEE International Conference on Robotics and Automation, ICRA 2004, vol. 5, pp. 4539–4544 (2004)

Improving the Forward Kinematics of Cable-Driven Parallel Robots Through Cable Angle Sensors

Xavier Garant, Alexandre Campeau-Lecours, Philippe Cardou$^{(\boxtimes)}$, and Clément Gosselin

Laboratoire de Robotique, Département de génie mécanique,
Université Laval, Quebec City, QC, Canada
xavier.garant.1@ulaval.ca,
{alexandre.campeau-lecours,pcardou,gosselin}@gmc.ulaval.ca

Abstract. This paper presents a sensor fusion method that aims at improving the accuracy of cable-driven planar parallel mechanisms (CDPMs) and simplifying the kinematic resolution. While the end-effector pose of the CDPM is usually obtained with the cable lengths, the proposed method combines the cable length measurement with the cable angle by using a data fusion algorithm. This allows for a resolution based on the loop closure equations and a weighted least squares method. The paper first presents the resolution of the forward kinematics for planar parallel mechanisms using cable angle only. Then, the proposed sensor fusion scheme is detailed. Finally, an experiment comparing the different procedures for obtaining the pose of the CDPM is carried out, in order to demonstrate the efficiency of the proposed fusion method.

Keywords: Cable-driven robot · Wire-driven robot · Planar parallel mechanism · Sensor fusion · Angular position sensor · Cable angle sensor · Measurement redundancy

1 Introduction

Cable-driven parallel mechanisms (CDPMs) have a proven track record in many different fields. Their architecture, allowing for large workspaces and relatively simple designs, has attracted much attention in the past two decades within academia and industry. This interest has resulted in multiple applications of CDPMs being developped, ranging from cable-driven cameras [4] to very large radio-telescopes [18], medical applications [17] and haptic devices [14]. Several researchers have also solved some of the computational problems associated with CDPMs. Related to the topic of this article, the problem of solving the direct kinematics of spatial fully constrained cable-driven parallel robots from six cable-length measurements alone has already been solved [9]. Indeed, it is equivalent to the problem of the direct kinematics of the Gough-Stewart platform. More recently, the problem of the forward displacement analysis of under-constrained

CDPMs was also tackled [3]. The nature of this problem is not purely kinematic, however, as it also involves equilibrium conditions. To the knowledge of the authors, no workable, general, real-time solution has ever been reported to this problem.

In this paper, we propose to improve the accuracy of CDPMs by the addition of cable angular sensors. Already, assessing the accuracy of these mechanisms has been the scope of some research [5,16]. Other researchers have proposed the use of additional length sensors for calibration purposes [15]. The strategic placement of angular sensors has also been discussed in [13], in order to solve the forward kinematics of CDPMs in closed-form. Previous studies on the implementation of cable angle sensors with the aim of improving accuracy have proposed a specific sensor design and a sensor fusion algorithm based on Kalman filtering [6].

The purpose of relying on additional angular sensor data for solving the forward kinematics is to increase the accuracy of the estimated CDPM pose, while also discriminating between different solutions. To do so, this paper proposes the use of the cable angle sensor first presented in [2,12], and shown in Fig. 1. The advantages of using this particular sensor architecture are discussed in [6]. It should be pointed out, however, that the accuracy of such a device is in the order of $1°$. The main source of error is the two slots that allow the cable to freely change direction. These slots must be somewhat wider than the cable diameter to avoid impeding its motion, which results in small backlash. Notice also that past experience has shown that the semi-circular arms of a cable angle sensor can preserve the rectilinear shape of the cable, provided that it is under sufficient tension [6]. Indeed, the semi-circular arms, which are already light, are balanced about their respective rotation axes, so that their weight has no effect on the cable they guide. They are also mounted on ball bearings to minimise friction. Their effect on the shape of the cable can thus be made negligible, for most practical intents.

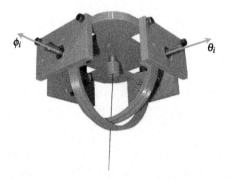

Fig. 1. Proposed angular sensor.

The structure of this paper is as follows. First, the geometric model of a generic planar CDPM is presented. Then, the forward kinematics with cable length measurement are briefly discussed, followed by the detailed resolution of the forward kinematics using cable angle measurement only. The proposed fusion method, using both cable length and angular position sensors, is then presented. Finally, an experiment is reported to compare the accuracy of the three different methods.

2 Geometric Model of a Planar Cable-Driven Parallel Robot

A cable-driven parallel robot consists of a mobile platform connected to reels on a fixed base by means of n cables acting in parallel. In the present case, a generic planar cable-driven parallel mechanism is considered. The geometric model of this mechanism is shown in Fig. 2. The position of the ith anchor point B_i is defined by vector $\mathbf{b}_i$ in the fixed reference frame, originating from O. The angle of the ith cable with respect to the X axis in the fixed reference frame is defined by θ_i. The origin O' of the moving reference frame, which corresponds to the position of the end effector, is defined by vector $\mathbf{t} = [x, y]^{\mathrm{T}}$ in the fixed reference frame. The orientation ϕ of the end effector is defined by the rotation matrix $\mathbf{Q}$. The point P'_i where the ith cable is attached to the end effector, in the moving reference frame, is defined by vector $\mathbf{p}'_i$ originating from O'. Finally, it must be noted that the effects of cable mass and elasticity are not considered in this model.

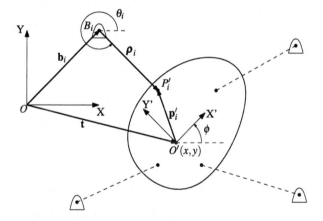

Fig. 2. The geometric model of a generic planar cable-driven parallel mechanism

3 Forward Kinematics with Cable-Length Measurements

The classical approach to solving the forward kinematics for planar parallel mechanisms relies on the measurement of the cable lengths. The length of the ith cable is given by the Euclidean norm of $\boldsymbol{\rho}_i$, the vector connecting B_i and P'_i, which yields

$$||\boldsymbol{\rho}_i||_2 = ||\mathbf{t} + \mathbf{Q}\mathbf{p}'_i - \mathbf{b}_i||_2, \ i = 1, 2, \ldots, n. \tag{1}$$

Computing (1) for each cable results in a nonlinear system of equations which must be solved to find the pose of the end effector. The analytical solution to this problem is known and results in a sixth order univariate polynomial [7,8], where each real root of the polynomial corresponds to a different assembly mode of the mechanism.

4 Forward Kinematics with Cable Angle Measurements

Another approach to solving the forward kinematics of planar parallel mechanisms involves using cable angle measurements. In order to obtain such a measurement, a cable angle sensor capable of measuring the cable angles about two orthogonal axes has been proposed in [2,12]. This section presents the forward kinematics solution by using the cable angles only, i.e., without any information on cable lengths. The loop closure equation for each cable is:

$$\mathbf{b}_i + \boldsymbol{\rho}_i - \mathbf{Q}\mathbf{p}'_i - \mathbf{t}_i = \mathbf{0}, \ i = 1, 2, \ldots, n, \tag{2}$$

where $\boldsymbol{\rho}_i = [\rho_i \cos\theta_i, \rho_i \sin\theta_i]^\mathrm{T}$. Let us introduce a new rotation matrix:

$$\hat{\mathbf{Q}}_i = \begin{bmatrix} \cos(-\theta_i) & -\sin(-\theta_i) \\ \sin(-\theta_i) & \cos(-\theta_i) \end{bmatrix}, \ i = 1, 2, \ldots, n. \tag{3}$$

We then multiply both sides of (2) by $\hat{\mathbf{Q}}_i$. Geometrically speaking, this transformation results in vector $\boldsymbol{\rho}_i$ being parallel to the X axis of the fixed reference frame. Thus, the new equation can be written as a set of two scalar equations:

$$(-P'_{i,x}\cos\phi + P'_{i,y}\sin\phi - x + B_{i,x})\cos\theta_i \\ + (-P'_{i,x}\sin\phi - P'_{i,y}\cos\phi - y + B_{i,y})\sin\theta_i + \rho_i = 0, \tag{4a}$$

$$(B_{i,y} - P'_{i,x}\sin\phi - P'_{i,y}\cos\phi - y)\cos\theta_i \\ + (-B_{i,x} + P'_{i,x}\cos\phi - P'_{i,y}\sin\phi + x)\sin\theta_i = 0, \tag{4b}$$

where ρ_i is absent from (4b). Consequently, we discard (4a) and obtain a new system of n equations where we can write the unknowns as a vector $\mathbf{x} = [x, y, \phi]^\mathrm{T}$. When the robot uses more than three cables, the system of equations is overdetermined, since the number of unknowns is always three. In the present case, the proposed method for solving this system is the least squares method, which is a typical approach in this regard. Let us write the problem as

$$\text{minimize } \frac{1}{2}\mathbf{f}^\mathrm{T}\mathbf{f},$$
$$\text{over } \mathbf{x}, \tag{5}$$

where **f**, the vector of residuals, is defined using Eq. (4b) as follows:

$$\mathbf{f}(\mathbf{x}) = \begin{bmatrix} (B_{1,y} - P'_{1,x}\sin\phi - P'_{1,y}\cos\phi - y)\cos(\theta_1) \\ +(-B_{1,x} + P'_{1x}\cos\phi - P'_{1,y}\sin\phi + x)\sin(\theta_1) \\ \vdots \\ (B_{n,y} - P'_{n,x}\sin\phi - P'_{n,y}\cos\phi - y)\cos\theta_n \\ +(-B_{nx} + P'_{n,x}\cos\phi - P'_{n,y}\sin\phi + x)\sin\theta_n \end{bmatrix}. \quad (6)$$

The condition for an extremum of $\frac{1}{2}\mathbf{f}^T\mathbf{f}$ is met when

$$\frac{\partial \frac{1}{2}\mathbf{f}^T\mathbf{f}}{\partial \mathbf{x}} = 0. \quad (7)$$

Equation (7) represents a set of three equations in three unknowns: x, y and ϕ. These three equations being linear in x and y, a resultant equation containing only ϕ can easily be obtained using a procedure such as the one used in [8]. The univariate resultant equation is then made algebraic by performing the tangent half-angle substitution: $\sin\phi = 2t/(1+t^2)$ and $\cos\phi = (1-t^2)/(1+t^2)$. Clearing the denominators and any factor $(1+t^2)$ in the remaining expression leads to a polynomial of degree five in t. This polynomial is easily solved, and its only real root corresponds to the minimum sought. The optimum values of ϕ, x and y are obtained by back-substitution.

5 Sensor Fusion

In the previous sections, the forward kinematic problem was solved by using either cable lengths or cable angles. In this section, sensor fusion algorithms are proposed to extract the most accurate estimates from all the available information.

5.1 Three Loop-Closure-Equation Components

Having information on cable angles allows the use of the loop closure equations for solving the forward kinematics, instead of using the equations described in Sect. 3. Thus, the most straightforward method to solve the forward kinematic problem with both cable length and cable angle measurements consists in using the X and Y components of Eq. (2) for one cable, along with either the X or Y component of Eq. (2) for any other cable. This results in a system of three equations and three unknowns, which can be solved symbolically. However, while being fast and simple, this solution does not use the full potential of every sensor in the mechanism, which may result in non-negligible inaccuracies in the estimation of the pose of the effector.

5.2 Weighted Fusion by Propagation of Variance

Another method for solving the forward kinematics problem consists in solving the the loop closure equations in Eq. (2) for all n cables, all at once. This can be done via a simple least square approach, which even yields a symbolic solution. With this approach, however, the same weight is used for all sensors, even though their accuracies may differ, which is a significant drawback. Indeed, it is important to adjust the weight of each sensor measurement in the equations according to its rated accuracy. Therefore, this section presents the symbolic solution to the sensor fusion problem, where weights are introduced to better reflect the relative uncertainty between the two types of measurements.

Let us rewrite the loop-closure equations defined in Eq. (2), in scalar form:

$$B_{i,x} + \rho_i \cos\theta_i - \cos\phi P'_{i,x} + \sin\phi P'_{i,y} - x = 0, \ i = 1, 2, \ldots, n \quad (8a)$$
$$B_{i,y} + \rho_i \sin\theta_i - \sin\phi P'_{i,x} - \cos\phi P'_{i,y} - y = 0, \ i = 1, 2, \ldots, n. \quad (8b)$$

We then express $\sin\phi$ and $\cos\phi$ in terms of the half-angle tangent $T = \tan(\phi/2)$, $\phi = 2\arctan(T)$, that is,

$$\sin(\phi) = \frac{2T}{1+T^2}, \quad \cos(\phi) = \frac{1-T^2}{1+T^2}. \quad (9)$$

Substituting (9) into (8a) and (8b) yields two quadratic polynomials in T:

$$B_{i,x} + \rho_i \cos\theta_i - P'_{i,x} - x + 2P'_{i,y}T + (B_{i,x} + \rho_i \cos\theta_i + P'_{i,x} - x)T^2 = 0, \quad (10a)$$
$$B_{i,y} + \rho_i \sin\theta_i - P'_{i,y} - y + 2P'_{i,x}T + (B_{i,y} + \rho_i \sin\theta_i + P'_{i,y} - y)T^2 = 0. \quad (10b)$$

We can then use (10a) and (10b) to define the residuals vector

$$\mathbf{f} = \begin{bmatrix} B_{i,x} + \rho_i \cos\theta_i - P'_{i,x} - x + 2P'_{i,y}T + (B_{i,x} + \rho_i \cos\theta_i + P'_{i,x} - x)T^2 \\ B_{i,y} + \rho_i \sin\theta_i - P'_{i,y} - y + 2P'_{i,x}T + (B_{i,y} + \rho_i \sin\theta_i + P'_{i,y} - y)T^2 \\ \vdots \\ B_{n,x} + \rho_n \cos\theta_n - P'_{n,x} - x + 2P'_{n,y}T + (B_{n,x} + \rho_n \cos\theta_n + P'_{n,x} - x)T^2 \\ B_{n,y} + \rho_n \sin\theta_n - P'_{n,y} - y + 2P'_{n,x}T + (B_{n,y} + \rho_n \sin\theta_n + P'_{n,y} - y)T^2 \end{bmatrix} \quad (11)$$

of a weighted least squares method

$$\text{minimize } K = (1/2)\mathbf{f}^T \mathbf{\Sigma_f}^{-1} \mathbf{f},$$
$$\text{over } \mathbf{x}. \quad (12)$$

According to the Gauss-Markov theorem [11], the weight matrix value leading to the best linear unbiased estimator (BLUE) of $\mathbf{x}$ is the inverse $\mathbf{\Sigma_f}^{-1}$ of the variance-covariance matrix of $\mathbf{f}$. In order to compute $\mathbf{\Sigma_f}$, we apply the principle of the propagation of uncertainties. First, let $\mathbf{z}$ be a vector containing the sensor values at any given time:

$$\mathbf{z} = [\theta_1, \ldots, \theta_n, \rho_1, \ldots, \rho_n]^\mathrm{T}. \tag{13}$$

Let us then express a linear approximation of $\mathbf{f}$ at point $\mathbf{z}$, using estimate $\hat{\mathbf{z}}$, as:

$$\mathbf{f}(\mathbf{z}) \approx \mathbf{f}(\hat{\mathbf{z}}) + \left.\frac{\partial \mathbf{f}}{\partial \mathbf{z}}\right|_{\hat{\mathbf{z}}} (\mathbf{z} - \hat{\mathbf{z}}). \tag{14}$$

Therefore, we can express $\boldsymbol{\Sigma}_\mathbf{f}$, the covariance matrix of $\mathbf{f}$, as

$$\boldsymbol{\Sigma}_\mathbf{f} = \mathrm{E}[(\mathbf{f} - \mathrm{E}[\mathbf{f}])(\mathbf{f} - \mathrm{E}[\mathbf{f}])^\mathrm{T}],$$

$$\boldsymbol{\Sigma}_\mathbf{f} = \mathrm{E}\left[\left.\frac{\partial \mathbf{f}}{\partial \mathbf{z}}\right|_{\hat{\mathbf{z}}} (\mathbf{z} - \mathrm{E}[\mathbf{z}])(\mathbf{z} - \mathrm{E}[\mathbf{z}])^T \left(\left.\frac{\partial \mathbf{f}}{\partial \mathbf{z}}\right|_{\hat{\mathbf{z}}}\right)^\mathrm{T}\right],$$

$$\boldsymbol{\Sigma}_\mathbf{f} = \left.\frac{\partial \mathbf{f}}{\partial \mathbf{z}}\right|_{\hat{\mathbf{z}}} \boldsymbol{\Sigma}_\mathbf{z} \left(\left.\frac{\partial \mathbf{f}}{\partial \mathbf{z}}\right|_{\hat{\mathbf{z}}}\right)^\mathrm{T},$$

where $\boldsymbol{\Sigma}_\mathbf{z}$ is the diagonal covariance matrix of $\mathbf{z}$, and an upright E denotes the statistical expectation. The values of $\boldsymbol{\Sigma}_\mathbf{z}$ are determined from theory or from previous experience and must reflect the expected accuracy of each sensor. Tuning these parameters has an impact on the final solution, since it gives more or less weight to each corresponding sensor. Notice also that this expression of $\boldsymbol{\Sigma}_\mathbf{f}$ requires an estimate of the current pose of the effector. This rough estimate can simply be obtained with the method presented in Sect. 5.1 or with a non-weighted least squares method.

Computing the inverse of $\boldsymbol{\Sigma}_\mathbf{f}$ yields a symmetric block-diagonal matrix, namely,

$$\boldsymbol{\Sigma}_\mathbf{f}^{-1} = \begin{bmatrix} \mathbf{M}_1 & 0 & \cdots & 0 \\ 0 & \mathbf{M}_2 & \cdots & 0 \\ \vdots & \vdots & \ddots & \vdots \\ 0 & 0 & \cdots & \mathbf{M}_n \end{bmatrix}, \quad \mathbf{M}_i \in \mathbb{R}^{2 \times 2}. \tag{15}$$

Thus, with the weight matrix derived, the objective function of (12) is now fully defined. The condition for an extremum of this function is met when

$$\frac{\partial K}{\partial \mathbf{x}} = 0. \tag{16}$$

Since the residual vector $\mathbf{f}$ is post-multiplied by its transpose, and $\boldsymbol{\Sigma}_\mathbf{f}^{-1}$ is the inverse of a covariance matrix, which is positive semi-definite, the problem (12) is convex [1]. Consequently, any local minimum is also a global minimum, which means that only one real solution exists for this problem.

K being scalar and $\mathbf{x} = [x, y, T]^\mathrm{T}$ being the vector of unknowns, this results in a system of three equations and three unknowns. Computing Eq. (16), we see that the derivatives with respect to x and y can be written in compact form as a linear system of two equations:

$$(1+T^2)\begin{bmatrix} \alpha_{1,1} & \alpha_{2,1} \\ \alpha_{2,1} & \alpha_{2,2} \end{bmatrix}\begin{bmatrix} x \\ y \end{bmatrix} = \begin{bmatrix} \beta_{1,1} & \beta_{1,2} & \beta_{1,3} \\ \beta_{2,1} & \beta_{2,2} & \beta_{2,3} \end{bmatrix}\begin{bmatrix} 1 \\ T \\ T^2 \end{bmatrix}. \quad (17)$$

We can then solve this system for x and y, that is

$$\begin{bmatrix} x \\ y \end{bmatrix} = \frac{1}{1+T^2}\begin{bmatrix} \alpha_{1,1} & \alpha_{2,1} \\ \alpha_{2,1} & \alpha_{2,2} \end{bmatrix}^{-1}\begin{bmatrix} \beta_{1,1} & \beta_{1,2} & \beta_{1,3} \\ \beta_{2,1} & \beta_{2,2} & \beta_{2,3} \end{bmatrix}\begin{bmatrix} 1 \\ T \\ T^2 \end{bmatrix} \quad (18)$$

where the α and β coefficients are constants. Substituting (18) in the remaining derivative in T yields, after simplification, a fifth order univariate polynomial in T:

$$\sum_{i=0}^{5} \gamma_i T^i = 0 \quad (19)$$

where the γ_i coefficients are functions of the sensor inputs, the weight matrix and the geometric parameters of the robot. Finally, finding the real root of Eq. (19) gives the value of T, which can then be substituted into Eqs. (18) and (9). Thus, the pose of the effector is fully defined.

Notice that the proposed algorithm is not iterative, except for the solution of Eq. (19), which is done using a standard eigenvalue algorithm. Although it is iterative, the computation of the eigenvalues of the 6×6 companion matrix resulting from Eq. (19) can be done reliably in microseconds using currently available algorithms and computers. Let us also point out that the rest of the algorithm is performed in a predetermined number of operations, although it requires an initial guess of the pose, which is only used to estimate the covariance matrix $\mathbf{\Sigma}_\mathbf{f}^{-1}$.

The implementation of the algorithm in Matlab runs in 6 ms on a laptop computer equipped with an Intel Core i7-2640M running 2.8 GHz and with 8 GB of RAM. These computation times could be reduced even more by implementing the algorithm in a lower-level programming language such as C.

6 Simulated Example

In this section, a simulation is performed to validate the effectiveness of the propagation of variance fusion method. The robot geometry used for this simulation is presented in Fig. 3. For simplicity, a three-cable configuration was chosen, even though the methods presented in this paper can be applied to n cables.

The simulation consists in comparing three sets of data corresponding to the end-effector pose, for given input parameters θ_i and ρ_i. The first data set is obtained by the forward kinematics solution using cable-length measurements only; The second data set is obtained from the forward kinematics using cable-angle measurements only; The third is obtained using a combination of both measurements through the sensor fusion algorithm described in Sect. 5.2.

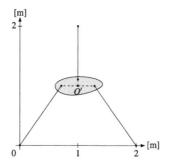

Fig. 3. Robot configuration used for the experiment.

The exact values of the parameters are first obtained by specifying the desired arbitrary poses of the effector and computing the inverse kinematics of the robot. Random noise from a normal distribution of zero mean and arbitrary variance is then added to θ_i and ρ_i. In the present case, the variance corresponds to an uncertainty on cable-length measurements of the order of one centimetre, and of the order of one degree for angular measurement. Finally, the pose of the effector is obtained using these values.

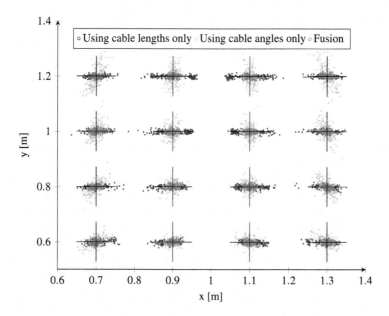

Fig. 4. Estimated position **t** of the effector for given ρ and θ parameters with added noise. True position marked with a black cross. True orientation (not shown in this figure) is $0°$. For the sensor fusion method, equal weights are given to ρ and θ parameters. The number of samples per position is 150 per method.

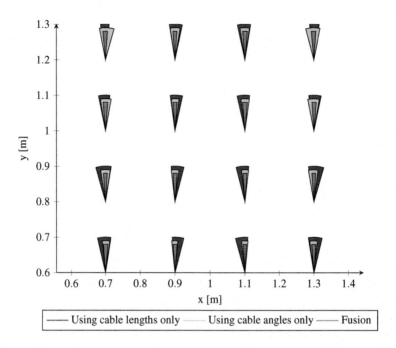

Fig. 5. Estimated orientation ϕ of the effector for given ρ and θ parameters with uncertainties of ± 1.5 cm and $\pm 1.15°$ on cable length and angles respectively. True orientation is $0°$ from vertical. The bounding sectors represent the maximum deviations in orientation for each position. For easier comparison, the error on the X,Y position is not shown. For the sensor fusion method, equal weights are given to ρ and θ parameters.

Figure 4 presents the resulting distribution of effector position for each method. While the data points from these methods are generally distributed in different directions, the fusion results in a distribution that is tightly centred on the true position. This is particularly true near the edges of the workspace, where the estimates computed from a single type of measurement exhibit larger errors. This presents a clear advantage over, for example, computing a simple average of the cable length solution and the angular position solution, as the resulting data would spread away from the true position between the two distributions. Therefore, the results in Fig. 4 indicate that the pose obtained through sensor fusion is more accurate in terms of position than the other two methods.

Figure 5 presents the maximum deviation in effector orientation for different points in the workspace, with uncertainties of ± 1.5 cm and $\pm 1.15°$ on cable length and angle respectively. Similar conclusions can be drawn from this figure. We observe that the bounding sector is consistently more narrow for the data obtained through sensor fusion. In other words, the maximum deviation in effector orientation is always smaller with this method. Moreover, while the maximum deviation values vary greatly with position for the cable length solution and angular solution, they are generally constant when using sensor fusion.

Table 1 summarizes the data from Fig. 4. According to these results, when computing the position of the effector with the sensor fusion method, a 62% reduction in average RMS error is observed. In addition, the maximum error is even more significantly impacted, at 75% reduction versus using cable angle only and 68% reduction versus using cable length only.

Table 1. Error between calculated and true effector X,Y position. RMS value is first computed for each position, then averaged over the number of positions.

	Average RMS error [cm]	Max error [cm]
Cable length only	2.73	11.58
Angular sensor only	2.75	13.86
Fusion	1.05	3.67

Table 2 presents the end effector orientation error. While Fig. 5 demonstrates that the maximum deviations are smaller with the sensor fusion method, we can conclude from this table that the average RMS error is also significantly lower. The data show a 61% reduction versus the cable length solution, and a 57% reduction versus the cable angle solution.

Table 2. Error between calculated and true effector orientation. RMS value is first computed for each position, then averaged over the number of positions.

	Average RMS error [°]	Max error [°]
Cable length only	5.30	19.43
Angular sensor only	4.89	29.88
Fusion	2.09	7.82

7 Conclusion

In this paper, angular sensors are combined with cable-length measurements in order to improve the accuracy of planar CDPMs. A method for solving the forward kinematics of such a mechanism only from cable angles was first presented. The proposed sensor fusion algorithm, based on the loop-closure equations and a weighted least squares method, was then detailed. A simulation was finally performed to show the effectiveness of this algorithm.

The results indicate an improved accuracy in terms of end-effector position and orientation. Moreover, the precision of the proposed method is generally constant throughout the workspace. The method also presents the advantage of yielding a single solution, discriminating between multiple forward kinematics solutions, which is not the case with classical methods based on a single type of sensors.

Further work will consist in generalizing the proposed data fusion algorithm to CDPMs with six degrees of freedom. The robustness of the proposed method to the choice of the a priori estimate $\hat{\mathbf{z}}$ could also be tested more extensively, since its effect on the final estimate of $\mathbf{z}$ was not investigated.

References

1. Boyd, S., Vandenberghe, L.: Convex Optimization. Cambridge University Press, New York (2004)
2. Campeau-Lecours, A., Foucault, S., Laliberte, T., Mayer-St-Onge, B., Gosselin, C.: A cable-suspended intelligent crane assist device for the intuitive manipulation of large payloads. IEEE/ASME Trans. Mechatron. **21**(4), 2073–2084 (2016). doi:10.1109/TMECH.2016.2531626
3. Carricato, M.: Direct geometrico-static problem of underconstrained cable-driven parallel robots with three cables. ASME J. Mech. Robot. **5**(3), 031008 (2013)
4. Cone, L.L.: Skycam: an aerial robotic camera system. Byte **10**(10), 122–132 (1985)
5. Nguyen, D.Q., Gouttefarde, M., Company, O., Pierrot, F.: On the simplifications of cable model in static analysis of large-dimension cable-driven parallel robots. In: 2013 IEEE/RSJ International Conference on Intelligent Robots and Systems, pp. 928–934. doi:10.1109/IROS.2013.6696461
6. Fortin-Côté, A., Cardou, P., Campeau-Lecours, A.: Improving cable-driven parallel robot accuracy through angular position sensors. In: IEEE/RSJ International Conference on Intelligent Robots and Systems, pp. 4350–4355. Daejeon, Korea (2016)
7. Gosselin, C., Merlet, J.P.: On the direct kinematics of planar parallel manipulators: special architectures and number of solutions. Mech. Mach. Theor. **29**(8), 1083–1097 (1994)
8. Gosselin, C., Sefrioui, J., Richard, M.J.: Solution polynomiale au problème de la cinématique directe des manipulateurs parallèles plans à 3 degrés de liberté. Mech. Mach. Theor. **27**(2), 107–119 (1992)
9. Husty, M.L.: An algorithm for solving the direct kinematics of general stewart-gough platforms. Mech. Mach. Theor. **31**(4), 365–379 (1996)
10. Jiang, Q., Kumar, V.: The direct kinematics of objects suspended from cables. In: ASME International Design Engineering Technical Conferences. Montreal, Canada (2010)
11. Kay, S.M.: Fundamentals of Statistical Signal Processing: Estimation Theory. Prentice Hall, Upper Saddle River (1993)
12. Lecours, A., Foucault, S., Laliberte, T., Gosselin, C., Mayer-St-Onge, B., Gao, D., Menassa, R.J.: Movement system configured for moving a payload in a plurality of directions (2015)
13. Merlet, J.P.: Closed-form resolution of the direct kinematics of parallel manipulators using extra sensor data. In: Proceedings IEEE International Conference in Robotics and Automation, pp. 200–204 (1993)
14. Morizono, T., Kurahashi, K., Kawamura, S.: Realization of a virtual sports training system with parallel wire mechanism. In: IEEE International Conference on Robotics and Automation, pp. 3025–3030. Albuquerque, NM, USA (1997)
15. Patel, A.J., Ehmann, K.F.: Calibration of a hexapod machine tool using a redundant leg. Int. J. Mach. Tools Manuf. **40**(4), 489–512 (2000). doi:10.1016/S0890-6955(99)00081-4

16. Riehl, N., Gouttefarde, M., Krut, S., Baradat, C., Pierrot, F.: Effects of non-negligible cable mass on the static behavior of large workspace cable-driven parallel mechanisms. In: 2009 IEEE International Conference on Robotics and Automation, pp. 2193–2198. doi:10.1109/ROBOT.2009.5152576
17. Surdilovic, D., Bernhardt, R.: String-man: a new wire robot for gait rehabilitation. In: IEEE International Conference on Robotics and Automation, pp. 2031–2036. New Orleans, LA, USA (2004)
18. Zi, B., Duan, B., Du, J., Bao, H.: Dynamic modeling and active control of a cable-suspended parallel robot. Mechatronics **18**(1), 1–12 (2008)

Direct Kinematics of CDPR with Extra Cable Orientation Sensors: The 2 and 3 Cables Case with Perfect Measurement and Ideal or Elastic Cables

Jean-Pierre Merlet(✉)

HEPHAISTOS Project, Université Côte d'Azur, Inria, Sophia-Antipolis, France
Jean-Pierre.Merlet@inria.fr

Abstract. Direct kinematics (DK) of cable-driven parallel robots (CDPR) based only on cable lengths measurements is a complex issue even with ideal cables and consequently even harder for more realistic cable models. A natural way to simplify the DK solving is to add sensors. We consider here sensors that give a partial or complete measurement of the cable direction at the anchor points and spatial CDPR with 2/3 cables and we assume that these measurements are exact. We provide a solving procedure and maximal number of DK solutions for an extensive combination of sensors while considering two different cables models: ideal and linearly elastic without deformation.

1 Introduction

We consider cable-driven parallel robot (CDPR) with 3 cables whose output point on the base is A_i and anchor point B_i on the platform. The known distance between B_i, B_j will be denoted d_{ij} and length of cable i will be denoted ρ_i. Solving the direct kinematics (DK) problem with only as input the ρ's is clearly an issue in parallel robotics. Although relatively well mastered for parallel robots with rigid legs, it is still an open issue for CDPR. Even if we assume ideal cable (with no elasticity and no deformation of the cable due to its own mass) the DK problem leads to a larger number of equations than in the rigid leg case [8] and consequently to solving problems [1,2,6,9–11,19], although finding all solutions is possible at the expense of a rather large computation time [5]. If we assume linearly elastic cables similar solving problem arise [15]. All the proposed DK algorithms exhibit a large computation time that prohibits their use in a real-time context. In this case fast and safe algorithms have been proposed [15,19]: still several DK solutions may exist even in a small neighborhood around the previous pose so that the proposed algorithms will fail.

An intuitive approach to avoid the non-unicity problem and to speed up the solving time of the DK is to add sensors that provide additional information on the cable beside the cable lengths, as already proposed for classical parallel robots [7,12,14,18]. A natural candidate will be to measure the cable tensions as they play an important role in the solving. Unfortunately force measurements are usually noisy and measuring these tensions on a moving platform submitted to various mechanical noises appears to be difficult [13,16]. Although several attempts have been made of integrating force sensing in CDPR, none of them have presented clear result about the reliability of the measurement.

In this paper we are considering another measurement possibility which consists in getting complete or partial information on the cable direction at the anchor points A. These measurement are, Fig. 1:

- the angle θ_V between the x axis and the vertical plane that includes the cable.
- the angle θ_H between the horizontal direction of the cable plane and the cable.

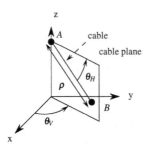

Fig. 1. Orientation sensors may provide the value of θ_V and/or θ_H

Realizing such measurement has already been considered: for example our CDPR MARIONET-Assist uses a simple rotating guide at A whose rotation is measured by a potentiometer in order to obtain the measurement of θ_V while our CDPR MARIONET-VR is instrumented with a more sophisticated cable guiding system which allows for the measurement of both θ_V and θ_H (Fig. 2). For measuring theses angles we may also consider a vision system as proposed in [4]. If ρ, θ_V, θ_H are known, then the location of B is fixed. If only ρ, θ_V are known, then B lies on a circle $\mathscr{C}_V$ centered at A which belong to the vertical cable plane. If only ρ, θ_H are known, then B lies on a horizontal circle $\mathscr{C}_H$ whose center U and radius can easily be calculated as function of ρ, θ_H. To characterize the sensor arrangement we will use the following notation:

- $\theta_V^j \theta_H^j$ indicates that the cable j has both θ_V, θ_H sensors.
- $\theta_V(H)^j$ indicates that the cable j has only $\theta_V(H)$ sensor.

We will also use $n\theta_V\theta_H$ to indicate that n cables have all both θ_V, θ_H sensors. Whenever needed xb_i, yb_i, zb_i will denote the coordinates of B_i while xa_i, ya_i, za_i are the coordinates of A_i. In some cases and angle α_i will appear and we define T_i as $tan(\alpha_i/2)$. We may have also to use the mechanical equilibrium equations:

$$\mathscr{F} = \mathbf{J}^{-\mathbf{T}} \tau \qquad (1)$$

Fig. 2. On the left the rotation guide of MARIONET-Assist which allows for the measurement of θ_V. On the right the system used on MARIONET-VR for the measurement of both θ_V and θ_H

where $\mathscr{F}$ is the external wrench applied on the platform, assumed here to be only the force applied by the gravity, $\mathbf{J}^{-T}$ is the transpose of the inverse kinematic Jacobian and τ the vector of the 3 tensions in the cable. Equations (1) are a set of 6 constraint equations. Furthermore if G denotes the center of mass of the platform there are constants l_i, k_i such that

$$\mathbf{OG} = l_1\mathbf{B_1B_2} + l_2\mathbf{B_1B_3} + l_3\mathbf{B_1B_2} \times \mathbf{B_1B_3} \qquad (2)$$
$$\mathbf{OB_3} = k_1\mathbf{B_1B_2} + k_2\mathbf{B_1G} + k_3\mathbf{B_1B_2} \times \mathbf{B_1G} \qquad (3)$$

Our objective is to consider an exhaustive set of sensor arrangements and number of sensors and for each of them to determine the computational effort that is required to solve the DK, together with an upper bound on the maximal number m of solutions that may be obtained. As we have redundant information we will consider in each case only one square system leading to a closed-form solution (thereby faster than the DK algorithm based only on cable lengths) and whenever possible one leading to a minimal number of solutions. The closed-form solution will be obtained through an elimination process leading to a univariate polynomial. Elimination may lead to a polynomial whose degree is higher than the minimal one: in this work we have tried to provide solution with the lowest degree but we cannot claim for minimality. For this preliminary, but exhaustive, work we will consider a spatial CDPR with only 2 and 3 cables. Furthermore we will assume that all measurements are exact, including the cable lengths. Clearly this assumption is not realistic but our purpose is to pave the way to a more complete analysis.

2 Ideal Cable

For an ideal cable the shape of the cable is the straight line between A and B and cable tension does not affect the length of the cable.

2.1 The 3 Ideal Cables Case

- **case $3\theta_V\theta_H$, 6 extra sensors:** this case is trivial as the measurements provide directly the coordinates of all three B and therefore a single solution of the DK.
- **case $\theta_V^1\theta_H^1 - \theta_V^2\theta_H^2 - \theta_V(H)^3$, 5 extra sensors:** in that case the locations of $B_1, B2$ are known. Consequently B_3 must lie on a circle C_3 lying in a plane that is perpendicular to B_1B_2 and whose center is located on the line B_1B_2, while B_3 is also located on the circle $\mathscr{C}_V^3$ Hence B_3 is located at the intersection of two circles, this leading to one of two solutions whose calculation involves solving a univariate quadratic polynomial. Note that the case $\theta_V^1\theta_H^1 - \theta_V^2\theta_H^2 - \theta_H^3$ is similar if we substitute $\mathscr{C}_V^3$ by $\mathscr{C}_H^3$.
- **case $\theta_V^1\theta_H^1 - \theta_V^2\theta_H^2$, 4 extra sensors:** as in the previous case B_3 lies on the circle C_3 and is also located on a sphere centered at A_3 with radius ρ_3. The intersection of this sphere with C_3 leads usually to 2 intersection points and involves solving a univariate quadratic polynomial. Hence the DK may have at most 2 solutions.
- **case $\theta_V^1\theta_H^1 - \theta_V^2 - \theta_V^3$, 4 extra sensors:** B_2 lies on a circle C_2 that is in a plane perpendicular to A_2B_1, whose center lies on the line A_2B_1 and whose radius may easily be calculated being given B_1, ρ_2, d_{12}. It lies also on the circle $\mathscr{C}_V^2$. Consequently there are two possible locations for B_2 that are obtained by solving a univariate quadratic polynomial. In the same manner B_3 lies on a circle that is perpendicular to B_1A_3 and whose center is located on this line while B_3 also belongs to $\mathscr{C}_V^3$, thereby leading to two possible locations for this point that are obtained by solving a univariate quadratic polynomial. Hence there may be at most 4 possible poses for the platform. Note that changing θ_V to θ_H for any of the cables 2 and 3 will lead to the same result.
- **case $\theta_V^1\theta_H^1 - \theta_V^2$, 3 extra sensors:** in that case B_1 is fixed and B_2 lies on the circle $\mathscr{C}_V^2$. At the same time B_2 lies on the sphere centered at B_1 with radius d_{12}. Consequently there are two possible locations for B_2 whose calculation amounts to solving a univariate quadratic polynomial. For each of these locations as seen in the previous sections there are up to 2 possible location for B_3. In summary there are up to four DK solutions that are obtained by solving two univariate quadratic polynomial. Note that the case $\theta_V^1\theta_H^1 - \theta_H^2$ is similar.
- **case $3 - \theta_V$, 3 extra sensors:** in that case each of the three B_i is constrained to lie on a known circle $\mathscr{C}_V^i$. The CDPR is therefore equivalent to a $3 - RS$ whose DK may lead to 16 solutions that are obtained by solving a 16th order univariate polynomial. The case $3 - \theta_H$ will be similar.
- **case $\theta_V^1\theta_H^1$, 2 extra sensors:** in that case B_1 has a fixed position while for $j = 2, 3$ B_j lies on a circle perpendicular to the the line B_1A_j whose center M_j lies on this line with a radius r_j than can easily be calculated. Hence $\mathbf{OB_j}$ may be written as $\mathbf{OM_j} + r_j \cos\alpha_j \mathbf{u_j} + r_j \sin\alpha_j \mathbf{v_j}$ where $\mathbf{u_j}, \mathbf{v_j}$ are two arbitrary unit vectors perpendicular to $\mathbf{B_1A_j}$ and perpendicular to each other while α_j is an unknown angle that parametrizes the location of B_j on its circle. A constraint is that $||B_2B_3|| = d_{23}$ but this provides only one constraint for the

2 unknowns α_2, α_3. We have therefore to look at the mechanical equilibrium equations that involve the 3 unknown tensions in the cable τ_j. Using the 3 first equations of the equilibrium (1) allows one to determine τ_1, τ_2, τ_3 as functions of α_2, α_3. Reporting this result in the last equation of the equilibrium enables us to obtain a second constraint on α_2, α_3. The 2 constraint equations are transformed into algebraic equations by using the Weierstrass substitution and calculating the resultant of these two equations leads to a univariate polynomial of degree 8, leading to up to 8 solutions for the DK.

- case $\theta_H^1 - \theta_H^2$, **2 extra sensors**: in that case B_1, B_2 are moving on the horizontal circles $\mathscr{C}_H^1, \mathscr{C}_H^2$. Hence we have $\mathbf{OB_j} = \mathbf{OU_j} + r_j \cos\alpha_j \mathbf{x} + r \sin\alpha_j \mathbf{y}$ for $j = 1, 2$. Then we have the constraint equations $||\mathbf{B_1B_2}||^2 = d_{12}^2$, $||\mathbf{B_1B_3}||^2 = d_{13}^2$, $||\mathbf{B_2B_3}||^2 = d_{23}^2$, $||\mathbf{A_3B_3}||^2 = \rho_3^2$ which is a set of 4 equations in the 5 unknowns $\alpha_1, \alpha_2, xb_3, yb_3, zb_3$. Hence the geometrical condition are not sufficient to determine the DK solution(s). The mechanical equilibrium equations (1) introduces three new unknowns τ_1, τ_2, τ_3 and 6 constraints. The 3 first equations of the mechanical equilibrium are linear in τ_1, τ_2, τ_3: solving this system leads to the 6th equation of the mechanical equilibrium, $||\mathbf{B_2B_3}||^2 - d_{23}^2 - ||\mathbf{A_3B_3}||^2 + \rho_3^2$ and $||\mathbf{B_1B_3}||^2 - d_{13}^2 - ||\mathbf{A_3B_3}||^2 + \rho_3^2$ being linear in xb_3, yb_3, zb_3. Consequently we have 3 linear equations in xb_3, yb_3, zb_3 that may be solved in these unknowns. It remain the equations $||\mathbf{B_1B_2}||^2 = d_{12}^2 (A)$ and the 4th and 5th equations of the mechanical equilibrium. These two later equations may be factored and have a common factor (B) whose cancellation will ensure that these 2 equations are satisfied. Then equations (A) and (B) are functions of the sine and cosine of α_1, α_2: using the Weierstrass substitution allows one to obtain 2 algebraic equations in T_1, T_2 whose resultant in T_2 is a univariate polynomial in T_1 of degree 12.

- case θ_H^1, **1 extra sensors**: this case is somewhat similar to the previous one: we have now as unknown $\alpha_1, xb_3, yb_3, zb_3$ and xb_2, yb_2, zb_2. with the additional constraint $||\mathbf{A_2B_2}||^2 = \rho_2^2$. As previously we solve the mechanical equilibrium equation to get τ_1, τ_2, τ_3 and the other constraints to obtain xb_3, yb_3, zb_3. We end up with a system of 4 equations $||\mathbf{B_1B_2}||^2 = d_{12}^2$, $||\mathbf{A_2B_2}||^2 = \rho_2^2$, $||\mathbf{A_3B_3}||^2 = \rho_3^2$ and the 4th equation of the mechanical equilibrium in the 4 unknowns $\alpha_1, xb_2, yb_2, zb_2$. The difference of the two first equations is linear in xb_2 and the last equation is linear in zb_2. Therefore 2 equations remain in the unknowns α_1, yb_2: the resultant in yb_2 leads to a polynomial in $T_1 = \tan(\alpha_1/2)$ which factors out in polynomials of degree 6, 8, 16 and 24.

Table 1 summarizes the previous results for the 3-cables case (the complexity indicates the degree of the polynomials that have to be solved). It must be noted that even a single sensor allows one to drastically reduce the computational effort to get all the DK solutions.

2.2 The 2 Ideal Cables Case

We should not forget that although the CDPR has 3 cables it may end up in a pose where only 2 cables are under tension, the remaining one being slack.

Table 1. For ideal cable: sensors arrangement, total number of sensors, complexity of the solving and maximal number of DK solution(s)

Case	Number of sensors	Complexity	Number of solution
$3\theta_V \theta_H$	6	1	1
$\theta_V^1 \theta_H^1 - \theta_V^2 \theta_H^2 - \theta_{V(H)}^3$	5	2	2
$\theta_V^1 \theta_H^1 - \theta_V^2 \theta_H^2$	4	2	2
$\theta_V^1 \theta_H^1 - \theta_{V(H)}^2 - \theta_{V(H)}^3$	4	2,2	4
$\theta_V^1 \theta_H^1 - \theta_{V(H)}^2$	3	2,2	4
$3 - \theta_{V(H)}$	3	16	16
$\theta_V^1 \theta_H^1$	2	8	8
$\theta_H(V)^1 - \theta_H(V)^1$	2	12	12
$\theta_H(V)^1$	1	6,8,16,24	54

Without losing generality we may assume that cable 1 and 2 are under tension and cable 3 is slack. A direct consequence is that the platform fully lies in the vertical plane that includes A_1, A_2, B_1, B_2 and G.

- **case** $\theta_V^1 \theta_H^1 - \theta_V^2 \theta_H^2$, 4 sensors: a necessary condition to have the platform in the vertical plane including A_1, A_2 is

$$(ya_2 - ya_1)/(xa_2 - xa_1) = \tan(\theta_V^1) = -\tan(\theta_V^2) \quad (4)$$

If this condition is fulfilled then the locations of B_1, B_2 are fixed. There are then 2 possible locations for G: one below $B_1 B_2$ (which is stable) and one above $B_1 B_2$ (unstable). By choosing an appropriate frame both locations may be determined by solving a linear equation. Using Eq. (3) we may determine the location of B_3 and check if $\rho_3 > ||\mathbf{A_3 B_3}||$ for confirming the slackness of cable 3.
- **case** $\theta_V^1 \theta_H^1 - \theta_V^2$, 3 sensors: we use Eq. (4) to check if A_1, A_2, B_1, B_2 may be in the same vertical plane (and this is the only use of θ_V^2). If this is so, then B_1 is in a fixed location, while B_2 belongs to a circle centered in B_1 with radius d_{12} and to a circle centered in A_2 with radius ρ_2. Hence there are two possible locations for B_2 that are obtained by solving a quadratic polynomial. The two possible location of B_3 for each location of B_2 are obtained using the same method as in the previous item for checking the slackness of cable 3.
- **case** $\theta_V^1 \theta_H^1$, 2 sensors: here we cannot check if A_1, A_2, B_1, B_2 are in the same vertical plane but we still may use the same method than in the previous item and we may obtain up to 4 solutions for the DK, two of them being unstable.
- **case** $\theta_V^1 - \theta_V^2$, 2 sensors: if condition (4) holds, then the CDPR becomes a planar CDPR with 2 cables and it is known that to obtain the DK solutions we will have to solve two univariate polynomials of degree 12 [10].
- **case** $\theta_H^1 - \theta_H^2$, 2 sensors: here we will assume that A_1, A_2, B_1, B_2 are in the same vertical plane. Being given the sensor measurements we are able to

get the location of B_1, B_2 in this plane, which will to check if the condition $||\mathbf{B_1 B_2}|| = d_{12}$ holds. If this is the case we may solve the DK by using the procedure described for the $\theta_V^1 \theta_H^1 - \theta_V^2 \theta_H^2$ case.
- **case θ_H^1, 1 sensor:** here again we will proceed under the assumption that A_1, A_2, B_1, B_2 are in the same vertical plane and use the procedure for the $\theta_V^1 \theta_H^1 - \theta_V^2$ case to obtain up to 4 DK solutions.
- **case θ_V^1, 1 sensor:** the sensor measurement allows to check if A_1, A_2, B_1 lie in the same vertical plane. If this is so we resort to the procedure for solving the planar 2-cable DK problem, i.e. solving two univariate polynomials of degree 12 [10].

3 Elastic Cable

The shape of the cable is still the straight line between A and B but the cable length and its length at rest ρ_r (which is the variable that is controlled and estimated from the winch motion) are related to the cable tension τ by:

$$\tau = k(\rho - \rho_r) \quad \text{if } \rho \geq \rho_r, \quad 0 \text{ otherwise} \tag{5}$$

where k is the known stiffness of the cable. There is no deformation of the cable whose shape is the straight line between A and B. The same measurement system as for the ideal cable may be implemented and we use the same notation for describing the sensor arrangement. The difference with the ideal case is that the measurement of both θ_V, θ_H is no more sufficient to determine the location of the B as the cable length is no more known (and so is the radius of the circles $\mathscr{C}_V, \mathscr{C}_H$).

- **case $3\theta_V \theta_H$, 6 extra sensors:**
  the 2 sensors on a given cable j provide the cable direction unit vector $\mathbf{u^j}$ and the three first equations of the mechanical equilibrium may be written as

$$\sum_{j=1}^{j=3} \mathbf{u}_x^j k(\rho_j - \rho_r^j) = 0 \quad \sum_{j=1}^{j=3} \mathbf{u}_y^j k(\rho_j - \rho_r^j) = 0 \quad \sum_{j=1}^{j=3} \mathbf{u}_z^j k(\rho_j - \rho_r^j) = mg$$

These 3 equations constitute a linear system in the ρ_j that can be solved to obtain these variables. We have then $\mathbf{OB_j} = \mathbf{OA_j} + \rho_j \mathbf{u_j}$ that allow to determine the unique pose of the platform.
- **case $2 - \theta_V \theta_H - \theta_V^3$, 5 extra sensors:**
  the ρ may be determined using the same method than in the previous case but they are now function of α_3, the angle used to define B_3 on its vertical circle $\mathscr{C}_V^3$. The constraint $||\mathbf{B_1 B_2}||^2 = d_{12}^2$ factors out in a polynomial of degree 2 and a polynomial of degree 4, leading to 6 possible DK solutions.
- **case $\theta_V^1 \theta_H^1 - \theta_V^2 \theta_H^2$, 4 extra sensors:**
  the unknowns are the 3 ρ and the three coordinates of B_3. The ρ can be determined by solving the first three equations of (1). We consider the 6th equation of the mechanical equilibrium (1) and the two constraints $||\mathbf{B_1 B_2}||^2 = d_{12}^2$,

$||\mathbf{A_3B_3}||^2 = \rho_3^2$. We compute in sequence the resultant with respect to xb_3, yb_3 of these 3 constraints to get a univariate polynomial in zb_3. This polynomial factors out in 3 polynomials of degree 4. Hence there are at most 12 DK solutions.

- **case** $\theta_V^1 \theta_H^1 - \theta_V(H)^2 - \theta_V(H)^3$, **4 extra sensors:**
  The ρ can be determined by solving the first three equations of (1) which are functions of α_2, α_3, the two angles that allow to determine the location of B_2, B_3 on the $\mathscr{C}_V^2, \mathscr{C}_V^3$ circles. The 6th equation of the mechanical equilibrium (1) factors out in 4 polynomials of degree (2,2) in T_2, T_3 and one polynomial of degree (4,4) in T_2, T_3, while $||\mathbf{B_1B_2}||^2 = d_{12}^2$ is a polynomial P of degree (4,8) in T_2, T_3. Taking all resultants in T_3 of all factors of the 6th equation of the mechanical equilibrium with P leads to 5 polynomials of degree 6, 6, 6, 12 and 12 in T_2.

- **case** $\theta_V(H)^1 - \theta_V(H)^2 - \theta_V(H)^3$, **3 extra sensors:**
  the unknowns here are the 3 angles α_i that define the location of the B_i on the vertical circle $\mathscr{C}_V$ and the ρ's which may be obtained by solving the first three equations of (1). The constraint $||\mathbf{B_1B_2}||^2 = d_{12}^2, ||\mathbf{B_1B_3}||^2 = d_{13}^2$ and the 6th equation of the mechanical equilibrium are functions of T_1, T_2, T_3. Successive resultants in T_1, T_2 leads to a univariate polynomial in T_3 which factors out in 6 polynomials of degree 72, one polynomial of degree 12, one of degree 24, 2 of degree 8 and two of degree 4. So an upper bound on the number of solutions is 492, a number which is most probably overestimated.

- **case** $\theta_V^1 \theta_H^1 - \theta_V(H)^2$, **3 extra sensors:**
  the unknowns are the 3 ρ, the 3 coordinates of B_3 and the angle α_2 that allow to define the position of B_2 on its vertical circle $\mathscr{C}_V^2$. We use the 3 first equations of the mechanical equilibrium (1) to determine xb_3, yb_3, zb_3. The 6th equation of the mechanical equilibrium, which is linear in ρ_1, will be used to calculate this unknown. The resultant R_1 of the constraints $||\mathbf{A_3B_3}||^2 - \rho_3^2$ and $||\mathbf{B_2B_3}||^2 = d_{23}^2$ in ρ_3 is a function of ρ_2, α_2. The constraint $||\mathbf{B_1B_2}||^2 = d_{12}^2$ is only function of α_2, ρ_2. The resultant of this equation and of R_1 in ρ_2 is only a function of α_2. Using the Weierstrass substitution this resultant factors out in 2 polynomials in T_2 of degree 12 and 40.

- **case** $\theta_V^1 \theta_H^1$, **2 extra sensors:**
  the unknowns are the 3 ρ and the 3 coordinates of B_2, B_3. The constraint equations are the 6 equations of the mechanical equilibrium (1), the two equations $||\mathbf{A_jB_j}||^2 - \rho_j^2$ for $j \in [2,3]$ and the 3 equations $||\mathbf{B_iB_j}||^2 = d_{ij}^2$ with $i, j > i \in [1,3], i \neq j$. We first use the 3 first equation of the mechanical equilibrium to determine xb_2, yb_2, zb_2. If we consider the difference between $||\mathbf{B_1B_3}||^2 = d_{13}^2$ and $||\mathbf{A_3B_3}||^2 - \rho_3^2$ and the 6th equation of the mechanical equilibrium we have a linear system in xb_3, yb_3. If we report the solution of this system into the remaining equations, then the 5th equation of the mechanical equilibrium is linear in zb_3. The remaining equations are now functions of ρ_1, ρ_2, ρ_3. Successive resultants in ρ_1, ρ_2 leads to a univariate polynomial in ρ_3 which factors out in polynomial of degree 162, 104, 68, 48, 22, 20, 8, 7 and 3, leading to a maximum of 442 solutions, a number which is most probably overestimated.

- case $\theta_V(H)^1 - \theta_V(H)^2$, **2 extra sensors:**
  the unknowns are the 3 ρ, the 2 angles α_1, α_2 that are used to determine the location of B_1, B_2 on their vertical circle $\mathscr{C}_V$ and the 3 coordinates of B_3. The constraint equationss are the 6 equations of the mechanical equilibrium (1), the equation $\|\mathbf{A_3B_3}\|^2 - \rho_3^2$ and the 3 equations $\|\mathbf{B_iB_j}\|^2 = d_{ij}^2$ with $i, j > i \in [1,3], i \neq j$. We first use the 3 first equation of the mechanical equilibrium to determine xb_3, yb_3, zb_3. The 6th equation of the mechanical equilibrium is linear in ρ_1. The resultant of $\|\mathbf{B_2B_3}\|^2 = d_{23}^2$ and of the 4th equation of the mechanical equilibrium with the constraint $\|\mathbf{B_1B_2}\|^2 = d_{12}^2$ allows one to obtain 2 equations free of ρ_2. The Weierstrass substitution is then used to obtain 2 polynomials P_1, P_2 in T_1, T_2 which factor out in several polynomials with $P_1 = \prod R_i$ and $P_2 = \prod S_j$. When considering the resultant of all possible combinations of P_i, Q_j we get polynomials in T_1 only of degree 936, 240, 112, 72, 8, 4, 4 and hence the maximum number of solutions is 1376. Trials have shown that the polynomials of degree 936, 240 may have real roots.

Table 2 summarizes the result for the 3 elastic cables case.

Table 2. For elastic cable: sensors arrangement, total number of sensors, complexity of the solving and maximal number of DK solution(s)

Case	Number of sensors	Complexity	Max number of solutions
$3\theta_V\theta_H$	6	1	1
$\theta_V^1\theta_H^1 - \theta_V^2\theta_H^2 - \theta_V(H)^3$	5	2,4	6
$\theta_V^1\theta_H^1 - \theta_V^2\theta_H^2$	4	4,4,4	12
$\theta_V^1\theta_H^1 - \theta_V(H)^2 - \theta_V(H)^3$	4	6,6,6,12,12	44
$\theta_V(H)^1 - \theta_V(H)^2 - \theta_V(H)^3$	3	$6 \times 72, 12, 24, 2 \times 8, 2 \times 4$	492
$\theta_V^1\theta_H^1 - \theta_V(H)^2$	3	40, 12	52
$\theta_V^1\theta_H^1$	2	162, 104, 68, 48, 22, 20, 8, 7, 3	442
$\theta_V(H)^1 - \theta_V(H)^2$	2	936, 240, 112, 72, 8, 4, 4	1376

3.1 The 2 Elastic Cables Case

- case $\theta_H^1 - \theta_H^2$, **2 extra sensors:** the unknowns are ρ_1, ρ_2 and the 2 first equations of the mechanical equilibrium are linear in these variables. The solution is unique for the planar CDPR but has 2 DK solutions for the spatial CDPR, see Sect. 2.2.
- case θ_H^1, **1 extra sensors:** the unknowns are ρ_1, ρ_2 and the 2 coordinates of B_2 in the CDPR plane. The 2 first equations of the mechanical equilibrium are used to determine these later unknowns. The third equation of the mechanical equilibrium becomes linear in ρ_1. After solving the constraint $\|\mathbf{A_2B_2}\|^2 = \rho_2^2$ becomes a polynomial of degree 4 in ρ_2.

4 Analysis and Uncertainty

As seen in Tables 1 and 2 the complexity of the calculation of the DK solution(s) and their maximal number increases very quickly as soon as the number of sensors is getting lower than 6 (in which case we get a single solution both for the ideal and elastic cables). Taking into account measurement uncertainty is not the purpose of this paper but our first trial with our measurement system (see Fig. 2) has shown that we cannot expect a high accuracy, especially when the cable tension is low. Furthermore the accuracy ΔB of the location of B based on the orientation sensors and assuming an exact measurement of ρ is $\Delta B = \rho \Delta \theta$ where $\Delta \theta$ is the sensor error. This implies that for large CDPR where ρ is much larger than 1 we may expect large error on the coordinates of the B. However it may be thought that the parallel structure may overall decrease this influence. To examine this point we have considered a simple planar CDPR with 2 cables connected at the same point. We assume that the ρ, θ are measured respectively with an accuracy $\pm \Delta \rho, \pm \Delta \theta$. For a given pose x_0, y_0 of the CDPR these uncertainties induce an error on the location of the CDPR and its real pose lies in a closed region around the nominal pose. To determine the border of this region we consider the poses $x_0 + r\cos(\alpha), y_0 + r\sin(\alpha)$ along a specific direction defined by the angle α, poses that are at a distance r from the pose x_0, y_0 For a given α a simple otimization procedure allows one to determine the maximum of r, i.e. the maximal positioning error that is compatible with $\Delta \rho, \Delta \theta$ along the direction defined by α. Starting from $\alpha = 0$ we increment α by a step of 5 degrees until we reach 360 degrees, giving us a reasonable approximation of the border of the region in which the CDPR will lie. The calculation of r at each α allows us to calculate a good approximation of the minimal, maximal and mean value for the maximal positioning error. We are thus able to calculate these variables as a function of $\Delta \theta$ for a fixed value of $\Delta \rho$. When $\Delta \theta$ is large the

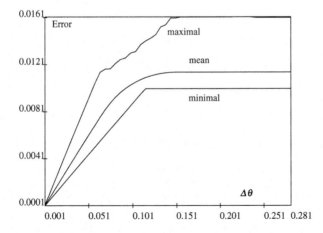

Fig. 3. Minimal, maximal and mean positioning error for $\Delta \rho = 0.01$ as a function of $\Delta \theta$ in degree

positioning error is just influenced by $\Delta\rho$ but when $\Delta\theta$ decreases their will be a switching point at which the maximal positioning error will start to decrease due to the influence of $\Delta\theta$. Hence this switching point indicates how accurate should be the measurement in $\Delta\theta$ in order to obtain a better accuracy than the one based on $\Delta\rho$ only. Figure 3 shows this function for the CDPR with $A_1 = (0,0)$, $A_2 = (10,0)$, the pose $x_0 = 5\sqrt{2}/2$, $y_0 = -5\sqrt{2}/2$ and $\Delta\rho = 0.01$. It may be seen that the switching point occurs around 0.1 degrees. Therefore the orientation measurement must be highly accurate to provide a better accuracy than the one obtained by using only the cable lengths.

5 Conclusion

As solving the DK of CDPR based only on the cable lengths is a complex task it is worth investigating how additional sensors may help this solving. Note that these additional sensor(s) may also be used for other tasks such as auto-calibration [3], identification [17] or workspace limit detection and consequently may be worth the limited additional cost. In this paper we have investigated sensors that provide partial or complete information on the cable orientation and have examined the effect of sensor number and arrangement on the DK solving for CDPR with 2 or 3 cables, ideal or elastic. This is a necessary work but also preliminary: it should be extended to CDPR with more than 3 cables. Furthermore we have assumed perfect sensor measurements which is an unrealistic hypothesis, and consequently the influence of the uncertainties on the DK solving has to be studied. We have shown on a 2dof planar CDPR that the uncertainty on the orientation sensor measurement must be very low to have an influence on the accuracy of the estimation of the DK solutions but this influence has to be studied in detail in more general cases. However cable orientation measurement, even with an uncertainty interval, may provide useful information for a numerical method solving the DK with the cable lengths only, allowing to safely eliminate possible DK solutions. Indeed some of these solutions may lead to angles that lie outside their measurement intervals and thus can be eliminated. Finally the use of extra orientation sensors has also to be investigated to manage redundantly actuated CDPR, singularity and sagging cables.

References

1. Abbasnejad, G., Carricato, M.: Real solutions of the direct geometrico-static problem of underconstrained cable-driven parallel robot with 3 cables: a numerical investigation. Meccanica **473**(7), 1761–1773 (2012)
2. Abbasnejad, G., Carricato, M.: Direct geometrico-static problem of underconstrained cable-driven parallel robots with n cables. IEEE Trans. Robot. **31**(2), 468–478 (2015)
3. dit Sandretto, J.A. et al.: Certified calibration of a cable-driven robot using interval contractor programming. In: Computational Kinematics, Barcelona, 12–15 May 2013 (2013)

4. Andreff, N., Dallej, T., Martinet, P.: Image-based visual servoing of a Gough-Stewart parallel manipulator using leg observations. Int. J. Robot. Res. **26**(7), 677–688 (2007)
5. Berti, A., Merlet, J.-P., Carricato, M.: Solving the direct geometrico-static problem of underconstrained cable-driven parallel robots by interval analysis. Int. J. Robot. Res. **35**(6), 723–739 (2016)
6. Berti, A., Merlet, J.-P., Carricato, M.: Solving the direct geometrico-static problem of the 3–3 cable-driven parallel robots by interval analysis: preliminary results. In: 1st International Conference on Cable-Driven Parallel Robots (CableCon), Stuttgart, 3–4 September 2012, pp. 251–268 (2012)
7. Bonev, I.A., et al.: A closed-form solution to the direct kinematics of nearly general parallel manipulators with optimally located three linear extra sensors. IEEE Trans. Robot. Autom. **17**(2), 148–156 (2001)
8. Bruckman, T., et al.: Wire robot part I, kinematics, analysis and design. In: Parallel Manipulators, New Developments, pp. 109–132. ITECH, April 2008
9. Carricato, M., Abbasnejad, G.: Direct geometrico-static analysis of underconstrained cable-driven parallel robots with 4 cables. In: 1st International Conference on Cable-Driven Parallel Robots (CableCon), Stuttgart, 3–4 September 2012, pp. 269–286 (2012)
10. Carricato, M., Merlet, J.-P.: Stability analysis of underconstrained cable-driven parallel robots. IEEE Trans. Robot. **29**(1), 288–296 (2013)
11. Carricato, M., Merlet, J.-P.: Direct geometrico-static problem of under-constrained cable-driven parallel robots with three cables. In: IEEE International Conference on Robotics and Automation, Shangai, 9–13 May 2011, pp. 3011–3017 (2011)
12. Han, K., Chung, W., Youm, Y.: New resolution scheme of the forward kinematics of parallel manipulators using extra sensor data. ASME J. Mech. Design **118**(2), 214–219 (1996)
13. Krauss, W., et al.: System identification and cable force control for a cable-driven parallel robot with industrial servo drives. In: IEEE International Conference on Robotics and Automation, Hong-Kong, 31 May–7 June 2014, pp. 5921–5926 (2014)
14. Merlet, J.-P.: Closed-form resolution of the direct kinematics of parallel manipulators using extra sensors data. In: IEEE International Conference on Robotics and Automation, Atlanta, 2–7 May 1993, pp. 200–204 (1993)
15. Merlet, J.-P.: On the real-time calculation of the forward kinematics of suspended cable-driven parallel robots. In: 14th IFToMM World Congress on the Theory of Machines and Mechanisms, Taipei, 27–30 October 2015 (2015)
16. Miermeister, P., Pott, A.: Auto calibration method for cable-driven parallel robot using force sensors. In: ARK, Innsbruck, 25–28 June 2012, pp. 269–276 (2012)
17. Ottaviano, E., Ceccarelli, M., Palmucci, F.: Experimental identification of kinematic parameters and joint mobility of human limbs. In: 2nd International Congress, Design and Modelling of mechanical systems, Monastir, 19–21 March 2007 (2007)
18. Parenti-Castelli, V., Di Gregorio, R.: Real-time computation of the actual posture of the general geometry 6–6 fully parallel mechanism using only two extra rotary sensors. ASME J. Mech. Design **120**(4), 549–554 (1998)
19. Pott, A.: An algorithm for real-time forward kinematics of cable-driven parallel robots. In: ARK, Piran, 28 June–1 July 2010, pp. 529–538 (2010)

Trajectory Planning and Control

Randomized Kinodynamic Planning for Cable-Suspended Parallel Robots

Ricard Bordalba[(✉)], Josep M. Porta, and Lluís Ros

Institut de Robòtica i Informàtica Industrial, CSIC-UPC, Barcelona, Spain
{rbordalba,porta,lros}@iri.upc.edu

Abstract. This paper proposes the use of a randomized kinodynamic planning technique to synthesize dynamic motions for cable-suspended parallel robots. Given two mechanical states of the robot, both with a prescribed position and velocity, the method attempts to connect them by a collision-free trajectory that respects the joint and force limits of the actuators, keeps the cables in tension, and takes the robot dynamics into account. The method is based on the construction of a bidirectional rapidly-exploring random tree over the state space. Remarkably, the technique can be used to cross forward singularities of the robot in a predictable manner, which extends the motion capabilities beyond those demonstrated in previous work. The paper describes experiments that show the performance of the method in point-to-point operations with specific cable-driven robots, but the overall strategy remains applicable to other mechanism designs.

1 Introduction

Cable-suspended parallel robots consist of a moving load hanging from a fixed base by means of cables. The load configuration can be changed by varying the cable lengths or anchor point locations, and gravity is typically used to maintain the cables under tension. As opposed to fully-constrained parallel cable-driven robots, cable-suspended robots are not redundantly actuated, and generally employ as many actuators as the number of degrees of freedom to be governed.

A fundamental issue in such robots is to guarantee that the cable tensions remain positive at all times. In this way, cable slackness is avoided, which allows the control of the load using proper tension adjustments. Traditionally, these robots have been used as robotic cranes, setting them to operate in static or quasi-static conditions, in which gravity is the sole source of cable tension [1,5]. While this simplifies the planning and control of the motions, it also confines them to the static workspace, which is a region limited by the footprint of the robot. More recently, inertia has also been proposed as another source of tension [11], extending the movement capacity to the dynamic workspace, i.e., the region that can be attained when load accelerations are allowed [3].

This research has been partially funded by project DPI2014-57220-C2-2-P.

© Springer International Publishing AG 2018
C. Gosselin et al. (eds.), *Cable-Driven Parallel Robots*,
Mechanisms and Machine Science 53, DOI 10.1007/978-3-319-61431-1_17

Using pendulum-like motions, for example, pick-and-place tasks between points well beyond the robot footprint can be planned [10,14].

In this new context, there is a strong need for an efficient planning technique that determines the force inputs required to move the robot between two mechanical states (both with a prescribed position and velocity). Such planner must avoid collisions of the robot with itself or with the environment while obeying the physical laws imposed by the motion equations, and the force and joint limits of the actuators. This problem, known as kinodynamic planning in the literature [16], is gaining attention in cable-driven robotics [9]. Early work in this regard includes planning methods for a remarkable mechanism like the Winch-bot, which can follow prescribed paths on a vertical plane with just a single actuator [8], or evolved architectures with additional manipulation abilities [18,22,23]. Because of their underactuation, however, these robots cannot control their pose exactly along the motion, which motivated the development of newer methods for fully-actuated designs. For instance, in [11] cyclic trajectories leaving the static workspace were given, and the approach was later extended to synthesize point-to-point trajectories for pick-and-place tasks [10,14]. Optimal control methods for Robocrane-type platforms were also provided in [2]. While such methods are remarkable, none of them were designed to avoid collisions. The methods in [10,14], moreover, rely on predefined trajectories, and they need some guidance to define intermediate waypoints when the start and goal configurations fail to be connected by such trajectories.

Particular solutions for specific robots are valuable, but it is the authors' belief that existing randomized techniques can solve the kinodynamic planning problem with great generality in cable-driven robots. The purpose of this paper is, precisely, to show that a recent method of this kind [7] can successfully cope with the kinematic, collision, and positive-tension constraints arising in such robots. The method is based on deploying an exploration tree over the state space and it is probabilistically complete, i.e., it finds a connecting trajectory whenever one exists and enough computing time is available. Remarkably, we show that the method can also be used to cross forward singularities in a predictable manner, which further extends the motion capabilities beyond those envisaged in earlier work [10,14]. The method is a generalization of a classic planning method [17]. Whereas the approach in [17] was suitable for mechanisms described by means of independent generalized coordinates, the one in [7] can also handle dependent coordinates coupled by kinematic constraints, which often arise in parallel mechanisms. After reviewing the contributions of [7,17] (Sect. 2), we show how the resulting technique can be applied to cable-driven robots (Sect. 3), and illustrate its performance on challenging problems (Sect. 4). We finally summarize the main strengths of the approach, and points deserving further attention (Sect. 5).

2 A Kinodynamic Motion Planner

The planning of dynamic motions typically takes place in the state space of the robot, i.e., the set $\mathscr{X}$ of kinematically-valid states $\boldsymbol{x} = (\boldsymbol{q}, \dot{\boldsymbol{q}})$, where $\boldsymbol{q}$ is a vector of n_q generalized coordinates describing the configuration of the robot, and $\dot{q}$ is the time derivative of q, which describes its velocity. The coordinates in q may be independent or not. In the former case, any pair $x = (q, \dot{q}) \in \mathbb{R}^{2n_q}$ is kinematically valid, and $\mathscr{X}$ becomes parametrically defined. The latter case is more complex. The configuration space (C-space) of the robot is the set $\mathscr{C}$ of points q that satisfy a system of n_e nonlinear equations

$$\boldsymbol{\Phi}(q) = 0 \tag{1}$$

encoding, e.g., loop-closure constraints, or geometric constraints due to nonminimal representations of $SO(3)$. As a result, the valid values of $\dot{q}$ are those that fulfill

$$\boldsymbol{\Phi}_q \, \dot{q} = 0, \tag{2}$$

where $\boldsymbol{\Phi}_q = \partial \boldsymbol{\Phi} / \partial q$. Then, $\mathscr{X}$ becomes a nonlinear manifold of dimension $d_{\mathscr{X}} = 2(n_q - n_e)$ generically, defined implicitly by Eqs. (1) and (2).

Irrespective of the form of $\mathscr{X}$, the motions must always be confined to a feasibility region $\mathscr{X}_{\text{feas}} \subseteq \mathscr{X}$ of collision-free states respecting joint and constraint force limits (such as tension positivities in cable-driven robots). Finally, the motions must also obey the dynamic equations of the robot, which can be written in the form

$$\dot{x} = g(x, u). \tag{3}$$

In this equation, $g(x, u)$ is an appropriate differentiable function, and u is a d-vector of actuator forces subject to lie in a bounded subset $\mathscr{U} \subset \mathbb{R}^d$. Then, given start and goal states, x_s and x_g, the kinodynamic planning problem consists in finding a time function $u(t)$ such that the system trajectory $x(t)$ determined by Eqs. (1)–(3) for $x(0) = x_s$, fulfills $x(t_f) = x_g$ for some time $t_f > 0$, and $u(t) \in \mathscr{U}$, $x(t) \in \mathscr{X}_{\text{feas}}$ for all $t \in [0, t_f]$.

The solution proposed in [17] assumes that the q coordinates are independent, so that Eqs. (1) and (2) need not be considered. The resulting planner looks for a solution by constructing a rapidly-exploring random tree (RRT) over $\mathscr{X}$. The RRT is rooted at x_s and it is grown incrementally towards x_g while staying inside $\mathscr{X}_{\text{feas}}$. Every tree node stores a feasible state $x \in \mathscr{X}_{\text{feas}}$, and every edge stores the action $u \in \mathscr{U}$ needed to move between the connected states. This action is assumed to be constant during the move. The expansion of the RRT proceeds by applying three steps repeatedly (Fig. 1, top-left). First, a state $x_{rand} \in \mathscr{X}$ is randomly selected; then, the RRT state x_{near} that is closest to x_{rand} is computed according to some metric; finally, a movement from x_{near} towards x_{rand} is performed by applying an action $u \in \mathscr{U}$ during a fixed time Δt. The movement from x_{near} towards x_{rand} is simulated by integrating Eq. (3) numerically, which yields a new state x_{new} that may or may not be in $\mathscr{X}_{\text{feas}}$. In the former case x_{new} is added to the RRT, and in the latter it is discarded. To test whether $x_{new} \in \mathscr{X}_{\text{feas}}$, x_{new} is checked for collisions by using standard algorithms [15], and the joint positions and constraint forces are computed to check whether they stay within bounds. The action u applied is typically chosen as the one from $\mathscr{U}$ that brings the robot closer to x_{rand}. One can either try all

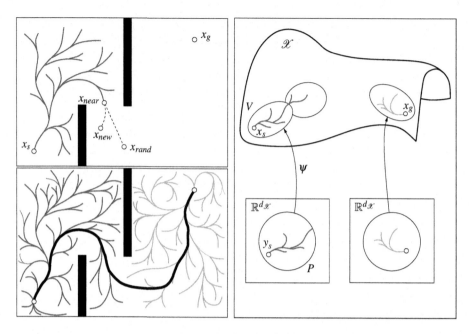

Fig. 1. Left-Top: Extension process of an RRT. Left-Bottom: A kinodynamic planning problem is often solved faster with a bidirectional RRT. Right: Construction of an RRT on an implicitly-defined state space manifold.

possible values in $\mathscr{U}$ (if it is a discrete set) or only those of n_s random points on $\mathscr{U}$ (if it is continuous). To force the RRT to extend towards $\bm{x}_g$, $\bm{x}_{rand}$ is set to $\bm{x}_g$ once in a while, stopping the whole process when a RRT leaf is close enough to $\bm{x}_g$. Usually, however, a solution trajectory can be found more rapidly if two RRTs respectively rooted at $\bm{x}_s$ and $\bm{x}_g$ are grown simultaneously towards each other (Fig. 1, left-bottom). The expansion of the tree rooted at $\bm{x}_g$ is based on the integration of Eq. (3) backward in time.

The previous strategy has proved to be effective in many situations, but in parallel robots the coordinates in $\bm{q}$ are often dependent. This fact complicates the generation of RRTs over $\mathscr{X}$, because there is no straightforward way to randomly select points $\bm{x} = (\bm{q}, \dot{\bm{q}})$ satisfying Eqs. (1) and (2), and the numerical integration of Eq. (3) easily drifts away from $\mathscr{X}$ when standard methods for ordinary differential equations are used. These two issues have been recently circumvented in [7] by constructing an atlas of $\mathscr{X}$ in parallel to the RRT.

An atlas is a collection of charts mapping $\mathscr{X}$ entirely, where each chart is a local diffeomorphism $\bm{\psi}$ from an open set $P \subseteq \mathbb{R}^{d_{\mathscr{X}}}$ of parameters to an open set $V \subset \mathscr{X}$ (Fig. 1, right). The V sets can be thought of as partially-overlapping tiles covering $\mathscr{X}$, in such a way that every $\bm{x} \in \mathscr{X}$ lies in at least one set V. Assuming that an atlas is available, the problem of sampling $\mathscr{X}$ boils down to generating random values $\bm{y}$ in the P sets, since these values can always be projected to $\mathscr{X}$ using $\bm{x} = \bm{\psi}(\bm{y})$. Also, the atlas allows the conversion of the vector field defined on $\mathscr{X}$ by Eq. (3) into one in the coordinate spaces P, which permits the integration of Eq. (3) using local coordinates [21]. As a result, the RRT motions satisfy Eqs. (1) and (2) by construction, eliminating any drift from $\mathscr{X}$ to machine precision.

As explained in [7], the construction of the atlas is incremental. The atlas is initialized with two charts covering x_s and x_g, respectively (Fig. 1, right). Then, these charts are used to pull the expansion of the RRT, which in turn adds new charts to the atlas as needed, until x_s and x_g become connected. To be able to construct the charts, the method requires $\mathscr{X}$ to be smooth, which implies that the robot cannot exhibit C-space singularities, i.e., points q for which Φ_q is rank deficient [4,6]. In practice, the exclusion of such singularities can be achieved by choosing appropriate mechanism dimensions, since generically Φ_q will be full rank. Another choice is to set joint limits excluding the presence of such singularities.

3 Application to a Cable-Suspended Robot

To apply the previous method to a specific cable-suspended robot we need to obtain Eqs. (1)–(3) and verify that Φ_q is full rank over the C-space. Moreover, to determine whether a given x belongs to $\mathscr{X}_{\text{feas}}$, note that the joint limits can be trivially checked, and we can use the methods in [15] to detect the collisions. Thus, we only need to provide a means to compute the cable tensions for each $x \in \mathscr{X}$. We next illustrate these points in the particular robot of Fig. 2.

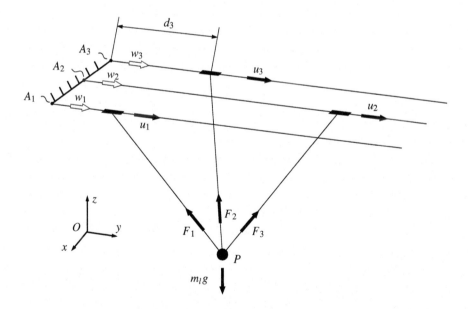

Fig. 2. A spatial 3-DOF cable-driven robot.

3.1 Kinematic Model

Consider a point mass suspended from three cables of fixed length ρ_i, $i = 1, 2, 3$. Each cable is connected to an actuated slider that can move along a guide line defined by a point $\boldsymbol{A}_i$, with position vector $\boldsymbol{a}_i$, and a unit vector $\boldsymbol{w}_i$, both given in a fixed frame O_{xyz} (Fig. 2). Although the guides are horizontal and parallel in this figure, they could take any direction in general. By changing the displacements $\boldsymbol{d} = [d_1, d_2, d_3]^\mathsf{T}$ of the sliders, the robot can control the position $\boldsymbol{P}$ of the point mass, with position vector $\boldsymbol{p} = [x, y, z]^\mathsf{T}$ in the mentioned frame. In this robot, it is natural to choose $\boldsymbol{q} = [x, y, z, d_1, d_2, d_3]^\mathsf{T}$ so that Eq. (1) becomes the system formed by

$$\rho_i^2 - \boldsymbol{c}_i^\mathsf{T} \boldsymbol{c}_i = 0, \tag{4}$$

for $i = 1, 2, 3$, where $\boldsymbol{c}_i = (\boldsymbol{a}_i + d_i \boldsymbol{w}_i) - \boldsymbol{p}$ is the vector from $\boldsymbol{P}$ to the ith slider position. By taking the partial derivatives of Eq. (4) with respect to $\boldsymbol{p}$ and $\boldsymbol{d}$ we obtain

$$\boldsymbol{\Phi}_q = \begin{bmatrix} 2\boldsymbol{c}_1^\mathsf{T} & -2\boldsymbol{c}_1^\mathsf{T} \boldsymbol{w}_1 & 0 & 0 \\ 2\boldsymbol{c}_2^\mathsf{T} & 0 & -2\boldsymbol{c}_2^\mathsf{T} \boldsymbol{w}_2 & 0 \\ 2\boldsymbol{c}_3^\mathsf{T} & 0 & 0 & -2\boldsymbol{c}_3^\mathsf{T} \boldsymbol{w}_3 \end{bmatrix}, \tag{5}$$

$$\underbrace{}_{\boldsymbol{\Phi}_p} \underbrace{}_{\boldsymbol{\Phi}_d}$$

which readily provides Eq. (2).

By inspection of the previous Jacobian, it is easy to see that a configuration $\boldsymbol{q}$ is a C-space singularity if, and only if, the three cables lie on a plane π orthogonal to the three guides. Certainly, if for a given $\boldsymbol{q}$ all cables lie on such a plane π, the subjacobian $\boldsymbol{\Phi}_d$ is null, and the subjacobian $\boldsymbol{\Phi}_p$ is rank deficient. This implies that all 3×3 minors of $\boldsymbol{\Phi}_q$ will vanish, so that $\boldsymbol{q}$ is a C-space singularity. Conversely, if not all cables lie on such a plane π, but they are still on a plane not orthogonal to the guides, $\boldsymbol{\Phi}_p$ is rank deficient but $\boldsymbol{\Phi}_d$ will be full rank. If the cables are not coplanar, $\boldsymbol{\Phi}_p$ is full rank. In any of the two situations, therefore, $\boldsymbol{\Phi}_q$ will be full rank.

In what follows, we shall assume that our robot does not exhibit C-space singularities. To ensure so, note that it suffices to choose cable lengths for which it is impossible to assemble the mechanism with all of its cables stretched and lying on the plane π just described.

The configurations in which $\boldsymbol{\Phi}_p$ is rank deficient are the so-called forward singularities of the mechanism [6]. In these singularities, the velocities of the actuators do not determine the velocity of the end effector if only Eq. (2) is considered. However, in the next section we shall see that, dynamically, the evolution of the mechanism is perfectly predictable across such singularities.

3.2 Dynamic Model

To formulate Eq. (3), we use the Euler-Lagrange equations with multipliers [12] which lead to a compact treatment of dynamics and are easily applicable to other cable-driven architectures. These equations take the form

$$\frac{d}{dt} \frac{\partial K}{\partial \dot{\boldsymbol{q}}} - \frac{\partial K}{\partial \boldsymbol{q}} + \frac{\partial U}{\partial \boldsymbol{q}} + \boldsymbol{\Phi}_q^\mathsf{T} \boldsymbol{\lambda} = \boldsymbol{\tau}, \tag{6}$$

where K and U are the expressions of the kinetic and potential energies of the robot, $\boldsymbol{\lambda}$ is a vector of n_e Lagrange multipliers, and $\boldsymbol{\tau}$ is the generalized force corresponding to the non-conservative forces applied on the system.

In the robot of Fig. 2, we assume that the cables have negligible mass, and let m_l and m_s refer to the mass of the moving load and the mass of each slider, respectively. By defining $\boldsymbol{M}_l = m_l \boldsymbol{I}_3$ and $\boldsymbol{M}_s = m_s \boldsymbol{I}_3$, where $\boldsymbol{I}_3$ is the 3×3 identity matrix, the kinetic energy of the robot is given by

$$K = \frac{1}{2}[\dot{\boldsymbol{p}}^\mathsf{T}\ \dot{\boldsymbol{d}}^\mathsf{T}]\begin{bmatrix} \boldsymbol{M}_l & 0 \\ 0 & \boldsymbol{M}_s \end{bmatrix}\begin{bmatrix} \dot{\boldsymbol{p}} \\ \dot{\boldsymbol{d}} \end{bmatrix} = \frac{1}{2}\dot{\boldsymbol{q}}^\mathsf{T}\boldsymbol{M}\dot{\boldsymbol{q}}, \tag{7}$$

where $\boldsymbol{M}$ is the so-called mass matrix, which is always symmetric and positive definite. The potential energy of the robot, on the other hand, is given by

$$U = m_l g z, \tag{8}$$

where g is the gravitational acceleration. By substituting Eq. (7) into Eq. (6), the Euler-Lagrange equations of our robot reduce to

$$\boldsymbol{M}\ddot{\boldsymbol{q}} + \boldsymbol{U}_{\boldsymbol{q}} + \boldsymbol{\Phi}_{\boldsymbol{q}}^\mathsf{T}\boldsymbol{\lambda} = \boldsymbol{\tau}. \tag{9}$$

where the term $\boldsymbol{U}_{\boldsymbol{q}}$ is given by the partial derivatives of Eq. (8), i.e.,

$$\boldsymbol{U}_{\boldsymbol{q}} = [0, 0, m_l g, 0, 0, 0]^\mathsf{T}. \tag{10}$$

Also, assuming for simplicity that all contacts are frictionless, and letting u_i denote the force exerted by the ith actuator, we have

$$\boldsymbol{\tau} = [0, 0, 0, u_1, u_2, u_3]^\mathsf{T}. \tag{11}$$

Since Eq. (9) is a system of n_q equations in $n_q + n_e$ unknowns (the values of $\ddot{\boldsymbol{q}}$ and $\boldsymbol{\lambda}$), we need extra equations to be able to solve for $\ddot{\boldsymbol{q}}$. These can be obtained by differentiating Eq. (2), which yields

$$\boldsymbol{\Phi}_{\boldsymbol{q}}\ddot{\boldsymbol{q}} - \boldsymbol{\xi} = \boldsymbol{0}, \tag{12}$$

where $\boldsymbol{\xi} = -(\boldsymbol{\Phi}_{\boldsymbol{qq}}\dot{\boldsymbol{q}})\dot{\boldsymbol{q}}$. Equations (9) and (12) can then be written as

$$\begin{bmatrix} \boldsymbol{M}(\boldsymbol{q}) & \boldsymbol{\Phi}_{\boldsymbol{q}}^\mathsf{T} \\ \boldsymbol{\Phi}_{\boldsymbol{q}} & 0 \end{bmatrix}\begin{bmatrix} \ddot{\boldsymbol{q}} \\ \boldsymbol{\lambda} \end{bmatrix} = \begin{bmatrix} \boldsymbol{\tau} - \boldsymbol{U}_{\boldsymbol{q}} \\ \boldsymbol{\xi} \end{bmatrix}. \tag{13}$$

Clearly, if $\boldsymbol{\Phi}_{\boldsymbol{q}}$ is full rank, i.e. there are no C-space singularities, the matrix on the left-hand side of Eq. (13) is invertible, even at forward singularities, and thus we can write

$$\ddot{\boldsymbol{q}} = \boldsymbol{f}(\boldsymbol{q}, \dot{\boldsymbol{q}}, \boldsymbol{u}) = \begin{bmatrix} \boldsymbol{I}_{n_q} & 0 \end{bmatrix}\begin{bmatrix} \boldsymbol{M}(\boldsymbol{q}) & \boldsymbol{\Phi}_{\boldsymbol{q}}^\mathsf{T} \\ \boldsymbol{\Phi}_{\boldsymbol{q}} & 0 \end{bmatrix}^{-1}\begin{bmatrix} \boldsymbol{\tau} - \boldsymbol{U}_{\boldsymbol{q}} \\ \boldsymbol{\xi} \end{bmatrix}. \tag{14}$$

To finally obtain Eq. (3), we transform Eq. (14) into a first-order ordinary differential equation using the change of variables $\dot{\boldsymbol{q}} = \boldsymbol{v}$, yielding

$$\dot{\boldsymbol{x}} = \begin{bmatrix} \dot{\boldsymbol{q}} \\ \dot{\boldsymbol{v}} \end{bmatrix} = \begin{bmatrix} \boldsymbol{v} \\ \boldsymbol{f}(\boldsymbol{q}, \boldsymbol{v}, \boldsymbol{u}) \end{bmatrix} = \boldsymbol{g}(\boldsymbol{x}, \boldsymbol{u}). \tag{15}$$

3.3 Tension Computation

Let $\boldsymbol{F}_i$ denote the force applied by the ith cable on the moving load (Fig. 2). Such a force can be written as $\boldsymbol{F}_i = \boldsymbol{c}_i F_i / \rho_i$, where F_i is the tension of the ith cable. We next see that the tensions F_i can be obtained from the Lagrange multipliers $\boldsymbol{\lambda}$. Note that Eq. (9) can be decomposed into

$$M_l \, \ddot{\boldsymbol{p}} = [0, 0, -m_l g]^\mathsf{T} - \boldsymbol{\Phi_p}^\mathsf{T} \boldsymbol{\lambda}, \tag{16}$$

$$M_s \, \ddot{\boldsymbol{d}} = [u_1, u_2, u_3]^\mathsf{T} - \boldsymbol{\Phi_d}^\mathsf{T} \boldsymbol{\lambda}, \tag{17}$$

which correspond, respectively, to Newton's 2nd law applied to the load and the sliders. Using Eq. (16), for instance, we see that the term $-\boldsymbol{\Phi_p}^\mathsf{T} \boldsymbol{\lambda}$ must be the resultant force applied by the cables $\boldsymbol{F}_c = \boldsymbol{F}_1 + \boldsymbol{F}_2 + \boldsymbol{F}_3$, because the other two terms are the weight of the load and the time derivative of its linear momentum. Thus we can say that $\boldsymbol{F}_c = -\boldsymbol{\Phi_p}^\mathsf{T} \boldsymbol{\lambda}$, or, using the value of $\boldsymbol{\Phi_p}$ in Eq. (5),

$$\boldsymbol{F}_c = 2\boldsymbol{c}_1 \lambda_1 + 2\boldsymbol{c}_2 \lambda_2 + 2\boldsymbol{c}_3 \lambda_3. \tag{18}$$

On the other hand, $\boldsymbol{F}_c$ can also be written as

$$\boldsymbol{F}_c = \frac{\boldsymbol{c}_1}{\rho_1} F_1 + \frac{\boldsymbol{c}_2}{\rho_2} F_2 + \frac{\boldsymbol{c}_3}{\rho_3} F_3, \tag{19}$$

and comparing Eqs. (18) and (19) we obtain $F_i = 2\rho_i \lambda_i$. We note this expression for F_i could also have been obtained by departing from Eq. (17) instead. Moreover, since the robot is assumed to be free from C-space singularities, the $\boldsymbol{\lambda}$ values are always determined by Eq. (13), implying that the tensions F_i will be determined too, even at forward singularities.

4 Experiments

The planner has been implemented in C, and it has been integrated into the CUIK Suite [20]. To illustrate its performance, we next show three experiments of increasing complexity. The first two experiments involve a planar version of the robot of Fig. 2, whereas the last experiment is three-dimensional. In all cases the mass of the load is 1 [kg], the mass of each slider is 0.1 [kg], and the force applied by the sliders is limited to the range $[-8, 8]$ [N], so that $\mathscr{U} = [-8, 8]^d$, with $d \in \{2, 3\}$. A bidirectional RRT is always constructed, and each time it is extended, $n_s = 25$ actions are randomly sampled from $\mathscr{U}$. Each of these actions is applied during $\Delta t = 0.5$ [s].

For each experiment, Table 1 summarizes the values of n_q, n_e, and $d_{\mathscr{X}}$, as well as the performance statistics on an iMac with an Intel i7 processor with 8 CPU cores running at 2.93 Ghz. The statistics include the number of samples and charts generated, and the planning time in seconds, all averaged over ten runs. The planner successfully connected the start and goal states in all runs. Finally, the table also indicates the execution time, t_f, for the trajectories of Fig. 3.

Collision-Free Kinodynamic Planning for Cable-Suspended Parallel Robots   203

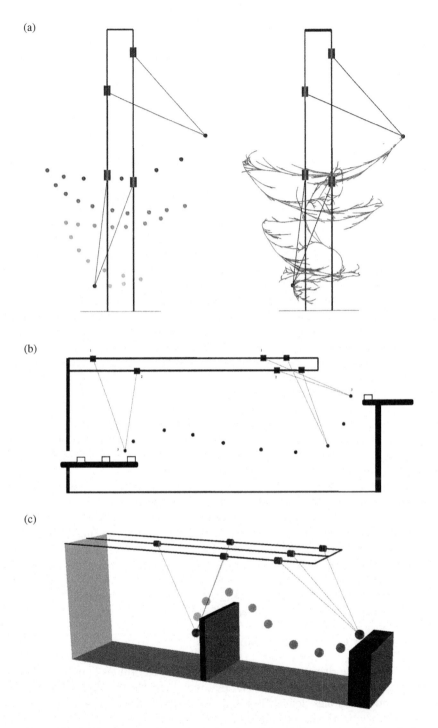

Fig. 3. Three planning problems on planar and spatial versions of the robot.

Table 1. Problem dimensions and performance statistics for the shown experiments.

Experiment	n_q	n_e	$d_\mathscr{x}$	No. of samples	No. of charts	Planning time [s]	t_f [s]
1	4	2	4	1928	283	12.6	10.1
2	4	2	4	20946	2622	140	7.1
3	6	3	6	24398	2244	234	8.4

Experiment 1: Moving in the dynamic workspace

In this example, the load is suspended from two cables, and the sliders move along vertical guides [Fig. 3(a)]. The cables and the sliders move on different planes and, thus, their collisions need not be checked. The distance between the two guides is 2 [m], and the cables' length is 8 [m]. The goal here is to move the load from a low position to a higher position, both in rest and outside the static workspace. The load has to oscillate along the trajectory in order to gain momentum and finally reach the goal. The smaller the allowable force on the motors, the larger the number of oscillations and the harder the planning problem. The bidirectional RRT created encompasses two trees rooted at the start and goal states, shown in red and green respectively. Note that although the robot has a limited static workspace (the region between the guides), including dynamics in the planning has increased the usable workspace substantially.

Experiment 2: Singularity crossing

We now consider the robot of Fig. 3(b), in which the load is suspended from two horizontal guides separated 1 [m] from each other. The lengths of the cables have been set to 6.6 and 8 [m], which allows them to align at 45° relative to the guides. The planning problem consists in finding a trajectory to move the load between the left and rightmost positions in the figure, assuming that the cables cannot collide with the guides nor with themselves. Note that the triangle 1-2-3 has a different orientation at the start and goal positions, so that the robot will have to cross a forward singularity to connect them. Although the inverse static problem is indeterminate in such a singularity, we have shown how both the tensions and the evolution of the robot remain dynamically determined (Sects. 3.2 and 3.3). The planner, as a result, has no trouble in computing the shown trajectory, which certainly crosses the singularity somewhere between the two configurations depicted on the right.

Experiment 3: Obstacle avoidance

Finally, a 3D cable-driven robot with three horizontal guides is used to demonstrate how obstacles are avoided. The distance between two consecutive guides is 3 [m], and the cables' length is 8 [m]. The robot moves from a rest position inside the static workspace to another position outside of it. Both positions are separated by a wall in the middle of the workspace, which has to be avoided during the move. The robot is able to overpass the obstacle and manages to reach the goal as seen in Fig. 3(c).

5 Conclusions

This paper has shown how a recent randomized kinodynamic planning technique can be applied to generate dynamic trajectories for cable-suspended parallel robots. Taking into account the system dynamics enlarges the robot workspace substantially, allowing to reach points further apart from the footprint of the supporting structure. Moreover, the joint consideration of obstacle avoidance, force and joint limits, positive tension constraints, and singularity crossings makes the planner applicable to challenging scenarios. The approach has been validated with experiments on particular architectures, but it remains applicable to other robot designs.

The trajectory directly returned by the planner is smooth in position, but not in velocity and acceleration. A point deserving further attention, thus, is the application of local optimization techniques to obtain twice-differentiable trajectories. Also, global optimization methods should be developed to obtain trajectories involving minimum-time or energy consumption [13,19]. Finally, efforts should be devoted to enhance the metric used to measure the distance between states, which is known to be a challenging task in all sampling-based kinodynamic planners.

References

1. Albus, J., Bostelman, R.V., Dagalakis, N.: The NIST Robocrane. J. Rob. Syst. **10**(5), 709–724 (1993). doi:10.1002/rob.4620100509
2. Bamdad, M.: Time-energy optimal trajectory planning of cable-suspended manipulators. In: Bruckmann, T., Pott, A. (eds.) Cable-Driven Parallel Robots, vol. 12, pp. 41–51. (2013). http://doi.org/10.1109/ICRoM.2014.6990907
3. Barrette, G., Gosselin, C.: Determination of the dynamic workspace of cable-driven planar parallel mechanisms. Trans. ASME-R-J. Mech. Des. **127**(2), 242–248 (2005). doi:10.1115/1.1830045
4. Bohigas, O., Manubens, M., Ros, L.: Singularities of non-redundant manipulators: a short account and a method for their computation in the planar case. Mech. Mach. Theory **68**, 1–17 (2013). doi:10.1016/j.mechmachtheory.2013.03.001
5. Bohigas, O., Manubens, M., Ros, L.: Planning wrench-feasiblemotions for cable-driven hexapods. IEEE Trans. Rob. **32**(2), 442–451 (2016). doi:10.1109/TRO.2016.2529640
6. Bohigas, O., Manubens, M., Ros, L.: Singularities of Robot Mechanisms: Numerical Computation and Avoidance Path Planning. Springer, Switzerland (2016). http://www.springer.com/us/book/9783319329208
7. Bordalba, R., Ros, L., Porta, J.M.: Kinodynamic planning on constraint manifolds. Institut de Robòtica i Infomàtica. CSIC-UPC, Technical report (2017)
8. Cunningham, D., Asada, H.H.: The Winch-bot: a cable-suspended, under-actuated robot utilizing parametric self-excitation. In: IEEE International Conference on Robotics and Automation, pp. 1844–1850 (2009). http://doi.org/10.1109/ROBOT.2009.5152378

9. Gosselin, C.: Cable-driven parallel mechanisms: state of the art and perspectives. Mech. Eng. Rev. 1(1), DSM0004 (2014). http://doi.org/10.1299/mer.2014dsm0004
10. Gosselin, C., Foucault, S.: Dynamic point-to-point trajectory planning of a two-DOF cable-suspended parallel robot. IEEE Trans. Rob. 30(3), 728–736 (2014). http://doi.org/10.1109/TRO.2013.2292451
11. Gosselin, C., Ren, P., Foucault, S.: Dynamic trajectory planning of a two-DOF cable-suspended parallel robot. In: IEEE International Conference on Robotics and Automation, pp. 1476–1481 (2012). http://doi.org/10.1109/ICRA.2012.6224683
12. Greenwood, D.T.: Advanced Dynamics. Cambridge University Press, Cambridge (2006). http://doi.org/10.1017/CBO9780511800207
13. Hauser, K., Zhou, Y.: Asymptotically optimal planning by feasible kinodynamic planning in a state-cost space. IEEE Trans. Rob. 32(6), 1431–1443 (2016). http://doi.org/10.1109/TRO.2016.2602363
14. Jiang, X., Gosselin, C.: Dynamic point-to-point trajectory planning of a three-DOF cable-suspended parallel robot. IEEE Trans. Rob. 32(6), 1550–1557 (2016). http://doi.org/10.1109/TRO.2016.2597315
15. Jiménez, P., Thomas, F., Torras, C.: 3D collision detection: a survey. Comput. Graph. 25(2), 269–285 (2001). doi:10.1016/S0097-8493(00)00130--8
16. LaValle, S.M.: Planning Algorithms. Cambridge University Press, New York (2006). http://planning.cs.uiuc.edu/
17. LaValle, S.M., Kuffner, J.J.: Randomized kinodynamic planning. Int. J. Rob. Res. 20(5), 378–400 (2001). http://doi.org/10.1177/02783640122067453
18. Lefrançois, S., Gosselin, C.: Point-to-point motion control of a pendulum-like 3-DOF underactuated cable-driven robot. In: IEEE International Conference on Robotics and Automation, pp. 5187–5193 (2010). http://doi.org/10.1109/ROBOT.2010.5509656
19. Li, Y., Littlefield, Z., Bekris, K.E.: Asymptotically optimal sampling-based kinodynamic planning. Int. J. Rob. Res. 35(5), 528–564 (2016). http://doi.org/10.1177/0278364915614386
20. Porta, J.M., Ros, L., Bohigas, O., Manubens, M., Rosales, C., Jaillet, L.: The cuik suite: analyzing the motion of closed-chain multibody systems. IEEE Rob. Autom. Mag. 21(3), 105–114 (2014). http://doi.org/10.1109/MRA.2013.2287462
21. Potra, F.A., Yen, J.: Implicit numerical integration for Euler-Lagrange equations via tangent space parametrization. J. Struct. Mech. 19(1), 77–98 (1991). http://doi.org/10.1080/08905459108905138
22. Zanotto, D., Rosati, G., Agrawal, S.K.: Modeling and control of a 3-DOF pendulum-like manipulator. In: IEEE International Conference on Robotics and Automation, pp. 3964–3969 (2011). http://doi.org/10.1109/ICRA.2011.5980198
23. Zoso, N., Gosselin, C.: Point-to-point motion planning of a parallel 3-DOF underactuated cable-suspended robot. In: IEEE International Conference on Robotics and Automation, pp. 2325–2330 (2012). http://doi.org/10.1109/ICRA.2012.6224598

Rest-to-Rest Trajectory Planning for Planar Underactuated Cable-Driven Parallel Robots

Edoardo Idá[1(✉)], Alessandro Berti[1], Tobias Bruckmann[2], and Marco Carricato[1]

[1] Department of Industrial Engineering (DIN), University of Bologna, Bologna, Italy
edoardo.ida@studio.unibo.it,
{alessandro.berti10,marco.carricato}@unibo.it
[2] Chair of Mechatronics, University of Duisburg-Essen, 47057 Duisburg, Germany
tobias.bruckmann@uni-due.de

Abstract. This paper studies the trajectory planning of underconstrained and underactuated planar cable-driven parallel robots in the case of rest-to-rest motions. For these manipulators, it is possible to control only a subset of the generalized coordinates describing the system. Furthermore, when an arbitrary motion is prescribed for a suitable subset of these coordinates, the lack of constraint on the others leads to the impossibility of bringing the system at rest in a prescribed time. As a consequence, the behavior of the system may not be stable, unless a suitable trajectory is planned. In this paper, a planar 3-DoFs robot suspended by two cables is studied, and a planning method for the trajectory of a reference point on the moving platform is proposed, so as to ensure that the assigned path is tracked accurately and the system is brought to a static condition in a prescribed time.

1 Introduction

Cable-driven parallel robots (CDPRs) employ cables in place of rigid-body extensible legs in order to control the end-effector pose. A CDPR is *fully constrained* if the end effector pose can be completely determined when actuators are locked and, thus, all cable length are assigned. Conversely, a CDPR is *underconstrained* if the end-effector preserves some freedoms once actuators are locked. This occurs either when the end-effector is controlled by a number of cables smaller than the number of degrees of freedom (DoF) that it possesses with respect to the base or when some cables become slack in a fully constrained robot [1]. In addition, if the number of actuators is less than the number of generalized coordinates needed to completely describe the manipulator, the robot is *underactuated* and thus inherently underconstrained as well.

The use of CDPRs with a limited number of cables is justified in several applications, in which the task to be performed allows for a limited number of controlled freedoms (only n DoFs may be governed by n cables) or a limitation of dexterity is acceptable in order to decrease complexity, cost, set-up time, likelihood of cable interference, etc. While a rich literature exists for fully-constrained CDPRs, little research has been conducted on under-constrained ones [1–7].

A major challenge in the analysis of these systems is the trajectory planning of the platform for point-to-point motions. In the case of full or redundant actuation the system can be shown to be "flat" [8] and the trajectory planning problem is completely algebraic [6]. When the platform is both underconstrained and the system underactuated, the flatness property does not necessarily hold and different techniques must be employed. This fact leads, for example, to the impossibility of bringing the platform to rest once the transition from the starting point to the ending point is completed.

In [9] a pendulum-like robot was proposed, consisting of a point mass suspended by a single cable. In order for this system to move outside a straight vertical line and perform point-to-point motions, non zero initial condition must be provided. In [10] a serial planar cable-driven mechanism was proposed whose mechanical architecture is such that the dynamic equations can be decoupled and the burden of trajectory planning is reduced. A planar 3 DoF CDPR suspended by 2 cables was proposed in [11], and the authors were able to generate point-to-point motions for the platform, even outside the static workspace, exploiting harmonic motion laws for the cable lengths; however, they were not able to impose constraints on the path to follow or on the transition time. An input-shaping filtering technique was then proposed in [12] for a planar CDPR and in [13] for a spatial model. Generic trajectories were proposed for which the platform oscillations were significantly reduced, but not eliminated, even in a simulation environment, mainly because of the approximation of the robot natural frequency used in the input shaper. In addition, the nature of the input shaping filter does not allow precise tracking of geometrical paths in point-to-point motions, since the nominal path is modified by the filter.

In this paper, an underactuated and underconstrained planar CDPR, whose mechanical model is identical to the one described in [11,12], is proposed. Our objective is to develop a trajectory-planning method suitable for a stationary setpoint transition when the transition path is assigned. A stationary setpoint transition refers to a point-to-point motion of the platform that is performed in a finite, precomputed time, between two static equilibrium poses. First, the dynamic model is developed. Then, a robust trajectory planning method is proposed which is based on the solution of a *Boundary Value Problem* (BVP) arising from the dynamics of the platform. Finally, the results of computer simulation and experimental tests on a prototype are briefly presented and discussed.

2 Kinematic Model

A planar 3-DoF underactuated CDPR consists of a mobile platform linked to the base by two cables, as shown in Fig. 1. The i-th cable connects point A_i on the base to point B_i on the platform. Each cable is actuated by a motorized winch. Since the platform is constrained by only two cables, it is underconstrained. Oxy is an inertial frame, whereas $Gx'y'$ is a coordinate system attached to the platform center of mass. The mutual orientation between the two frames is described by angle θ. Cables are considered nonextensible and massless.

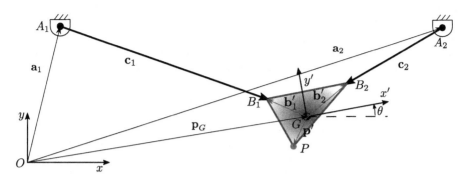

Fig. 1. Geometric Model of a general CDPR with two cables

This assumption does not introduce noticeable errors, if cables are light and have high Young modulus.

The coordinates of a generic point P on the platform, as well as of B_1 and B_2, are described by vectors $\mathbf{p}'$, $\mathbf{b}'_1$ and $\mathbf{b}'_2$ in the moving frame. $\mathbf{p}$, $\mathbf{b}_1$ and $\mathbf{b}_2$ denote, instead, the coordinates of the same points in the base frame. The position vectors of points A_1 and A_2 in the base frame are $\mathbf{a}_1$ and $\mathbf{a}_2$. The i-th cable length is $\mathbf{c}_i$, with its length being $|\mathbf{c}_i| = \rho_i$ ($i = 1, 2$). The array $\boldsymbol{\rho} = [\rho_1, \rho_2]^T$ contains the actuated variables of the system. The mass of the platform is m and its mass moment of inertia with respect to G is I_G. Finally, the configuration of the robot is described by $\mathbf{q} = [x_G, y_G, \theta]^T$, where $\mathbf{p}_G = [x_G, y_G]^T$.

The geometrical constraints that the cables impose on the platform are:

$$\mathbf{c}_i^T \mathbf{c}_i - \rho_i^2 = 0, \quad i = 1, 2 \tag{1}$$

where the vector $\mathbf{c}_i$ connecting the attachment points of the i-th cable is:

$$\mathbf{c}_i = \mathbf{b}_i - \mathbf{a}_i = \mathbf{p}_G + \mathbf{R}(\theta)\mathbf{b}'_i - \mathbf{a}_i, \quad \mathbf{R}(\theta) = \begin{bmatrix} \cos(\theta) & -\sin(\theta) \\ \sin(\theta) & \cos(\theta) \end{bmatrix} \tag{2}$$

Differentiating Eqs. (1) and (2) with respect to time yields:

$$\mathbf{c}_i^T \dot{\mathbf{c}}_i - \rho_i \dot{\rho}_i = 0, \quad i = 1, 2 \tag{3}$$

$$\dot{\mathbf{c}}_i = \dot{\mathbf{p}}_G + \mathbf{E}\mathbf{R}\mathbf{b}'_i \dot{\theta} \tag{4}$$

The antisymmetric matrix $\mathbf{E}$ is defined in the planar case by:

$$\dot{\mathbf{R}}(\theta) = \mathbf{E}\mathbf{R}(\theta)\dot{\theta}, \quad \mathbf{E} = \begin{bmatrix} 0 & -1 \\ 1 & 0 \end{bmatrix} \tag{5}$$

The Jacobian matrix is found by expressing Eqs. (3)–(4) expressed in compact form as:

$$\mathbf{J}\dot{\mathbf{q}} - \boldsymbol{\nu} = \mathbf{0} \tag{6}$$

with:

$$\mathbf{J} = \begin{bmatrix} \mathbf{c}_1^T & \mathbf{c}_1^T \mathbf{E}\mathbf{R}\mathbf{b}'_1 \\ \mathbf{c}_2^T & \mathbf{c}_2^T \mathbf{E}\mathbf{R}\mathbf{b}'_2 \end{bmatrix}, \quad \boldsymbol{\nu}(t) = \begin{bmatrix} \rho_1 \dot{\rho}_1 \\ \rho_2 \dot{\rho}_2 \end{bmatrix} \tag{7}$$

3 Dynamic Model

Due to the underactuated nature of the system, there is no way to predict the evolution of the system without taking into account the dynamics of the platform. If gravity is considered as the only external force acting on the platform and the tension in each cable is denoted by T_i, the dynamic equilibrium can be formulated as:

$$\mathbf{M}\ddot{\mathbf{q}} + \mathbf{J}^T \boldsymbol{\tau} - \mathbf{f} = \mathbf{0} \qquad (8)$$

where:

$$\mathbf{M} = \begin{bmatrix} m & 0 & 0 \\ 0 & m & 0 \\ 0 & 0 & I_G \end{bmatrix}, \quad \boldsymbol{\tau} = \begin{bmatrix} T_1/\rho_1 \\ T_2/\rho_2 \end{bmatrix}, \quad \mathbf{f} = \begin{bmatrix} 0 \\ mg \\ 0 \end{bmatrix}$$

In the case of an underactuated system, the generalized coordinates can be partitioned in *actuated* and *unactuated*, depending on the task that the platform is required to perform. As an example, we may consider the task of positioning the end-effector center of mass, whose coordinates are thus the actuated ones. In this case, Eq. (8) can be rewritten as:

$$\mathbf{M}_a \ddot{\mathbf{p}}_G + \mathbf{J}_a^T \boldsymbol{\tau} - \mathbf{f}_a = \mathbf{0} \qquad (9)$$

$$I_G \ddot{\theta} + \mathbf{J}_u^T \mathbf{T} = 0 \qquad (10)$$

where:

$$\mathbf{M}_a = \begin{bmatrix} m & 0 \\ 0 & m \end{bmatrix}, \quad \mathbf{J}_a = \begin{bmatrix} \mathbf{c}_1^T \\ \mathbf{c}_2^T \end{bmatrix}, \quad \mathbf{f}_a = \begin{bmatrix} 0 \\ mg \end{bmatrix}, \quad \mathbf{J}_u = \begin{bmatrix} \mathbf{c}_1^T \mathbf{E} \mathbf{R} \mathbf{b}_1' \\ \mathbf{c}_2^T \mathbf{E} \mathbf{R} \mathbf{b}_2' \end{bmatrix}$$

Equation (9) can then be solved for vector $\boldsymbol{\tau}$, thus yielding:

$$\boldsymbol{\tau} = -\mathbf{J}_a^{-T} \mathbf{M}_a (\ddot{\mathbf{p}}_G - \mathbf{g}) \qquad (11)$$

where $\mathbf{g}$ is the gravitational acceleration. Substituting (11) in (10) results in a differential equation of second order in the unactuated coordinate θ:

$$I_G \ddot{\theta} - \mathbf{J}_u^T \mathbf{J}_a^{-T} \mathbf{M}_a (\ddot{\mathbf{p}}_G - \mathbf{g}) = 0 \qquad (12)$$

This differential equation is sometimes referred to as the second-order nonholonomic constraint [14] arising from the underconstrained nature of the system. The platform generalized coordinates must fulfill this differential equation at any time, in the same way they must satisfy the geometrical constraints in Eq. (1).

4 Rest-to-Rest Trajectory Planning

The problem of rest-to-rest trajectory planning for an underactuated mechanical system is referred to as a transition problem between stationary setpoints in system theory [15]. The aforementioned transition has been proven to be possible, in most cases, if a pre-actuation or post-actuation phase is considered for the system at end [16]. This essentially leads to a theoretical impossibility to bring the system at rest in a finite time T and provides no prediction on the uncontrolled coordinates behavior, which can lead to the system instability (i.e. oscillatory behavior), thus strongly limiting practical applications.

In [17], a new methodology was proposed for the trajectory design of *Single Input Single Output* (SISO) systems and in [18] the same methodology was extended to *Multi Input Multi Output* (MIMO) systems. The result of bringing the system to a stationary position in the prescribed time T is accomplished thanks to a substantial modification of a nominal trajectory that the actuated coordinates are supposed to track. This is often not desirable (when not dangerous) in industrial applications involving robots, because of possible interference with obstacles.

In this section, a novel method is proposed that avoids geometric modification of the path to track, focusing instead on a specific design of the *motion law* that describes how the platform moves along the path itself. The design of such a motion law relies on the solution of a *Boundary Value Problem* (BVP), which is formulated as the problem of finding a solution to the differential Eq. (12), with constraints on position, velocity and acceleration at start and end positions.

4.1 Formulation of the Problem

When we refer to the trajectory planning of a generic manipulator, the goal is to define both the geometric path of a reference point on the end-effector and the orientation of the latter. In the case of an underactuated planar CDPR supported by two cables, however, it is not possible to assign both the geometric path and the orientation of the platform, due to the underactuated nature of the system. For the sake of simplicity, in the following we will consider the case in which the geometric path of the platform center of mass G is prescribed. It is convenient to consider a parametric representation of the path to track, such as $\mathbf{p}_G = \mathbf{p}_G(u)$. The parameter u is referred to as the *motion law*, which describes how the curve is tracked by G and it is usually a function of time $u = u(t)$, with initial and final conditions $u(0) = 0$ and $u(T) = 1$. The composition $(\mathbf{p}_G \circ u) = \mathbf{p}_G(u(t))$ is what we refer to as the *trajectory*, that is, the time function which describes the evolution of vector $\mathbf{p}_G(t)$.

The design criteria for the motion law are various and application specific. However, when the manipulator is completely or redundantly actuated, the problem of a stationary point change always leads to the solution of a system of linear equations emerging from the fulfillment of some boundary conditions and the necessity of a continuous and differentiable function. A polynomial motion law is often sufficient to satisfy continuity and differentiability up to a predetermined order in the start and end points. An easy way to devise such polynomials is to use the so-called *transition polynomials* [19] of degree $2r + 1$:

$$u(t) = \sum_{i=r+1}^{2r+1} a_i \left(\frac{t}{T}\right)^i \qquad t \in [0, T] \tag{13}$$

where the coefficients a_i's do not depend on the task at hand and are given by:

$$a_i = \frac{(-1)^{i-r-1}(2r+1)!}{i \cdot r!(i-r-1)!(2r+1-i)!} \quad (14)$$

The index r stands for the maximum order of derivation up to which the continuity of the polynomial is required. In the case of an underactuated system, though, this approach is generally not sufficient to bring the platform to rest in the end position. This is due to the fact that the time evolution of the uncontrolled coordinate is the result of the evolution of the system, which is subject to the inertial effect due to the geometric path and the chosen motion law. In order to achieve the desired result, the nonholonomic constraint (12) must be considered in the planning phase. The stationary constraints that the platform has to meet in the start and end points are then considered as boundary conditions (BCs) for the differential equation, thus leading to a BVP. This problem has no solution if the complete trajectory of the actuated coordinates is assigned. On the other hand, if modifications are allowed on the path or the motion law, the problem may admit a solution.

In order to numerically solve the differential Eq. (12), we first express it in (first-order) state-space representation:

$$\dot{\mathbf{x}} = \mathbf{f}(\mathbf{x}, \mathbf{p}_G, \ddot{\mathbf{p}}_G) \quad (15)$$

$$\mathbf{x} = \begin{bmatrix} \theta \\ \dot{\theta} \end{bmatrix}, \quad \mathbf{f}(\mathbf{x}, \mathbf{p}_G, \ddot{\mathbf{p}}_G) = \begin{bmatrix} \dot{\theta} \\ \mathbf{J}_u^T \mathbf{J}_a^{-T} \mathbf{M}_a(\ddot{\mathbf{p}}_G - \mathbf{g})/I_G \end{bmatrix} \quad (16)$$

For a rest-to-rest trajectory planning, the BCs can be expressed as:

$$\mathbf{x}(0) = \begin{bmatrix} \theta_0(\mathbf{p}_G(0)) \\ 0 \end{bmatrix}, \quad \mathbf{x}(T) = \begin{bmatrix} \theta_T(\mathbf{p}_G(T)) \\ 0 \end{bmatrix} \quad (17)$$

where $[\mathbf{p}_G(0), \theta_0(\mathbf{p}_G(0))]^T$ and $[\mathbf{p}_G(T), \theta_T(\mathbf{p}_G(T))]^T$ are stable equilibrium configurations of the system, obtained as in [1].

Equation (15) has dimension 2 and can only match 2 out of the 4 BCs considered in (17). One way to provide a solution to the problem is to consider a number of additional scalar parameters $k_1, \ldots, k_h$ (called *free parameters*) equal to the difference between the number of boundary conditions and the dimension of the differential problem, so that $\mathbf{p}_G = \mathbf{p}_G(\mathbf{k}, t)$. In the case at hand, $h = 2$ and $\mathbf{k} = [k_1, k_2]^T$. The free parameters k_1 and k_2 are calculated so as conditions (17) are satisfied.

4.2 Modification of the Motion Law

If we do not wish to modify the geometric path of the trajectory, the only allowed modification concerns the motion law, that is, $\mathbf{p}_G = \mathbf{p}_G(u(\mathbf{k}, t))$ [20]. One way to design such a modified motion law u, so that the actuated coordinates can meet the start and end conditions prescribed by the task, is to consider the composition $u(\mathbf{k}, t) = (\tilde{u} \circ \gamma)(\mathbf{k}, t)$, where the analytical expression of $\tilde{u}(\gamma)$ is:

$$\tilde{u}(\gamma) = \sum_{i=r+1}^{2r+1} a_i \gamma^i(\mathbf{k}, t) \qquad \gamma(\mathbf{k}, t) \in \mathbb{R} \tag{18}$$

where a_i is still expressed as in (14) and the function $\gamma(\mathbf{k}, t)$ is arbitrary, as long as it satisfies the requirements:

$$\gamma(\mathbf{k}, 0) = 0, \qquad \gamma(\mathbf{k}, T) = 1 \qquad \forall \mathbf{k} \tag{19}$$

In the case under examination, $r = 3$ is also chosen, so that continuity of jerk can be imposed in the start and end positions. This way, any discontinuity in the cable tensions $\boldsymbol{\tau}$ is avoided, thus eliminating a different potential source of residual oscillation. As an example, $\gamma(\mathbf{k}, t)$ may be designed as a polynomial of order 3:

$$\gamma(\mathbf{k}, t) = \alpha t + k_1 t^2 + k_2 t^3, \qquad \alpha = \frac{1 - k_1 T^2 - k_2 T^3}{T} \tag{20}$$

Accordingly, the velocity of point G can be expressed as:

$$\dot{\mathbf{p}}_G = \frac{\partial \mathbf{p}_G}{\partial u} \frac{\partial u}{\partial \gamma} \frac{\partial \gamma}{\partial t} = \mathbf{p}'_G u^* \dot{\gamma} \tag{21}$$

where $(\cdot)'$ denotes the partial derivative with respect to u and $(\cdot)^*$ the partial derivative with respect to γ. The acceleration vector is consequently:

$$\ddot{\mathbf{p}}_G = \mathbf{p}''_G (u^* \dot{\gamma})^2 + \mathbf{p}'_G (u^* \ddot{\gamma} + u^{**} \dot{\gamma}^2) \tag{22}$$

4.3 Solution of the BVP with Free Parameters

In order to determine the vector of free parameters $\mathbf{k}$, the BVP expressed by Eq. (15) with BC (17) must be numerically solved. A number of algorithms are proposed in the literature and even implemented in commercial softwares, such as the *bvp4c* and *bvp5c* routines available in any MATLAB distribution [21]. These algorithms are finite-difference codes that implement a collocation formula [22] and, thus, require a suitable set-up in order to work efficiently and find a solution within a reasonable tolerance. However, even in this case, there is still no guarantee of success.

Alternatively, the problem can be formulated as a combination of an Initial Value Problem (IVP) followed by the solution of a system of nonlinear equations. Provided an initial guess for the unknown parameter vector $\mathbf{k}$, which can be as easy as the zero vector, the system of nonlinear equations is formulated as:

$$\tilde{\mathbf{x}}_i(\mathbf{k}_i, T) - \mathbf{x}(T) = \mathbf{0}, \qquad \mathbf{x}(T) = \begin{bmatrix} \theta_T(\mathbf{p}_G(T)) \\ \mathbf{0} \end{bmatrix} \tag{23}$$

where the vector $\tilde{\mathbf{x}}_i(\mathbf{k}_i, T)$ is evaluated at every iteration as the end point of the IVP defined by:

$$\dot{\mathbf{x}} = \mathbf{f}(\mathbf{x}, \mathbf{p}_G(\mathbf{k}_i, t), \ddot{\mathbf{p}}_G(\mathbf{k}_i, t)), \qquad \mathbf{x}(0) = \begin{bmatrix} \theta_0(\mathbf{p}_G(0)) \\ 0 \end{bmatrix} \qquad (24)$$

Once a solution $(\mathbf{k}, \mathbf{x}(t))$ is found, the trajectory $\mathbf{p}_G(\mathbf{k}, t)$ can be computed. By considering the calculated $\mathbf{x}(t)$, the cable's length is found according to (1) as:

$$\rho_i = \sqrt{\mathbf{c}_i^T \mathbf{c}_i}, \qquad i = 1, 2 \qquad (25)$$

It should be noted that no explicit constraints on cable tensions or motor torques are considered in this work other than cable positive tension, which is taken into account during the integration of Eq. (24).

5 Simulation and Experimental Results

The presented trajectory-planning method has been implemented in a MATLAB code in order to determine the trajectory of the actuated coordinates and to compute the cable lengths that are used as position setpoints in a open-loop control scheme. The geometrical and inertial parameters of the planar robot are chosen according to the architecture of the prototype that is used for the experimentation:

$$\mathbf{b}'_1 = \begin{bmatrix} -0.177 \\ 0.271 \end{bmatrix} \text{m} \qquad \mathbf{b}'_2 = \begin{bmatrix} 0.177 \\ 0.271 \end{bmatrix} \text{m}$$

$$\mathbf{a}_1 = \begin{bmatrix} -1.150 \\ 1.930 \end{bmatrix} \text{m} \qquad \mathbf{a}_2 = \begin{bmatrix} 1.150 \\ 1.930 \end{bmatrix} \text{m}$$

$$m = 7.510 \,\text{Kg} \qquad I_G = 0.357 \,\text{Kg} \cdot \text{m}^2 \qquad g = 9.807 \,\text{m/s}^2$$

As an example, a horizontal line path will be considered for the transition of the center of mass $\mathbf{p}_G$ between 4 stationary points, which will be executed one after the other, with a pause of 1 s between each one of them. The effects of residual oscillations of the platform, caused by non-equilibrium final conditions, are thus amplified and easily detectable. The stationary points are evaluated according to the geometrico-static analysis as:

$$\mathbf{q}_0 = \begin{bmatrix} 0 \\ 0.495 \\ 0 \end{bmatrix} \quad \mathbf{q}_1 = \begin{bmatrix} 0.4 \\ 0.495 \\ 0.153 \end{bmatrix} \quad \mathbf{q}_2 = \begin{bmatrix} -0.4 \\ 0.495 \\ -0.153 \end{bmatrix} \quad \mathbf{q}_3 = \begin{bmatrix} 0 \\ 0.495 \\ 0 \end{bmatrix}$$

and the time of the transitions are set to:

$$T_1 = 1\text{s} \qquad T_2 = 2\text{s} \qquad T_3 = 1\text{s}$$

In the case of a line, the parametrization of the s-th path ${}^s\mathbf{p}_G$ may be simply set as:

$${}^s\mathbf{p}_G(u) = \mathbf{p}_{G,i} + (\mathbf{p}_{G,i+1} - \mathbf{p}_{G,i})u \qquad (26)$$

The solution of the BVP gives the value of the unknown constant parameters $^s\mathbf{k}$:

$$^1\mathbf{k} = \begin{bmatrix} -1.4496 \\ 0.9847 \end{bmatrix} \qquad ^2\mathbf{k} = \begin{bmatrix} -0.0603 \\ 0.0201 \end{bmatrix} \qquad ^3\mathbf{k} = \begin{bmatrix} -1.4496 \\ 0.9847 \end{bmatrix}$$

as well as a prediction of the orientation of the platform (Fig. 2a). Consequently, it is possible to evaluate the cable lengths at every time step (Fig. 2b).

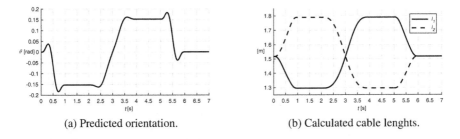

(a) Predicted orientation. (b) Calculated cable lenghts.

Fig. 2. Planning results

5.1 Simulation

In order to test the proposed trajectory planning method, a simulation was first performed by way of the commercial software MSC ADAMS, which makes it possible to model cable-driven mechanisms by employing the Cable-Toolbox. The test of a planar system consistent with the mathematical model considered in this paper and with the available prototype was possible by considering a 4 cable actuation, in which two pairs of cables are parallel to each other and whose actuation laws are coincident. In addition, we introduced a flexible body model for cables to test the robustness of our trajectory planing in a more realistic scenario. In the MSC ADAMS Cable-Toolbox the cables elastic behavior can be modeled based on Hooke's law for springs. Our prototype is equipped with 6 mm diameter Dyneema Pro cables, whose Young Modulus is $E = 115$ GPa.

Simulation results show that oscillations in the stationary positions are eliminated even without any feedback control loop on the pose of the end-effector, as it is shown in Fig. 3, where the proposed planning method for $u(t)$ is compared to the one expressed by Eq. (13).

5.2 Experimental Results

The proposed trajectory was tested on a prototype consistent with the ADAMS' model. In order to prove the effectiveness of the method, an open loop control scheme was employed, so that only the computed lengths of the cables are fed to the actuators. The experimentation showed a remarkable reduction of oscillations in each stationary point in comparison to a movement whose motion law was planned according to Eq. (13). The result, along with other geometrical

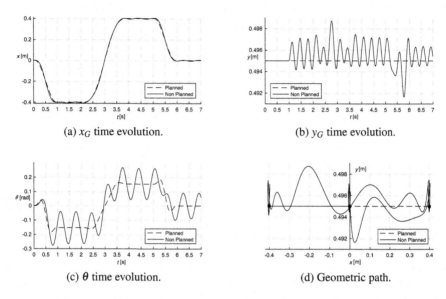

Fig. 3. Simulation results showing a comparison between a trajectory planned according to (18) and a non-planned trajectory resulting from (13)

trajectories (such as circular and parabolic arcs), can be observed in a videoclip available at [23]. The residual oscillations of the platform, which can not be damped because of the open loop control scheme, are mainly caused by the pulley system. In fact, the model implemented for the trajectory planning takes into account neither pulley friction nor the cable sliding into the pulley grooves. Numerical details about the experimental results are omitted due to space limitations and will be provided in a future enhanced version of this contribution.

6 Conclusions

This paper presented the rest-to-rest trajectory planning of planar underactuated CDPRs. The problem was formulated as a boundary value problem (BVP) with free parameters. A novel technique was proposed, which introduces no geometric modification of the path that a reference point on the platform has to follow. The methodology was first implemented in MATLAB and the resulting trajectories were tested both on a commercial simulation environment, MSC ADAMS, and experimentally on a prototype. Experimental results are satisfactory. Even without a feedback control loop on the platform pose, the platform shows a very limited oscillatory behavior, which is easily damped by the system internal friction. In the near future the same planning method will be applied to a spatial CDPR with a refined geometric model that takes into account pulley geometry.

References

1. Carricato, M., Merlet, J.P.: Stability analysis of underconstrained cable-driven parallel robots. IEEE Trans. Robot. **29**(1), 288–296 (2013)
2. Abbasnejad, G., Carricato, M.: Real solutions of the direct geometrico-static problem of under-constrained cable-driven parallel robots with 3 cables: a numerical investigation. Meccanica **47**(7), 1761–1773 (2012)
3. Carricato, M.: Direct geometrico-static problem of underconstrained cable-driven parallel robots with three cables. J. Mech. Robot. **5**(3), 031008 (2013)
4. Carricato, M.: Inverse geometrico-static problem of underconstrained cable-driven parallel robots with three cables. J. Mech. Robot. **5**(3), 031002 (2013)
5. Fattah, A., Agrawal, S.K.: On the design of cable-suspended planar parallel robots. J. Mech. Design **127**(5), 1021–1028 (2005)
6. Heyden, T., Woernle, C.: Dynamics and flatness-based control of a kinematically undetermined cable suspension manipulator. Multibody Syst. Dyn. **16**(2), 155–177 (2006)
7. Michael, N., Kim, S., Fink, J., Kumar, V.: Kinematics and statics of cooperative multi-robot aerial manipulation with cables. In: ASME 2009 International Design Engineering Technical Conferences and Computers and Information in Engineering Conference, pp. 83–91, January 2009
8. Fliess, M., Lévine, J., Martin, P., Rouchon, P.: Flatness and defect of non-linear systems: introductory theory and examples. Int. J. Control **61**(6), 1327–1361 (1995)
9. Cunningham, D., Asada, H.H.: The winch-bot: a cable-suspended, under-actuated robot utilizing parametric self-excitation. In: IEEE International Conference on Robotics and Automation, ICRA 2009, pp. 1844–1850, May 2009
10. Zanotto, D., Rosati, G., Agrawal, S.K.: Modeling and control of a 3-DOF pendulum-like manipulator. In: 2011 IEEE International Conference on Robotics and Automation (ICRA), pp. 3964–3969. IEEE, May 2011
11. Zoso, N., Gosselin, C.: Point-to-point motion planning of a parallel 3-dof underactuated cable-suspended robot. In: 2012 IEEE International Conference on Robotics and Automation (ICRA), pp. 2325–2330. IEEE, May 2012
12. Park, J., Kwon, O., Park, J.H.: Anti-sway trajectory generation of incompletely restrained wire-suspended system. J. Mech. Sci. Technol. **27**(10), 3171–3176 (2013)
13. Hwang, S.W., Bak, J.H., Yoon, J., Park, J.H., Park, J.O.: Trajectory generation to suppress oscillations in under-constrained cable-driven parallel robots. J. Mech. Sci. Technol. **30**(12), 5689–5697 (2016)
14. Oriolo, G., Nakamura, Y.: Control of mechanical systems with second-order nonholonomic constraints: Underactuated manipulators. In: Proceedings of the 30th IEEE Conference on Decision and Control, 1991, pp. 2398–2403. IEEE, December 1991
15. Isidori, A.: Nonlinear Control Systems. Springer Science & Business Media, New York (2013)
16. Seifried, R.: Dynamics of Underactuated Multibody Systems: Modeling, Control and Optimal Design, vol. 205. Springer Science & Business Media, Berlin (2013)
17. Graichen, K., Hagenmeyer, V., Zeitz, M.: A new approach to inversion-based feedforward control design for nonlinear systems. Automatica **41**(12), 2033–2041 (2005)
18. Graichen, K., Zeitz, M.: Feedforward control design for finite-time transition problems of non-linear MIMO systems under input constraints. Int. J. Control **81**(3), 417–427 (2008)

19. Piazzi, A., Visioli, A.: Optimal noncausal set-point regulation of scalar systems. Automatica **37**(1), 121–127 (2001)
20. Bobrow, J.E., Dubowsky, S., Gibson, J.S.: Time-optimal control of robotic manipulators along specified paths. Int. J. Robot. Res. **4**(3), 3–17 (1985)
21. Shampine L.F. , Kierzenka J., Reichelt M.W.: Solving boundary value problems for ordinary differential equations in Matlab with bvp4c. https://www.mathworks.com/matlabcentral/answers/uploaded_files/1113/bvp_paper.pdf
22. Ascher, U.M., Petzold, L.R.: Computer Methods for Ordinary Differential Equations and Differential-Algebraic Equations, vol. 61. Siam, Philadelphia (1998)
23. https://www.dropbox.com/sh/dkitevyyjpujk5t/AAAwxeJq4jaidStCH6LAIlA9a?dl=0

Dynamically-Feasible Elliptical Trajectories for Fully Constrained 3-DOF Cable-Suspended Parallel Robots

Giovanni Mottola[1(✉)], Clément Gosselin[2], and Marco Carricato[1]

[1] Department of Industrial Engineering, University of Bologna, Bologna, Italy
{giovanni.mottola3,marco.carricato}@unibo.it
[2] Département de Génie Mécanique, Université Laval, Québec, Canada
gosselin@gmc.ulaval.ca

Abstract. A cable suspended robot can be moved beyond its static workspace while keeping all cables in tension, by relying on end-effector inertia forces. This allows the robot capabilities to be extended by choosing suitable dynamical trajectories. In this paper, we study 3D elliptical motions, which are the most general case of spatial sinusoidal oscillations, for a robot with a point-mass end-effector and an arbitrary base architecture. We find algebraic conditions that define the range of admissible frequencies for feasible trajectories; furthermore, we show that, under certain conditions, a special frequency exists, which allows arbitrarily large oscillations to be reached. We also study transition trajectories that displace the robot from an initial state of rest (within the static workspace) to the elliptical trajectory, and vice versa.

1 Introduction

Cable-suspended parallel robots (CSPRs) are fully constrained if they are supported by a number of taut cables greater than or equal to the degrees of freedom (DOF) of the moving platform; they are underconstrained otherwise. This paper focuses on the former, so that all platform freedoms can be controlled. CSPRs are often assumed to work in quasi-static conditions: in such cases, the Static Equilibrium Workspace (SEW), where the robot can be brought to static equilibrium, coincides with the robot footprint [1]. Recently, the dynamic workspace of fully constrained CSPRs was investigated. The dynamic workspace is the set of poses that can be reached in a controlled way, while maintaining positive tensions in cables, by exploiting inertial effects [2].

The first work in this direction concerned a 2-DOF robot controlled by two cables [3]. Here, parametric periodic trajectories were introduced and, for each trajectory, a range of motion frequencies that ensure global feasibility was found; in this case the feasibility of a trajectory can be verified *ex ante*, without solving the inverse dynamic problem. A special motion frequency was also found, akin to the natural frequency of a simple pendulum, which always belongs to the aforementioned range. Such results were extended to 3-DOF spatial robots with a point-mass end-effector [4] and 3-DOF planar robots [5], for periodic trajectories and point-to-point motions [6, 7]. These studies expand the applicability of CSPRs, allowing them to move beyond the SEW.

In this paper, we study generic elliptical dynamic trajectories, for a spatial 3-DOF robot with a point-mass platform (Sect. 2). We find a special frequency that allows arbitrarily large periodic motions to be realized (Sect. 3), together with a range of trajectory frequencies that guarantees positive cable tensions (Sect. 4). In our study, cable exit points on the base are arbitrarily placed, while [4, 7] consider them on a *horizontal equilateral triangle;* more recent papers (see [8, 9]) consider cable exit points lying on a generic *horizontal* triangle. Reference [9] deals with general circular trajectories in 3D space, finding a smaller range of admissible frequencies compared to the range emerging from our study: for a horizontal circle (the case studied in [8]), the two ranges coincide and the frequency bounds in [9] are strict. Trajectories such as those in [4, 8, 9] can be found as special cases of the ones presented here.

We also study transition motions (Sect. 5) from a dynamic condition to a state of rest (or vice versa). These can be used to recover a CSPR after a cable failure (see [10] for an application) by bringing it back in the SEW and slowing it down.

2 Dynamic Model

A general periodic spatial sinusoidal trajectory Γ is defined by three sinusoidal functions along the three coordinate axes of a fixed reference frame:

$$\mathbf{p} = \begin{bmatrix} x \\ y \\ z \end{bmatrix} = \mathbf{p}_c + \mathbf{p}_d = \mathbf{p}_c + \begin{bmatrix} x_A \sin(\omega t + \varphi_x) \\ y_A \sin(\omega t + \varphi_y) \\ z_A \sin(\omega t + \varphi_z) \end{bmatrix} \quad (1)$$

where:

- $\mathbf{p}_C = [x_c, y_c, z_c]^T$ is the position of center C and $\mathbf{p}_d$ is the displacement from C;
- x_A, y_A, z_A are the amplitudes of oscillation;
- $\varphi_x, \varphi_y, \varphi_z$ are the phase angles;
- ωt defines the position of P along Γ, with ω being the motion frequency;

It may be proven that the trajectory thus described is always an ellipse. Note that such a trajectory includes, as special cases, circles and line segments, either horizontal, vertical or oblique: such special cases were dealt with in [4, 8, 9].

We consider a spatial CSPR, whose end-effector is a point-mass P with position $\mathbf{p} = [x, y, z]^T$ with respect to the fixed frame. The position vectors of the cable exit points A_i on the frame are denoted as $\mathbf{a}_i = [x_{ai}, y_{ai}, z_{ai}]^T$, for $i = 1, 2, 3$ (Fig. 1). In Fig. 1, for the sake of simplicity $O \equiv A_1$. For future convenience, we define

$$\mathbf{v}_{ci} = [x_{Cai}, y_{Cai}, z_{Cai}]^T = \mathbf{a}_i - \mathbf{p}_c \; (i \in \{1, 2, 3\}) \quad (2)$$

$$\mathbf{v}_{jk} = [x_{ajk}, y_{ajk}, z_{ajk}]^T = \mathbf{a}_k - \mathbf{p}_j \; (i \in \{1, 2, 3\}) \quad (3)$$

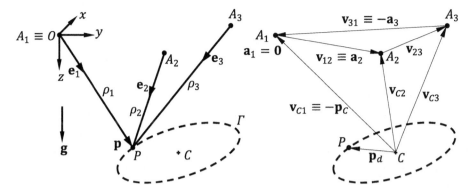

Fig. 1. A3-DOF spatial CSPR following a spatial elliptical trajectory.

$$\lambda_i = [\lambda_{xi}, \lambda_{yi}, \lambda_{zi}]^T = \mathbf{v}_{Cj} \times \mathbf{v}_{Ck} \qquad (4)$$

In Eq. (4), the indexes are as follows:

$$\begin{cases} i = 1 \rightarrow j = 3, k = 2 \\ i = 2 \rightarrow j = 1, k = 3 \\ i = 3 \rightarrow j = 2, k = 1 \end{cases} \qquad (5)$$

We also define the cable length $\rho_i = \|\mathbf{p} - \mathbf{a}_i\|$ and the unit vector $\mathbf{e}_i = (\mathbf{p} - \mathbf{a}_i)/\rho_i$, for $i = 1, 2, 3$.

The forces on the end-effector are the cable tensions $\tau_i \mathbf{e}_i$, the inertia force $-m\ddot{\mathbf{p}}$ and gravity $m\mathbf{g}$ (with $\mathbf{g} = [0, 0, g]^T$, directed along the positive z axis). Cable mass and elasticity are neglected. The dynamic model of the robot is then given by

$$-\sum_{i=1}^{3} \tau_i \mathbf{e}_i + m\mathbf{g} - m\ddot{\mathbf{p}} = 0 \qquad (6)$$

By defining the matrix $\mathbf{M} = [\mathbf{e}_1, \mathbf{e}_2, \mathbf{e}_3]$, Eq. (6) may be rewritten as

$$m(\mathbf{g} - \ddot{\mathbf{p}}) = \mathbf{M}[\tau_1, \tau_2, \tau_3]^T \qquad (7)$$

If the base anchor points are numbered clockwise (looking along the z axis) and P is below the plane Π passing through A_1, A_2 and A_3, it can be shown that $\det(\mathbf{M}) < 0$; in such a case, tensions τ_i are positive if and only if ([4], Eqs. (17)–(19))

$$\begin{cases} \mu_1 := [\mathbf{p} \times (\mathbf{a}_2 - \mathbf{a}_3) + \mathbf{a}_2 \times \mathbf{a}_3]^T (\ddot{\mathbf{p}} - \mathbf{g}) > 0 \\ \mu_2 := [\mathbf{p} \times (\mathbf{a}_3 - \mathbf{a}_1) + \mathbf{a}_3 \times \mathbf{a}_1]^T (\ddot{\mathbf{p}} - \mathbf{g}) > 0 \\ \mu_3 := [\mathbf{p} \times (\mathbf{a}_1 - \mathbf{a}_2) + \mathbf{a}_1 \times \mathbf{a}_2]^T (\ddot{\mathbf{p}} - \mathbf{g}) > 0 \end{cases} \qquad (8)$$

The plane Π has equation

$$a \cdot x + b \cdot y + c \cdot z + d = 0 \qquad (9)$$

where coefficients a, b, c, d can be derived from coordinates x_{ai}, y_{ai}, z_{ai} as

$$a = \begin{vmatrix} 1 & y_{a1} & z_{a1} \\ 1 & y_{a2} & z_{a2} \\ 1 & y_{a3} & z_{a3} \end{vmatrix} \quad b = \begin{vmatrix} x_{a1} & 1 & z_{a1} \\ x_{a2} & 1 & z_{a2} \\ x_{a3} & 1 & z_{a3} \end{vmatrix}$$
$$c = \begin{vmatrix} x_{a1} & y_{a1} & 1 \\ x_{a2} & y_{a2} & 1 \\ x_{a3} & y_{a3} & 1 \end{vmatrix} \quad d = -\begin{vmatrix} x_{a1} & y_{a1} & z_{a1} \\ x_{a2} & y_{a2} & z_{a2} \\ x_{a3} & y_{a3} & z_{a3} \end{vmatrix} \qquad (10)$$

Here, c is twice the signed area of triangle T_{xy}, formed by points (x_{ai}, y_{ai}). With the anchor points numbered clockwise, the signed area is <0, so that c is <0 too.

If P has to remain below Π, no intersection can occur between Γ and Π. We then substitute Eq. (1) in Eq. (9), thus obtaining (after some simplifications)

$$P + Q \sin(\omega t) + R \cos(\omega t) = 0 \qquad (11)$$

where:

$$\begin{cases} P = & ax_C + by_C + cz_C + d \\ Q = & ax_A \cos \varphi_x + by_A \cos \varphi_y + cz_A \cos \varphi_z \\ R = & ax_A \sin \varphi_x + by_A \sin \varphi_y + cz_A \sin \varphi_z \end{cases} \qquad (12)$$

By introducing $t_n = \tan(\omega t/2)$ and using the tangent half-angle formulae, Eq. (11) becomes a quadratic equation in t_n. For no intersections to occur between Γ and Π, this equation must have no solutions and thus a negative discriminant, namely

$$4(Q^2 + R^2 - P^2) < 0 \qquad (13)$$

This condition has to be checked *ex ante* for trajectory Γ to be feasible.

3 Natural Frequency

In [3, 4] a special frequency was determined that allows arbitrarily large periodic motions with positive cable tensions to be realized. This "special" frequency is expressed as $\omega_n = \sqrt{g/z_0}$, where $z_0 = z_C$ is the vertical coordinate of the trajectory center. We show that a similar frequency also exists for the system studied here.

In analogy to [7], we set tension τ_i in Eq. (6) to be proportional to cable length ρ_i; this can be achieved by a suitable actuator control. The dynamic equations become

$$-\sum_{i=1}^{3} k_i(\mathbf{p} - \mathbf{a}_i) + m\mathbf{g} = m\ddot{\mathbf{p}} \qquad (14)$$

where k_i are virtual cable stiffnesses. We conveniently express Eq. (1) as

$$\mathbf{p} = \mathbf{p}_C + \mathbf{c}\cos(\omega t) + \mathbf{s}\sin(\omega t) = \mathbf{p}_C + \mathbf{p}_d \qquad (15)$$

This elliptical trajectory is a general case of the harmonic motion introduced in Eqs. (11) and (12) of [7]. Substituting Eq. (15) in (14) we obtain

$$m[\mathbf{g} + \omega^2(\mathbf{p} - \mathbf{p}_C)] = \sum_{i=1}^{3} k_i(\mathbf{p} - \mathbf{a}_i) \qquad (16)$$

This equation becomes simpler if the terms containing $\mathbf{p}$ cancel out. This means choosing $\omega = \omega_n$ such that $m\omega_n^2 = k_1 + k_2 + k_3 = K$, so that Eq. (16) becomes:

$$m\mathbf{g} - K\mathbf{p}_C = -\sum_{i=1}^{3} k_i \mathbf{a}_i \qquad (17)$$

and ω_n is the natural frequency of the second order ODE in (14). If $k_i > 0$ and (17) is satisfied, the cable tensions are positive, since $\tau_i = k_i \rho_i$ (where clearly $\rho_i \geq 0$).

Trajectory Γ may then be realized with $\tau_i = k_i \rho_i$ only if (17) holds. Equation (17) gives a linear system of three equations in the unknowns k_i, whose solution is

$$[k_1 \ k_2 \ k_3] = \frac{-mg}{ax_C + by_C + cz_C + d}[\lambda_{z1} \ \lambda_{z1} \ \lambda_{z1}] \qquad (18)$$

The natural frequency ω_n is thus:

$$\omega_n^2 = \frac{K}{m} = -\frac{g(\lambda_{z1} + \lambda_{z2} + \lambda_{z3})}{ax_C + by_C + cz_C + d} = \frac{gc}{ax_C + by_C + cz_C + d} \qquad (19)$$

Since ω_n^2 must be positive and c is negative (cf. Sect. 2), it must be $ax_C + by_C + cz_C + d < 0$. It follows from Eq. (18) that $k_i > 0$ infers $\lambda_{zi} > 0$. This requires that the projection of point C on the $x - y$ plane must lie within the triangle T_{xy} defined in Sect. 2. Point C must then lie inside the SEW (which is the convex hull of the fixed attachment points, plus all points below this region [1]).

These results generalize those reported in [4, 8, 9], where the authors assumed points A_i to lie on a horizontal plane and Γ to be a circular trajectory.

4 Generic frequency

Periodic motions may also be realized along Γ with frequencies ω that are distinct from ω_n. By substituting Eq. (1) in Eq. (8), each μ_i can be expressed as a sum of sines and cosines, plus a constant term:

$$\mu_i = C_i \cos(\omega t) + D_i \sin(\omega t) + E_i, \quad i \in \{1, 2, 3\} \qquad (20)$$

With

$$C_i = C_{i,v}\omega^2 + C_{i,c}$$
$$D_i = D_{i,v}\omega^2 + D_{i,c} \qquad (21)$$
$$E_i = g\lambda_{zi}$$

$$\begin{aligned}
C_{i,v} &= x_A\lambda_{xi}\sin(\varphi_x) + y_A\lambda_{yi}\sin(\varphi_y) + z_A\lambda_{zi}\sin(\varphi_z) \\
D_{i,v} &= x_A\lambda_{xi}\cos(\varphi_x) + y_A\lambda_{yi}\cos(\varphi_y) + z_A\lambda_{zi}\cos(\varphi_z) \\
C_{i,c} &= g\left[x_A y_{akj}\sin(\varphi_x) - y_A x_{akj}\sin(\varphi_y)\right] \\
D_{i,c} &= g\left[x_A y_{akj}\cos(\varphi_x) - y_A x_{akj}\cos(\varphi_y)\right]
\end{aligned} \qquad (22)$$

where j,k depend on i as per Eq. (5). Coefficients C_i, D_i and E_i are constant for a given trajectory: C_i and D_i are linear functions of ω^2, while E_i only depends on g and the position of C. Equation (20) is akin to the ones obtained in [4], for the special cases of circular trajectories on vertical or horizontal planes.

The extreme values for each μ_i are

$$\mu_{i,1} = \sqrt{C_i^2 + D_i^2} + E_i, \quad \mu_{i,2} = -\sqrt{C_i^2 + D_i^2} + E_i \qquad (23)$$

Having $\mu_i > 0$ is equivalent to having extrema that are all positive. Notice that, if $E_i = \lambda_{zi} g \leq 0$, then $\mu_{i,2}$ is bound to be less than zero and there can be no values of ω satisfying the constraint that μ_i must be positive at all times. This implies $\lambda_{zi} > 0$ (which is the same condition found in Sect. 3).

Since $\mu_{i,2} < \mu_{i,1}$, it suffices to check that $\mu_{i,2}$ is >0, namely $E_i > \sqrt{C_i^2 + D_i^2}$. Since both sides of this inequality are nonnegative, we may square them and obtain

$$E_i^2 > C_i^2 + D_i^2 \Rightarrow \left(C_{i,v}\omega^2 + C_{i,c}\right)^2 + \left(D_{i,v}\omega^2 + D_{i,c}\right)^2 - g^2\lambda_{zi}^2 < 0 \qquad (24)$$

We finally obtain three biquadratic inequalities, namely

$$\alpha_i \omega^4 + 2\beta_i \omega^2 + \gamma_i < 0 \qquad (25)$$

The expressions for $\alpha_i, \beta_i, \gamma_i$ are found by expanding Eq. (24):

$$\begin{aligned}
\alpha_i &= C_{i,v}^2 + D_{i,v}^2 \\
\beta_i &= C_{i,c}C_{i,v} + D_{i,c}D_{i,v} \\
\gamma_i &= C_{i,c}^2 + D_{i,c}^2 - g^2\lambda_{zi}^2
\end{aligned} \qquad (26)$$

In refs. [4, 8, 9], for circular trajectories and points A_i located at the same height, it was found that there is a range of values of ω for which the trajectory is feasible. We expect the same to happen here, provided that there is at least one value of ω that satisfies the constraint in (25): this is in fact the case, as seen in Sect. 3.

The biquadratic inequality (25) can be solved by setting $\omega^2 = w$. As a sum of squares, α_i is always nonnegative, so the parabola defined by $\alpha_i w^2 + 2\beta_i w + \gamma_i = \zeta$ in the $w\zeta$ plane is convex (the degenerate case $\alpha_i = 0$ will be dealt with later on).

Depending on the sign of $\Delta_i = \beta_i^2 - \alpha_i \gamma_i$, we have two cases:

- $\Delta_i \leq 0$: inequality (25) has no solutions;
- $\Delta_i > 0$: there are $w_{i,min}, w_{i,max}$ such that, if $w \in [w_{i,min}, w_{i,max}]$, then (25) holds. These two values are generally different for the three cables.

If the conditions in Sect. 3 are fulfilled (i.e. C is in the SEW and the trajectory is below plane Π), we know that (25) admits at least one solution, $\omega = \omega_n$; therefore, all Δ_i's must be positive. When Δ_i is positive, $w_{i,min}$ and $w_{i,max}$ are given by:

$$w_{i,min} = \left(-\beta_i - \sqrt{\Delta_i}\right)/\alpha_i, \quad w_{i,max} = \left(-\beta_i + \sqrt{\Delta_i}\right)/\alpha_i \quad (27)$$

Since $\omega^2 = w$, we can have three cases:

1. If both $w_{i,min}$ and $w_{i,max}$ are > 0, we simply require $\omega \in \left[\sqrt{w_{i,min}}, \sqrt{w_{i,max}}\right]$.
2. If $w_{i,min} \leq 0$ and $w_{i,max} > 0$, then we require $\omega \in \left]0, \sqrt{w_{i,max}}\right]$. This means that, as ω approaches zero, the end-effector moves quasi-statically. In this case the inertial force is negligible and the trajectory is entirely in the SEW.
3. If $w_{i,min}$ and $w_{i,max}$ are ≤ 0, no values of ω allow trajectory Γ to be realized.

We exclude $\omega = 0$ as a solution, since in this case the end-effector does not move.

Finally, we end up with 3 constraints on ω (one for each cable), each one of which is of type $\omega \in \left[\sqrt{w_{i,min}}, \sqrt{w_{i,max}}\right]$ or $\omega \in \left]0, \sqrt{w_{i,max}}\right]$, depending on whether $w_{i,min}$ is > 0 or not. To ensure positive cable tensions, we must choose ω such that all three constraints are satisfied, namely

$$\omega_{min} = \sqrt{\max\{\max\{w_{i,min}\}, 0\}}, \quad \omega_{max} = \sqrt{\min\{w_{i,max}\}} \quad (28)$$

where $w_{i,min}$ and $w_{i,max}$ are known from Eq. (27). Note that these quantities can be determined from explicit algebraic formulae.

If the conditions set in Sect. 3 hold (namely, C is in the SEW and Γ lies below plane Π) and $\omega_{min} < \omega < \omega_{max}$, the trajectory Γ is feasible.

It is worth emphasizing that the conditions presented here are strict: if $\omega = \omega_{min}$ or $\omega = \omega_{max}$, at least one cable tension reaches zero during a period (while still remaining nonnegative). The conditions found in [9], instead, are sufficient, but not strictly necessary: the range for ω given in [9] is, therefore, smaller.

We have yet to address the special case where one of the α_i's is zero. Since each α_i is the sum of two squares, it can be zero if and only if both squared terms are zero, namely $C_{i,v} = D_{i,v} = 0$. This is a linear homogeneous system of two equations in three unknowns x_A, y_A, z_A, whose solutions are

$$\begin{bmatrix} x_A \\ y_A \\ z_A \end{bmatrix} = f_1 \cdot \begin{bmatrix} \lambda_{yi}\lambda_{zi}\sin(\varphi_{yz}) \\ \lambda_{zi}\lambda_{xi}\sin(\varphi_{zx}) \\ \lambda_{xi}\lambda_{yi}\sin(\varphi_{xy}) \end{bmatrix} \qquad (29)$$

where $f_1 \in \,]0, +\infty[$ is an arbitrary scalar. From the definitions (26), it is clear that $C_{i,v} = D_{i,v} = 0$ implies $\beta_i = 0$. Equation (25) then requires $\gamma_i < 0$: this is the only condition to check and it does not depend on ω. It can be proved that, in this case, Γ lies on the plane connecting the center C with points j and k (with $i \neq j, i \neq k$).

Another special case to consider is when points A_i are at the same height: $z_{a,1} = z_{a,2} = z_{a,3} = z_a$. In this case, one finds from (10) that $a = b = 0$ and $d = -cz_a$, so the special frequency becomes $\omega_n = \sqrt{g/(z_C - z_a)}$. Substituting these values in Eqs. (20–22) we find, after simplification, $\mu_i = \lambda_{zi}[g + z_A\omega_n^2 \sin(\omega_n t + \varphi_z)]$. μ_i is then automatically > 0, provided that the maximum acceleration along z, $\max\{|\ddot{z}|\} = z_A\omega_n^2$, is not greater than g, as one would expect [10].

5 Transition Trajectories

Along the harmonic trajectories defined by Eq. (1), the velocity $\dot{\mathbf{p}}$ and acceleration $\ddot{\mathbf{p}}$ are always nonzero, so the robot cannot be in a state of rest. To actually implement such trajectories we need to bring the robot, initially at rest, in a dynamic state; similarly, we also need to stop the robot by bringing it back to rest.

For such a purpose, we developed transition trajectories defined by

$$\mathbf{p} = \mathbf{p}_C + U(\xi)\begin{bmatrix} x_A \sin(\omega t + \varphi_x) \\ y_A \sin(\omega t + \varphi_y) \\ z_A \sin(\omega t + \varphi_z) \end{bmatrix} \qquad (30)$$

where $U(\xi)$ is a function of class C^2.

Here, U is defined in terms of the a dimensional variable $\xi = t/T$, with T being the total time required for the transition motion. Notice that Eq. (30) corresponds to Eq. (1) where the amplitudes of oscillation are no longer kept constant.

Having defined the derivatives of U as $V(\xi) = dU/d\xi$, $W(\xi) = d^2U/d\xi^2$, we impose the boundary conditions: $U(0) = 0$, $U(1) = 1$ and $V(0) = V(1) = W(0) = W(1) = 0$. In this way, the robot starts at $t = \xi = 0$ in $\mathbf{p} = \mathbf{p}_C$ with $\dot{\mathbf{p}} = \ddot{\mathbf{p}} = 0$ and then moves with growing amplitudes along the three axes; when the transition is complete, the trajectory can blend into the final ellipse Γ. We also require $V(\xi) \geq 0$ for all $\xi \in [0, 1]$, so that $U(\xi)$ is monotonically increasing.

With different boundary conditions, Eq. (30) can also define a transition trajectory that slows down the robot after it has been moving along an ellipse Γ, so that it finally stops in the center of Γ. By taking $U(0) = U_0$, $U(1) = U_1$, the trajectory in (30) can also be used to connect two ellipses Γ_0, Γ_1 having the same center and phase angles but different amplitudes (so that Γ_1 is Γ_0 scaled by a factor U_1/U_0). Here, we will only consider the first case (with $U(0) = 0$ and $U(1) = 1$), as the other two can be studied in the same way. Substituting Eq. (30) into (8) we find

$$\mu_i = q_{i,W}\frac{W(\xi)}{T^2} + q_{i,V}\frac{V(\xi)}{T} + q_{i,UV}\frac{U(\xi)V(\xi)}{T} + \underbrace{q_{i,U}U(\xi) + E_i}_{\mu_{i,0}} \quad (31)$$

where the coefficients q_i are defined as follows:

$$q_{i,W} = -\lambda_i \cdot \mathbf{p}_d, \quad q_{i,V} = -2\lambda_i \cdot \dot{\mathbf{p}}_d$$
$$q_{i,UV} = 2\omega\left(x_{akj}y_A z_A \sin\varphi_{yz} + x_A y_{akj} z_A \sin\varphi_{zx} + x_A y_A z_{akj} \sin\varphi_{xy}\right) \quad (32)$$
$$q_{i,U} = C_i \cos(\omega t) + D_i \sin(\omega t)$$

with C_i and D_i defined as in Eq. (21).

The minimum value of $q_{i,U}$ is $q_{i,U,min} = -\sqrt{C_i^2 + D_i^2}$. For ξ going from 0 to 1, U monotonically increases from 0 to 1 (since we assumed $V(\xi) \geq 0$) so that its maximum value is $U_{max} = 1$. From Eq. (31), if $T \to \infty$, $\mu_i \to q_{i,U}U(\xi) + E_i =: \mu_{i,0}$.

Clearly, a lower bound for the minimum value of $\mu_{i,0}$ is $\mu_{i,0,LB} = q_{i,U,min}U_{max} + E_i = -\sqrt{C_i^2 + D_i^2} + E_i = \mu_{i,2}$; this was already shown to be ≥ 0 in Sect. 4, as long as Γ is feasible. All this shows that, by taking a sufficiently large, T, $\mu_i \approx \mu_{i,0} > \mu_{i,0,LB} > 0$, so that the cable tensions are positive throughout the transition.

It would be interesting to find the minimum value of T that guarantees a feasible movement. For this we should differentiate Eq. (31) with respect to t and set the result to 0 to find the extrema $\mu_{i,max}, \mu_{i,min}$ of μ_i, then find the minimum values of T (for $i = 1, 2, 3$) such that $\mu_{i,min}$ is positive. This leads to complex equations, whose numerical solution seems unsuitable for real-time applications.

An alternative is to set a lower bound for the minimum value that μ_i can take. This requires to find the extreme values of $q_{i,W}$ and $q_{i,V}$, which depend on t. It can be proved that such extrema are $q_{i,W,e} = \max\{|q_{i,W}|\} = \|\Phi\|$ and $q_{i,V,e} = \max\{|q_{i,V}|\} = 2\omega\|\Phi\|$, where the magnitude of Φ is given by

$$\|\Phi\| = \left[\left(x_A \lambda_{xi} \sin\varphi_x + y_A \lambda_{yi} \sin\varphi_y + z_A \lambda_{zi} \sin\varphi_z\right)^2 + \left(x_A \lambda_{xi} \cos\varphi_x + y_A \lambda_{yi} \cos\varphi_y + z_A \lambda_{zi} \cos\varphi_z\right)^2\right]^{1/2} \quad (33)$$

We also define the extreme values $V_e = \max\{|V(\xi)|\}$, $W_e = \max\{|W(\xi)|\}$ and $(UV)_e = \max\{|U(\xi) \cdot V(\xi)|\}$ for $\xi \in [0, 1]$, which depend on the choice of $U(\xi)$.

The minimum possible value of Eq. (31) is then

$$\mu_{i,LB} = -q_{i,W,e}\frac{W_e}{T^2} - q_{i,V,e}\frac{V_e}{T} + q_{i,UV}\frac{(UV)_e}{T} + \mu_{i,0,LB} \quad (34)$$

if $q_{i,UV} < 0$, and

$$\mu_{i,LB} = -q_{i,W,e}\frac{W_e}{T^2} - q_{i,V,e}\frac{V_e}{T} + \mu_{i,0,LB} \tag{35}$$

otherwise. If $\mu_{i,LB}$ from Eq. (34) or (35) is positive, then μ_i is always positive.

Multiplying the right hand sides (RHS's) of Eqs. (34 and 35) by T^2, they can be both expressed as $M_i(T) = \mu_{i,cnst} + \mu_{i,T}T + \mu_{i,T^2}T^2 > 0$, where $\mu_{i,cnst} = -q_{i,W,e}W_e$, $\mu_{i,T^2} = \mu_{i,0,LB}$ and the linear coefficient is either $\mu_{i,T} = -q_{i,V,e}V_e$ or $\mu_{i,T} = q_{i,UV}(UV)_e - q_{i,V,e}V_e$ depending on the sign of $q_{i,UV}$.

We have already shown that $\mu_{i,T^2} = \mu_{i,0,LB} = \mu_{i,2} > 0$ (if Γ is feasible); it then follows that, if there are solutions $T_{i,min}, T_{i,max}$ to $M_i(T) = 0$, then $M_i(T) < 0$ for $T \in [T_{i,min}, T_{i,max}]$, and $M_i(T) > 0$ otherwise. By definition, $\mu_{i,cnst} < 0$, so that $M_i(0) < 0$; this means $T_{i,min} < 0 < T_{i,max}$. Then, in order to have positive μ_i along the trajectory, we have simply to ensure that

$$T > T_{i,max} = \frac{-\mu_{i,T} + \sqrt{\Delta_{i,\mu}}}{2\mu_{i,T^2}} \tag{36}$$

with $\Delta_{i,\mu} = \mu_{i,T}^2 - 4\mu_{i,T^2}\mu_{i,cnst}$. Finally, we find a sufficient (albeit not necessary) condition on T that ensures the feasibility of the transition trajectory, by taking

$$T > \max\{T_{1,max}, T_{2,max}, T_{3,max}\} \tag{37}$$

One potential drawback of the transition trajectory described above appears when ω is equal to one of the limits $\omega_{min}, \omega_{max}$ of the admissible range, as defined in Eq. (28). In this case, ω^2 is either equal to $w_{i,min}$ or $w_{i,max}$ (for one $i \in \{1,2,3\}$) and the minimum value $\mu_{i,2}$ of the corresponding μ_i becomes zero, so that one of the cable tensions τ_i reaches zero at one position (while still remaining positive elsewhere). If $\mu_{i,0,LB} = \mu_{i,2} = 0$, there are no values of T for which the RHS's of Eqs. (34 and 35) are > 0, since all terms except $\mu_{i,0,LB}$ are negative: this corresponds to one of the $T_{i,max}$ from Eq. (36) tending to $+\infty$ when $\omega \to \omega_{min}$ or $\omega \to \omega_{max}$.

One exception occurs when $\omega_{min} = 0$; this happens when one $w_{i,min}$ is negative. ω_{min} then is not given by the requirement of positive cable tensions, but by the condition $\omega > 0$ (see again Eq. (28)). In this case all $T_{i,max}$ from Eq. (36) remain bounded even as $\omega \to \omega_{min}$. This however happens only if Γ is entirely in the SEW (see Sect. 4), a case for which transition trajectories have less practical interest.

The conditions found above are not strictly necessary, so that the T given by (37) is actually overestimated: numerical experimentation shows that T is reasonably close to the actual minimum required for feasibility when ω is safely in the middle of the admissible range and progressively less so when moving close to its limits.

6 Simulations

Several simulations were performed, to test the results shown in this paper. In some simulations, we used a rigid body model with massless cables having infinite stiffness, thus reproducing exactly the model presented here. An example is shown in Fig. 2, where we compared the range of admissible frequencies obtained from the method in [9] with the range given in Eq. (28).

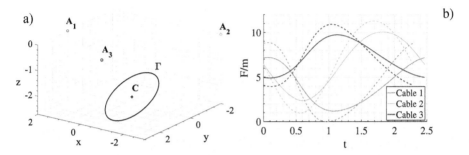

Fig. 2. The trajectory Γ is defined by $\mathbf{p}_C = [-1, 1, -2]^T$, $\mathbf{a}_1 = [2, 1, 0]^T$, $\mathbf{a}_2 = [-3, -2, 0]^T$, $\mathbf{a}_3 = [-1, 3, 0]^T$, with cable exit points being at the same height. Γ lies on a plane normal to $\mathbf{n} = [1, 2, 3]^T$, is circular and has radius $R = 1.2$ (length units are arbitrary).

The frequencies obtained with the method in [9] are $\omega_{min} = 1.548$, $\omega_{max} = 2.55$, while our method gives $\omega'_{min} = 1.387$, $\omega'_{max} = 2.75$: the first range is strictly contained in the second. The cable tensions (divided by the mass of the end-effector) are shown in Fig. 2b: the continuous lines correspond to $\omega = \omega_{max}$, the dashed ones to $\omega = \omega'_{max}$ (in which case one tension reaches zero).

In other simulations, we introduced a flexible body model, to test the robustness of the obtained results. Cable elasticity was introduced with a model based on Hooke's law for springs [10]: $\tau_i = \max\{(k_s/\rho_{i,l}) \cdot (\rho_i - \rho_{i,l}), 0\}$. Here, $\rho_{i,l}$ is the free length of the i-th cable (which depends on the rotation of the i-th motorized winch), $k_s = ES$ is the axial rigidity, E is Young's modulus and S is the cross-sectional area of the cable.

Such simulations were developed by solving the direct dynamic problem for the robot under consideration. The results of these simulations are not shown here due to space constraints. However, they all confirmed that the theoretical results obtained in the paper may hold for a real robot.

7 Conclusions and Future Work

We found a general class of spatial dynamical trajectories that allow a cable-suspended parallel robot (CSPR) to move beyond its static workspace while keeping all cables in tension. We also found transition trajectories that allow the robot to move from the aforementioned trajectories to rest, and vice versa. The conditions that ensure the trajectory feasibility can be verified in a few milliseconds and, thus, are compatible with real-time applications.

The elliptical trajectories studied in this paper expand and generalize the results previously obtained in [4, 9]. Such trajectories provide more design freedoms and appear to be more flexible for practical applications. We plan to expand this work by combining these spatial trajectories into piecewise movements, thus generalizing previous work on point-to-point motions for CSPRs [6, 7].

Acknowledgments. The authors wish to express their gratitude to Ms. Xiaoling Jiang for her relevant comments on the paper.

References

1. Riechel, A.T., Ebert-Uphoff, I.: Force-feasible workspace analysis for underconstrained point-mass cable robots. In: Proceedings of the IEEE International Conference on Robotics and Automation, pp. 4956–4962 (2004)
2. Barrette, G., Gosselin, C.: Determination of the dynamic workspace of cable-driven planar parallel mechanisms. ASME J. Mech. Des. **127**(2), 242–248 (2005)
3. Gosselin, C., Ren, P., Foucault, S.: Dynamic trajectory planning of a two-DOF cable-suspended parallel robot. In: Proceedings of the IEEE International Conference on Robotics and Automation, pp. 1476–1481 (2012)
4. Gosselin C.: Global planning of dynamically feasible trajectories for three-DOF spatial cable-suspended parallel robots. In: Proceedings of the 1st International Conference on Cable-Driven Parallel Robots, pp. 3–22 (2012)
5. Jiang, X., Gosselin, C.: Dynamically feasible trajectories for three-DOF planar cable-suspended parallel robots. In: Proceedings of the ASME IDETC (2014)
6. Gosselin, C., Foucault, S.: Dynamic point-to-point trajectory planning of a two-DOF cable-suspended parallel robot. IEEE Trans. Robot. **30**(3), 728–736 (2014)
7. Jiang, X., Gosselin, C.: Dynamic point-to-point trajectory planning of a three-DOF cable-suspended parallel robot. IEEE Trans. Robot. **32**(6), 1550–1557 (2016)
8. Zhang, N., Shang, W.: Dynamic trajectory planning of a 3-DOF under-constrained cable-driven parallel robot. Mech. Mach. Theory **98**, 21–35 (2016)
9. Zhang, N., Shang, W., Cong, S.: Geometry-based trajectory planning of a 3-3 cable-suspended parallel robot. IEEE Trans. Robot. **33**(2), 484–491 (2016)
10. Berti, A., Gouttefarde, M., Carricato, M.: Dynamic recovery of cable-suspended parallel robots after a cable failure. In: Proceedings of the ARK, pp. 337–344 (2016)

Dynamic Transition Trajectory Planning of Three-DOF Cable-Suspended Parallel Robots

Xiaoling Jiang[✉] and Clément Gosselin

Université Laval, Québec, QC G1V 0A6, Canada
xiaoling.jiang.1@ulaval.ca, gosselin@gmc.ulaval.ca

Abstract. This paper proposes a dynamic transition trajectory planning technique for three-degree-of-freedom (three-DOF) cable-suspended parallel robots (CSPRs). This trajectory is designed to connect multiple target trajectories beyond the static workspace in sequence with different starting points, as well as having the ability of starting from/ending with a resting position, while ensuring continuity up to the acceleration level. Two consecutive target trajectories are involved in the transition trajectory by using proper time functions, such that a goal trajectory is gradually reached by approaching the amplitude parameters and frequencies from those of a source trajectory. Additionally, each transition is based on the optimization of THE departure point from its source trajectory and a minimum time for the transition to its goal trajectory. An example is provided to demonstrate the novel trajectory-planning technique. The robot is requested to start from the state of rest, merge into two consecutive ellipses, a straight line and a circle in sequence and then go back to the state of rest.

1 Introduction

Cable-driven parallel robots (CDPRs) possess some desirable features and have therefore attracted the interest of researchers. For this type of robots, dynamic constraints must be taken into account in trajectory planning in most cases, due to the fact that cables can only pull but cannot push on the end-effector [1–3].

More recently, dynamic trajectories for point-mass cable-suspended parallel robots (CSPRs, a class of CDPRs) that can extend beyond the static workspace were generated in [4,5]. A series of periodic trajectories were defined parametrically as analytic functions of time, which produces the algebraic inequalities of cable tensions based on the dynamic equations. As a result, global constraints that can guarantee positive cable tensions were obtained and there is no need to verify or impose tension constraints locally along the trajectory.

In [6], a family of periodic trajectories were directly obtained from the integration of the dynamic equations, with no restriction on the amplitude of oscillations. Linear, circular, and elliptical trajectories were produced by oscillations in different directions. The results obtained there generalized the harmonic trajectories obtained in [4,5]. A three-DOF CSPR was modelled as a linear system of three mass-springs, with the ratio between cable force and length being the equivalent spring stiffness. The integrated periodic trajectories are responses/motions of the linear system, which are always feasible since the spring forces of the linear mass-spring system are positive at all times. In [7], such periodic trajectories for the translational component of three-DOF planar CSPRs were also obtained, with constant positive cable force-length ratios. The trajectory planning scheme proposed in [6,7] simplifies the cable tension calculation and more importantly, it reveals the fundamental properties of CSPRs. Since these dynamic trajectories can extend beyond the static workspace of the robot, they can potentially be used for many new CSPR applications.

Researches on planning periodic trajectories for three-DOF point-mass CSPRs that can go beyond the static workspace can also be found in some other work. In [8], a user friendly control method was proposed to extend and control the trajectories designed in [4,5]. In [9], a geometry-based approach was introduced, with linear and circular trajectories designed in the position-acceleration phase plane.

Although specific trajectories can be generated separately, in a real application the robot is required to move from one trajectory to the next in order to automatically chain multiple pre-generated trajectories with different starting points. Therefore, the goal of this work is to plan such feasible transition trajectories with positive cable tensions at all times. In particular, the transition trajectory is designed to connect any two or multiple periodic trajectories, obtained in [6], in sequence. Since these target trajectories can extend beyond the static workspace of the robot and do not include points representing the state of rest, the transition trajectories must be capable of extending beyond the static workspace, as well as having the ability to start from/end with a resting position. In the proposed scheme, the source and goal trajectories are blended with a transition trajectory that increasingly adapts the parameters to those of the goal trajectory, as illustrated in Fig. 1. The transition trajectory begins as the robot leaves the source trajectory at point *start* and ends as the robot merges into the goal trajectory at point *arrival*.

The rest of the paper is arranged as follows. In Sect. 2, the kinematic and dynamic models of the mechanism are presented. Section 3 introduces the novel transition trajectory planning technique. Section 4 provides example trajectories and results.

2 Kinematic and Dynamic Modelling

A spatial three-DOF CSPR with three cables and a point mass end-effector is considered here. All cables are assumed to be massless and inelastic and to extend as straight lines from the supporting pulleys to the end-effector.

A fixed reference frame $\mathscr{F}_O$ with origin O is defined on the base of the robot, as illustrated in Fig. 2. The z axis of the fixed reference frame points upwards, opposite to the direction of gravity. Point B_j, with $j = 1, 2, 3$, is fixed and corresponds to the location where cable j exits its pulley or eyelet. Below,

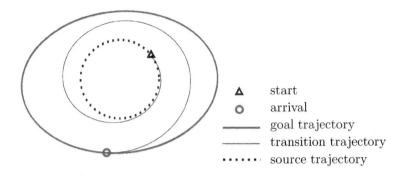

Fig. 1. Blending source and goal trajectories with a transition trajectory that increasingly adapts the parameters to those of the goal trajectory.

indices j should be assumed to go from 1 to 3 and a plural form of an item h_j is denoted $\{h_j\}_1^3$, unless specified otherwise. The vector $\mathbf{b}_j$ connects origin O to point B_j.

The inverse kinematic equations can be written as

$$\rho_j = \sqrt{(\mathbf{p} - \mathbf{b}_j)^T (\mathbf{p} - \mathbf{b}_j)}, \tag{1}$$

where ρ_j is the length of the jth cable and $\mathbf{p} = [x, y, z]^T$ is the position vector of the end-effector. Also, $\mathbf{e}_j$ is defined as a unit vector in the direction of the jth cable and oriented from the pulley to the end-effector. One has

$$\mathbf{e}_j = \frac{1}{\rho_j} (\mathbf{p} - \mathbf{b}_j). \tag{2}$$

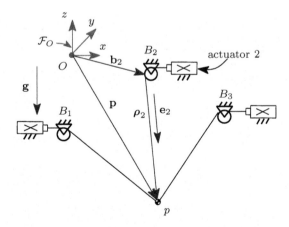

Fig. 2. A spatial three-DOF CSPR with point-mass.

The dynamic model is built by considering the force balance on the point mass end-effector, which yields

$$\sum_{j=1}^{3}(-f_j \mathbf{e}_j) + m\mathbf{g} = m\ddot{\mathbf{p}}, \tag{3}$$

where f_j is the tension in the jth cable, m is the mass of the end-effector, $\ddot{\mathbf{p}}$ is its acceleration, and $\mathbf{g} = [0, 0, -g]^T$ is the vector of gravitational acceleration. The dynamic Eq. (3) can be considered as a system of three linear equations in three unknowns (tensions f_j): one can explicitly solve for f_j, according to a specific architecture.

3 Transition Trajectory Planning

Transition trajectories that can automatically chain multiple feasible target trajectories in sequence with different starting points are developed. This trajectory is generated by combining two consecutive target trajectories, considered as source and goal trajectories for each transition, with a proper time function. As a result, the goal trajectory is gradually reached by changing parameters from that of the source trajectory. As illustrated in Fig. 3, the robot follows a source trajectory since its arrival and until the departure for a goal trajectory while starting the transition at a proper starting point. It is obvious that the preparation stage for transition is guaranteed to be feasible, since the robot is following the source trajectory during which cable tensions remain positive at all times.

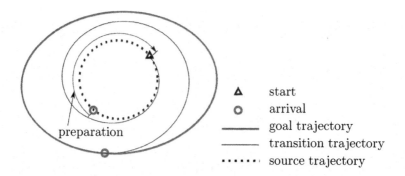

Fig. 3. Transition trajectory planning with preparation stage.

Mathematically, such a procedure for the preparation stage is given as

$$\mathbf{p}(t) = \mathbf{p}_i(t), \quad 0 \le t \le \delta_i, \tag{4}$$

and for the transition given as

$$\mathbf{p}(t) = \mathbf{p}_i(t) + U(\tau)\mathbf{s}_i(t), \quad \delta_i \le t \le T_i + \delta_i, \quad 0 \le \tau \le 1, \tag{5}$$

where $\mathbf{p}_i(t)$ represents trajectory i, $\mathbf{s}_i(t) = \mathbf{p}_{i+1}(t-\delta_i) - \mathbf{p}_i(t)$, $U(\tau) \in [0,1]$, with

$$U(0) = 0, \quad U(1) = 1, \tag{6}$$

is an increasingly monotonous time function with $\tau = (t-\delta_i)/T_i$ whereas δ_i, a time offset, is the departure time, T_i is the total transitioning time from $\mathbf{p}_i(t)$ to $\mathbf{p}_{i+1}(t)$, and there is no time offset for $\mathbf{p}_{i+1}(t)$. Therefore, the transition trajectory depends only on two parameters, the departure time (δ_i) from the source trajectory and the total transition time (T_i) until arrival at the goal trajectory. The values of the two parameters can be adjusted to obtain optimal transition trajectories.

Taking the first three derivatives of the transition trajectory (5) with respect to time, yields

$$\begin{aligned}\dot{\mathbf{p}}(t) &= V(\tau)\mathbf{s}_i(t) + \dot{\mathbf{p}}_i(t) + U(\tau)\dot{\mathbf{s}}_i(t),\\ \ddot{\mathbf{p}}(t) &= W(\tau)\mathbf{s}_i(t) + 2V(\tau)\dot{\mathbf{s}}_i(t) + \ddot{\mathbf{p}}_i(t) + U(\tau)\ddot{\mathbf{s}}_i(t),\\ \dddot{\mathbf{p}}(t) &= J(\tau)\mathbf{s}_i(t) + 3W(\tau)\dot{\mathbf{s}}_i(t) + 3V(\tau)\ddot{\mathbf{s}}_i(t) + \dddot{\mathbf{p}}_i(t) + U(\tau)\dddot{\mathbf{s}}_i(t).\end{aligned} \tag{7}$$

where $V(\tau)$, $W(\tau)$, $J(\tau)$ are the first three derivatives of $U(\tau)$ with respect to time. Thus, the continuity of $\mathbf{p}(t)$ can be preserved up to the acceleration level with (6) and the following constraints on $V(\tau)$ and $W(\tau)$, namely

$$V(0) = V(1) = 0, \quad W(0) = W(1) = 0. \tag{8}$$

Indeed, according to (6) and (8), the first two rows of (7) become

$$\begin{aligned}\dot{\mathbf{p}}(\delta_i) &= \dot{\mathbf{p}}_i(\delta_i), \quad \dot{\mathbf{p}}(T_i+\delta_i) = \dot{\mathbf{p}}_{i+1}(T_i),\\ \ddot{\mathbf{p}}(\delta_i) &= \ddot{\mathbf{p}}_i(\delta_i), \quad \ddot{\mathbf{p}}(T_i+\delta_i) = \ddot{\mathbf{p}}_{i+1}(T_i).\end{aligned}$$

Additionally, in some cases where the continuity of the jerk is required to yield smooth accelerations/cable forces of the robot, i.e.,

$$\dddot{\mathbf{p}}(\delta_i) = \dddot{\mathbf{p}}_i(\delta_i), \quad \dddot{\mathbf{p}}(T_i+\delta_i) = \dddot{\mathbf{p}}_{i+1}(T_i), \tag{9}$$

the corresponding constraints can also be found. Substituting (6) and (8) into the last row of (7) yields the value of the jerk at both ends of a transition trajectory, which is

$$\begin{aligned}\dddot{\mathbf{p}}(\delta_i) &= J(0)\mathbf{s}_i(\delta_i) + \dddot{\mathbf{p}}_i(\delta_i),\\ \dddot{\mathbf{p}}(T_i+\delta_i) &= J(1)\mathbf{s}_i(T_i+\delta_i) + \dddot{\mathbf{p}}_{i+1}(T_i).\end{aligned} \tag{10}$$

Thus, according to (10), the requirement for respecting (9) can be readily obtained, namely

$$J(0) = J(1) = 0. \tag{11}$$

In summary, transition trajectories with continuity up to the acceleration level can be obtained according to (5), with proper time function $U(\tau)$ while respecting (6) and (8). Moreover, for preserving the continuity of the jerk, an additional requirement (11) should also be respected.

4 Example

An example is given to demonstrate the transition trajectory technique proposed above. In this example, the robot transitions between the feasible periodic trajectories that were generalized in [6]. Two elementary time functions $U(\tau)$ are applied to generate the transition trajectory defined in (5), including the standard 5-th degree polynomial function and the 7-th degree polynomial function. Then, an optimal combination of the parameters δ_i and T_i are obtained, such that the cable tensions remain positive along the trajectory, and the execution time is as small as possible.

4.1 Specific Architecture

In order to provide physical insight, the specific robot architecture used in [6] is briefly recalled. In this architecture, the cable attachment points $\{\mathbf{b}_j\}_1^3$ form an equilateral triangle with circumradius R and are located at the same height, i.e., in a horizontal plane. One then has

$$\mathbf{b}_j = R\left[\cos\beta_j, \sin\beta_j, 0\right]^T, \quad \beta_j = \frac{2\pi(j-1)}{3}. \tag{12}$$

Substituting the above geometric parameters into (3), one can explicitly solve for f_j yielding

$$f_j = -\frac{m\rho_j}{3Rz}f'_j > 0, \tag{13}$$

where $z < 0$ and

$$\begin{aligned}
f'_1 &= (2x+R)(g+\ddot{z}) - 2z\ddot{x} > 0, \\
f'_2 &= (-x+\sqrt{3}y+R)(g+\ddot{z}) + z(\ddot{x}+\sqrt{3}\ddot{y}) > 0, \\
f'_3 &= (-x-\sqrt{3}y+R)(g+\ddot{z}) + z(\ddot{x}-\sqrt{3}\ddot{y}) > 0.
\end{aligned} \tag{14}$$

The above inequalities represent the constraints to be satisfied in order to ensure that the cables remain under tension. Since the first factor $-\frac{m\rho_j}{3Rz}$ of (13) is always positive under the assumption that the platform remains *suspended*, i.e., $z < 0$, inequalities (13) are automatically satisfied when inequalities (14) are satisfied. Therefore, if inequalities (14) are satisfied at all points of a given trajectory, it can be guaranteed that the cables will remain under tension throughout the trajectory. These conditions are necessary and sufficient.

4.2 Time Function $U(\tau)$

Two elementary time functions $U(\tau)$ based on polynomials are applied to generate the transition trajectory defined in (5). The standard 5-th degree polynomial trajectory which can ensure the continuity up to the acceleration level, proposed in [10], is first introduced. This time function and its first three derivatives are recalled, namely

$$U(\tau) = 6\tau^5 - 15\tau^4 + 10\tau^3, \quad V(\tau) = (30\tau^4 - 60\tau^3 + 30\tau^2)/T_i,$$
$$W(\tau) = (120\tau^3 - 180\tau^2 + 60\tau)/T_i^2, \quad J(\tau) = (360\tau^2 - 360\tau + 60)/T_i^3, \tag{15}$$

which meets the constraints given in (6) and (8). However, there exists discontinuities of the jerk at both ends of the trajectory since

$$J(0) = J(1) = 60/T_i^3, \tag{16}$$

which implies that the minimization of the trajectory time T_i increases the jerk.

Then, the 7-th degree polynomial trajectory with third derivative continuity, proposed in [10], is applied as an alternative. This time function and its first three derivatives are recalled, namely

$$\begin{aligned}
U(\tau) &= -20\tau^7 + 70\tau^6 - 84\tau^5 + 35\tau^4, \\
V(\tau) &= (-140\tau^6 + 420\tau^5 - 420\tau^4 + 140\tau^3)/T_i, \\
W(\tau) &= (-840\tau^5 + 2100\tau^4 - 1680\tau^3 + 420\tau^2)/T_i^2, \\
J(\tau) &= (-4200\tau^4 + 8400\tau^3 - 5040\tau^2 + 840\tau)/T_i^3,
\end{aligned} \tag{17}$$

which satisfies the constraints given in (6) and (8). It can be readily seen from the latter time function that the jerk is smooth at both ends of the trajectory since (11) is respected.

4.3 Transition Between Periodic Trajectories

Based on (4) and (5), transition with its preparation stage between a sequence of periodic trajectories is demonstrated. Periodic trajectories obtained in [6] are briefly recalled. The CSPR is equivalent to a passive mechanical system, obtained by replacing the actuator-cable units with linear constant-stiffness springs. Applying (2), its dynamic Eq. (3) is written as

$$m\ddot{\mathbf{p}} + \sum_{j=1}^{3} k_j \mathbf{p} - \mathbf{w} - m\mathbf{g} = 0, \tag{18}$$

where $k_j = f_j/\rho_j$, a constant, is the stiffness of the jth equivalent spring, and

$$\mathbf{w} = \sum_{j=1}^{3} k_j \mathbf{b}_j = m[v_1, v_2, 0]^T.$$

Periodic trajectories are then directly integrated from (18), namely

$$\mathbf{p} = \mathbf{p}_d + \mathbf{p}_s \tag{19}$$

with

$$\mathbf{p}_s = [x_s, y_s, z_s]^T = [-\frac{v_1}{\omega_n^2}, -\frac{v_2}{\omega_n^2}, -\frac{g}{\omega_n^2}]^T, \quad \omega_n^2 = \sum_{j=1}^{3} k_j/m,$$

and $\mathbf{p}_d = \mathbf{c}\cos\omega_n t + \mathbf{s}\sin\omega_n t$, with constant $\mathbf{c} = [\mu_{xc}, \mu_{yc}, \mu_{zc}]^T$ and $\mathbf{s} = [\mu_{xs}, \mu_{ys}, \mu_{zs}]^T$. Geometrically speaking, $\mathbf{p}_s$ represents the central static equilibrium position, and the *unbounded* $\mathbf{c}$ and $\mathbf{s}$ are design parameters that determine the shape of the motion, implying that these trajectories can extend beyond the static workspace. When trajectories of the real robot are designed as in (19) with parameters $\mathbf{p}_s$ located in the static workspace, it can be guaranteed that the tensions in the cables will remain positive.

In the example, the robot is requested to start from the state of rest, merge into two consecutive ellipses, a straight line and a circle in sequence and then go back to the state of rest. All these target trajectories can be represented by (19) by selecting proper parameters $\mathbf{p}_s$, $\mathbf{c}$, and $\mathbf{s}$. For each transition, the robot moves along a source trajectory for one turn since its arrival, continues following this trajectory while preparing for the next transition within the first phase, then starts the transition to reach a goal trajectory at a proper departure point. The transition with its preparation stage is designed in (5) and (4).

Optimal values of the phase $\delta_{start} = \omega_n \delta_i$ that corresponds to the moment when the robot leaves trajectory i (the source trajectory) and the minimum time T_i for the transition from trajectory i to $i+1$ are fixed by sweeping the (δ_{start}, T_i) plane, as shown in Fig. 4. Points marked by a small circle represent the optimal values of δ_{start} and $T_{\min}$ for each transition. The strategy is as follows. For a given value of T_i and δ_{start}, the constraint of positive cable tensions during a transition trajectory is numerically verified through a discretization of time. Each point on the curves in Fig. 4 represents a feasible minimum T_i for each δ_{start}. Then, the global minimum value of $T_{\min}$ for all the values of δ_{start} in one period can be readily obtained.

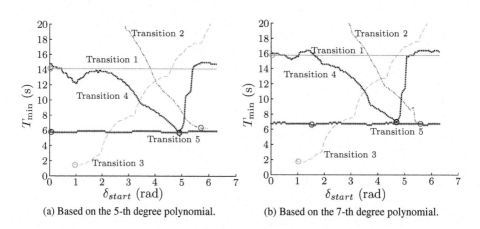

Fig. 4. Optimized $T_{\min}$ and δ_{start}.

Applying the optimized δ_{start} and $T_{\min}$, the example trajectory guarantees continuous and positive cable tensions, as is confirmed in Fig. 5. In this figure, a small triangle represents the moment at which the end-effector starts a transition

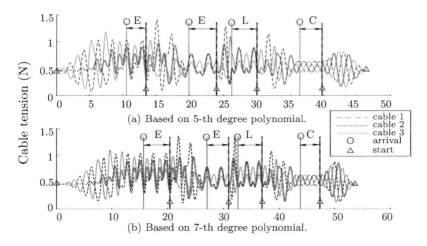

Fig. 5. Cable Tensions during the trajectory (E: Ellipse, L: Straight Line, C: Circle).

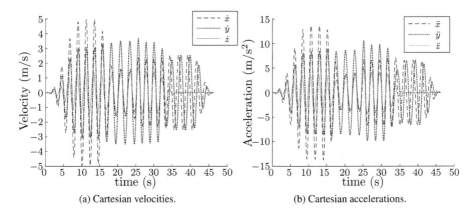

Fig. 6. Velocities and accelerations of the trajectory using the 5th degree polynomial.

while a small circle indicates the moment when the end-effector arrives and merges into the steady-state target trajectory. Additionally, the transition is able to start from/end with a resting position, as shown in Fig. 6. The advantage of using the 5-th degree polynomial is that the transition time is smaller than when using the 7-th degree polynomial.

The complete trajectory, using the 5-th degree polynomial for its transitions, is illustrated in Fig. 7. The transition trajectories using the 7-th degree polynomial are similar and thus the corresponding trajectory is not presented due to space limitations. As demonstrated in Fig. 7, transition trajectories gradually go beyond the static workspace to join target trajectories which are located beyond the static workspace. Points marked by small triangles represent points at which the end-effector starts the transition and leaves the source trajectory while points

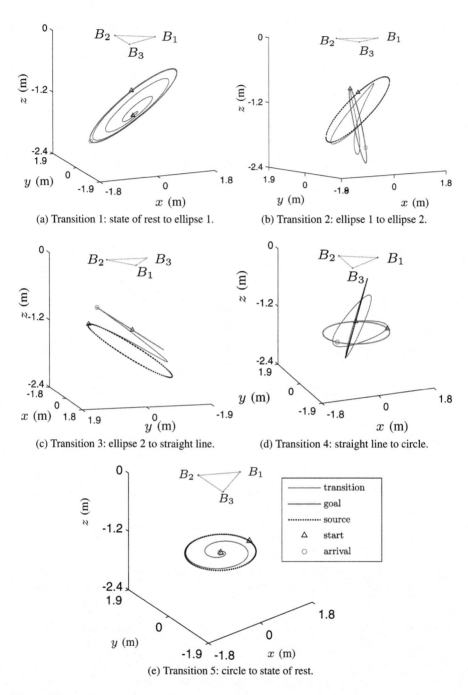

Fig. 7. Transition trajectories between periodic trajectories using the 5-th degree polynomial.

marked by small circles represent the points at which the end-effector arrives at and merges into the steady-state target trajectory.

5 Conclusion

The contribution of this paper lies in the definition of transition trajectories that can be used to chain multiple dynamic trajectories beyond a robot's static workspace in sequence, with continuity up to the acceleration level. The approach is relatively simple but effective: a target trajectory is gradually reached by approaching the amplitude and frequency of its parameters from that of its preceding trajectory, using proper time functions. Transition trajectories are obtained by optimizing the time offset of a source trajectory, corresponding to the departure time for a goal trajectory, and minimum execution time, as long as the cable tensions remain positive.

An example trajectory is performed by applying the novel technique to a specific mechanism. Periodic trajectories obtained in the literature are chosen as target trajectories. Using the transition trajectory proposed, the robot is able to smoothly start from the state of rest, merge into two consecutive ellipses, a straight line and a circle in sequence and then go back to the state of rest. In future work, time offsets will be added to both the source and goal trajectories such that optimal points at which the robot leaves for and lands on the goal trajectory can be obtained based on the minimization of the transitioning time or other criteria.

Acknowledgements. This work was supported by the Natural Sciences and Engineering Research Council of Canada (NSERC), by the Canada Research Chair Program and by the China Scholarship Council (CSC) through a scholarship to the first author.

References

1. Barnett, E., Gosselin, C.: Time-optimal trajectory planning of cable-driven parallel mechanisms for fully specified paths with G1-discontinuities. ASME J. Dyn. Syst. Meas. Control **137**(7), 071007 (2015)
2. Fahham, H.R., Farid, M.: Minimum-time trajectory planning of spatial cable-suspended robots along a specified path considering both tension and velocity constraints. Eng. Optim. **42**(4), 387–402 (2010)
3. Bamdad, M.: Time-energy optimal trajectory planning of cable-suspended manipulators. In: Cable-Driven Parallel Robots, pp. 41–51. Springer (2013)
4. Gosselin, C., Ren, P., Foucault, S.: Dynamic trajectory planning of a two-dof cable-suspended parallel robot. In: IEEE International Conference on Robotics and Automation, pp. 1476–1481 (2012)
5. Gosselin, C.: Global planning of dynamically feasible trajectories for three-dof spatial cable-suspended parallel robots. In: Cable-Driven Parallel Robots, pp. 3–22. Springer (2013)
6. Jiang, X., Gosselin, C.: Dynamic point-to-point trajectory planning of a three-dof cable-suspended parallel robot. IEEE Trans. Robot. **32**(6), 1–8 (2016a)

7. Jiang, X., Gosselin, C.: Trajectory generation for three-degree-of-freedom cable-suspended parallel robots based on analytical integration of the dynamic equations. ASME J. Mech. Robot. **8**(4), 041001 (2016)
8. Schmidt, V., Kraus, W., Ho, W.Y., Seon, J., Pott, A., Park, J.O., Verl, A.: Extending dynamic trajectories of cable-driven parallel robots as a novel robotic roller coaster. In: ISR/Robotik 2014; 41st International Symposium on Robotics, pp. 1–7 (2014)
9. Zhang, N., Shang, W.: Dynamic trajectory planning of a 3-dof under-constrained cable-driven parallel robot. Mech. Mach. Theor. **98**, 21–35 (2016)
10. Gosselin, C., Hadj-Messaoud, A.: Automatic planning of smooth trajectories for pick-and-place operations. ASME J. Mech. Des. **115**(3), 450–456 (1993)

Transverse Vibration Control in Planar Cable-Driven Robotic Manipulators

Mitchell Rushton[✉] and Amir Khajepour

University of Waterloo, Waterloo, Canada
{mmrushto,a.khajepour}@uwaterloo.ca

Abstract. This paper presents an analysis of the possibility of using cables to eliminate out-of-plane vibrations in planar cable-driven parallel robotic manipulators. A new control strategy is presented with a complete stability analysis. The results show it is theoretically possible to stabilize planar cable-driven systems in non-planar directions using cables alone. The developed controller is shown to be effective both in simulation and experimentally at attenuating transverse platform vibrations.

1 Introduction

Cable-driven parallel robots (CDPR) consist of a rigid platform suspended by a number of elastic cables [1]. This design has a number of considerable advantages over traditional robotic manipulators, which consist of solely rigid links. Cables are relatively lightweight and inexpensive and the reduced inertial load enables CDPRs to command very high accelerations [2]. While a well-known drawback of parallel manipulators is their limited work space, CDPRs are able to span distances upwards of 500 m [3].

The numerous benefits of CDPRs create a significant potential for use in large-scale industrial applications, such as high-speed automated warehousing. However, a number of technical challenges exist that complicate manipulator design configurations and control strategies. The most significant of these challenges is the fact that cables are only able to transmit force while under tension. This uniaxial force condition requires CDPRs to be redundantly actuated in order to be fully constrained by the use of internal antagonistic cable forces [4]. A second major issue is that the stiffness of the cables is much lower than a comparable rigid mechanism. As a result, end effector vibrations are a potential hazard that require active compensation during highly dynamic motions to ensure manipulator safety and accuracy.

An interesting and particularly challenging class of CDPRs are those limited to motion within a plane. In the planar case, most, if not all, of the cable internal forces lie within the motion plane. As a result, the translational stiffness in the planar normal direction is very low and creates a potential for large displacements that are uncontrollable using traditional continuous control methods [5]. While the use of redundant cables and design optimization can help maximize the overall manipulator stiffness, the potential for displacements in the planar normal direction remains problematic [6].

To the best of the authors' knowledge at the time of writing this paper, with two exceptions, all previous studies on the control of planar CDPRs have completely neglected to consider the dynamics in non-planar directions [7–9]. In [10,11], it was shown that persistent out-of-plane vibrations in a planar CDPR can effectively be eliminated by adding additional actuators to the mobile platform. The purpose of this study is to investigate the possibility of using cables alone for eliminating out-of-plane disturbances and whether there is any practical benefit to using such an approach.

2 Physical Model

A cable-driven parallel robot (CDPR) consists of a rigid platform suspended and actuated using a number of elastic cables. Consider the two-cable CDPR presented in Fig. 1. For the sake of simplifying the proceeding analysis, platform rotations have been ignored.

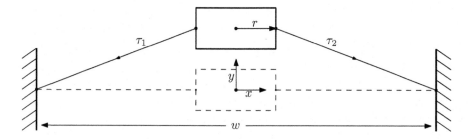

Fig. 1. 2-cable parallel robotic manipulator

For each cable, define a vector c_i as:

$$c_i = a_i - b_i \tag{1}$$

where a_i and b_i are respectively the platform and frame mount points for the ith cable. The length of each cable can then be found by taking the magnitude of c_i:

$$l_i = \|c_i\| \tag{2}$$

Additionally, a unit vector pointing along each cable can be obtained as

$$\hat{c}_i = \frac{c_i}{l_i} \tag{3}$$

Each cable can be modeled as a linear spring with some stiffness k_i for the ith cable. Using such a model, the tension in each cable can be determined as

$$\tau_i = -k_i \left(l_i - \delta l_i \right) \tag{4}$$

where δl_i is the unstretched cable length and considered as an input to the system. Combined with the cable direction vector defined in (3), the force applied on the platform by each cable is found to be

$$F_{c,i} = \hat{c}_i \cdot \tau_i \tag{5}$$

The x and y accelerations can then be calculated by summing the individual forces produced by the n cables:

$$m_p \begin{bmatrix} \ddot{x} \\ \ddot{y} \end{bmatrix} = \sum_{i=1}^{n} F_{c,i} \tag{6}$$

where m_p is represents the mass of the platform and, in this particular case, $n = 2$. Assuming there are no additional external forces, the dynamics of the platform are fully described by (6).

In [12], Behzadipour and Khajepour present an analysis of the platform and cable stiffness and show that the stiffness of each cable in the transverse direction is a function of the cable tension. Additionally, they show that increasing the cable tensions always leads to an increase in the platform stiffness in all directions. With that in mind, one can consider only the transverse platform dynamics (y direction) as

$$m_p \ddot{y} = -k_y y \tag{7}$$

where k_y is the combined stiffness of the two cables projected along the y axes.

3 Controller Design

Using the model for the out-of-plane dynamics of the platform as defined in (7), the goal is to design a controller that is able to dissipate any induced vibrations in the y direction and maintain a constant position within the equilibrium plane (where $y = 0$). A major difficulty in designing such a controller is the fact that cables are only able to transmit force uniaxially. As a result, cable forces projected along the y-axis always point toward the equilibrium plane.

One potential control strategy to eliminate vibrations using only the effect of the cables is to increase cable tensions when the platform is moving away from its equilibrium plane and decrease the cable tensions when the platform is moving toward its equilibrium plane. This task can be achieved by the following switching controller:

$$\delta l_i = \delta l_{i,eq} - \text{sign}(y)\,\text{sign}(\dot{y}) K_{sw} \tag{8}$$

where $\delta l_{i,eq}$ corresponds to the unstretched cable length of the ith cable when the platform is resting within the plane and K_{sw} is a gain term that can be used to adjust the rate of convergence. In order to analyze the stability of this new closed-loop system, let us first construct a candidate Lyapunov function based on the energy of the system:

$$V = \frac{1}{2}m_p \dot{y}^2 + \frac{1}{2}k_y y^2 \qquad (9)$$

By the law of conservation of energy, it is known that V is constant, assuming that there is no change in k_y or m_p. Since V is constant, $\dot{V} = 0$. Therefore, for a given k_y, the continuous dynamics of the system are marginally stable. It still remains, however, to examine the effect of the discrete switching terms and the effect of the discontinuity in the continuous dynamics on the overall system stability.

If for all possible switching events, occurring at some time t_{sw}, $V(t_{sw})^+ \leq V(t_{sw})^-$, then it can be concluded that the closed-loop system is globally asymptotically stable. Recall that k_y is a function of cable tensions and that k_y increases when cable tensions are similarly increased. The maximum value for k_y is therefore obtained when cable tensions are at a maximum, or in other words, when cable unstretched lengths are at a minimum. Conversely, the minimum value of k_y is obtained when cable unstretched lengths are at a maximum.

Discontinuities occur when either y or $\dot{y}$ changes sign, in other words, when the system is either in a state of purely kinetic or potential energy. Consider the value of V immediately before and after y changes sign, when the system is in a state of purely kinetic energy:

$$\begin{aligned} V(t_{sw,y})^- &= \frac{1}{2}m_p\,\dot{y}^2 + \frac{1}{2}k_{y,min}\,0^2 \\ V(t_{sw,y})^+ &= \frac{1}{2}m_p\,\dot{y}^2 + \frac{1}{2}k_{y,max}\,0^2 \end{aligned} \qquad (10)$$

It is obvious from (10) that $V(t_{sw,y})^+ = V(t_{sw,y})^-$ for all possible instances of $t_{sw,y}$. Consider now the case where $\dot{y}$ changes sign and the system is in a state of purely potential energy:

$$\begin{aligned} V(t_{sw,\dot{y}})^- &= \frac{1}{2}m_p\,0^2 + \frac{1}{2}k_{y,max}\,y^2 \\ V(t_{sw,\dot{y}})^+ &= \frac{1}{2}m_p\,0^2 + \frac{1}{2}k_{y,min}\,y^2 \end{aligned} \qquad (11)$$

So long as $k_{y,max} \geq k_{y,min}$, it is always the case that $V(t_{sw,\dot{y}})^+ \leq V(t_{sw,\dot{y}})^-$. It is therefore shown that for all possible switching events generated by using the switching controller presented in (8), $V(t_{sw})^+ \leq V(t_{sw})^-$. Thus, it can be concluded that the closed-loop system is globally asymptotically stable and can be used for eliminating out-of-plane translational displacements.

4 Results

The controller presented in Sect. 3 was implemented and tested in simulation and on an experimental test setup to verify and study its effectiveness and potential for real applications. The procedure used for evaluating the controller performance in both the simulated and experimental cases was to start the platform from some arbitrary nonzero initial condition and allow the controller to attempt to drive y back to the origin. The test would be repeated with the controller both active and inactive in order to compare the rate of convergence and observe what, if any, effect the controller has on the transverse platform dynamics.

The preceding subsections summarize the results obtained in simulation and experiment respectively.

4.1 Simulation

The simulation study was performed using Simulink and the nonlinear 2-cable robot model described in Sect. 2. The specific model parameters that were used for the simulation are presented in Table 1, where m_p represents the mass of the platform and k corresponds to the cable stiffness constants. w and r are two additional parameters that define the geometry of the robot, as shown in Fig. 1.

Table 1. Simulated cable robot model parameters

Parameter	Value
m_p	1 kg
k	1000 N/m
w	1 m
r	0.1 m

The control inputs to the system as described in the model of Sect. 2 are the unstretched lengths of the two cables. As such, the feedback controller developed in Sect. 3 is applied directly. At the start of simulation, the platform is given an initial position of $x = 0$, $y = 50$ mm. The equilibrium cable unstretched lengths used were $\delta l_{eq} = 0.38$ m for both cables. The simulation was repeated under identical conditions for various values of the controller switching gain, K_{sw}. The simulation results are summarized in Fig. 2.

From Fig. 2 it can be seen that when the switching controller is inactive ($K_{sw} = 0$) and the cable unstretched lengths are held constant, the platform oscillates perpetually. This is the expected result as there is no inherent damping considered in the model. With the controller enabled ($K_{sw} = 0.1$) the oscillations in y are swiftly attenuated, eventually bringing the platform to rest at its equilibrium plane.

4.2 Experimental

The procedure used for evaluating the controller of Sect. 3 in simulation was repeated on the experimental test setup shown in Fig. 3. The setup is a highly-stiff 12 cable planar cable-driven parallel robot, driven by 4 motors. The interested reader may refer to [13,14] for a detailed discussion of the mechanism and its operating principles.

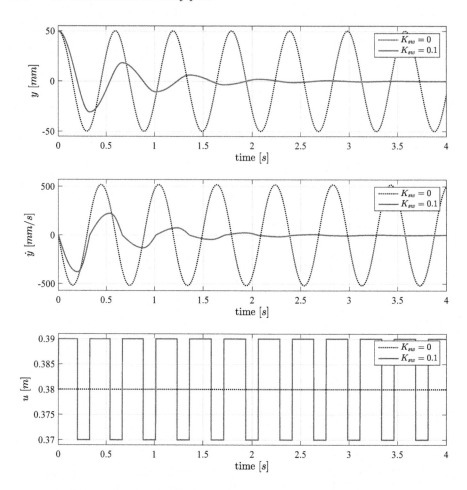

Fig. 2. Simulation of a 2-cable robot with switching controller

Fig. 3. Experimental setup

In contrast to the model developed in Sect. 2 for a generic 2-cable robot, the low-level control system for the experimental setup has been organized such that the bottom cables are operated in force control mode. As such, the switching controller of Sect. 3 has been rearranged to be in terms of cable tensions rather than cable unstretched lengths. The top cables are operated in position control mode, however, only the bottom cables have been used for vibration control in this particular experiment in order to simplify the implementation.

In an attempt to soften the shocks imposed on the system by the discontinuities in the control signal, the sign function was replaced with the following approximation: $sign(x) \approx \frac{x}{|x|+\varepsilon}$, where $\varepsilon \ll 1$. The resulting form of the transverse vibration controller, after having been modified to fit the requirements of the experimental test setup, is then found to be

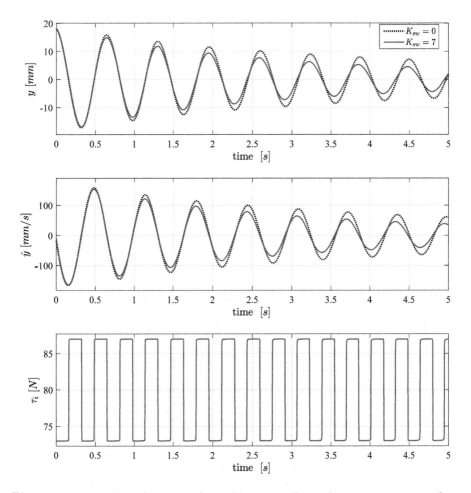

Fig. 4. Experimental performance of switching controller at damping transverse vibrations

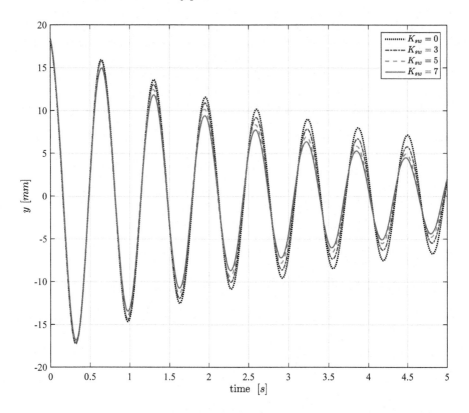

Fig. 5. Comparison of various positive values for the controller switching gain, K_{sw}

$$\tau_i = \tau_{i,eq} - \frac{y}{|y| + \varepsilon_y} \frac{\dot{y}}{|\dot{y}| + \varepsilon_{\dot{y}}} K_{sw} \qquad (12)$$

The controller parameters used for this specific implementation were $\varepsilon_y = \varepsilon_{\dot{y}} = 10^{-5}$ and $\tau_{eq} = 80N$ for both the bottom cables. Various values of K_{sw} were considered, as will be seen below. Within the plane, the platform was held at the centre of it's workspace.

At the start of the tests, the platform was released from an initial position of $y = 18$ mm. The observed performance of the controller is summarised in Fig. 4. The estimations for y and $\dot{y}$ were obtained by the use of an accelerometer, mounted on the mobile platform, and a Kalman filter based on previously obtained system identification data.

As can be seen in Fig. 4, the switching controller does indeed offer an improvement over the natural system damping at attenuating out-of-plane vibrations. The tests were repeated multiple times with various values of K_{sw}. The results are presented in Fig. 5. As predicted by the analysis in Sect. 3, increasing the value of K_{sw} is shown to improve the rate of convergence. The maximum value

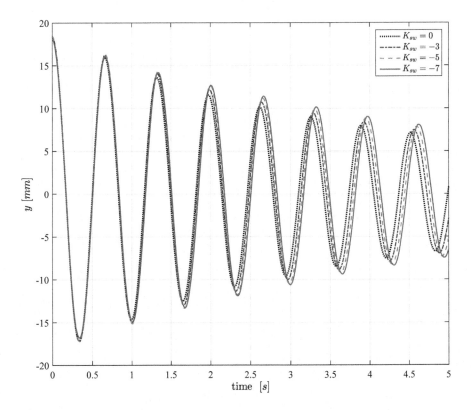

Fig. 6. Comparison of various negative values for the controller switching gain, K_{sw}

of K_{sw} used was limited to 7 N because it was feared that beyond that, the impulsive loading on the cable drive system may damage the test setup.

In order to verify that the observed increase in the rate of convergence of y is caused by the switching controller working correctly, and not some other phenomenon, the tests were repeated using various negative values of K_{sw}. From the analysis in Sect. 3, if negative values of K_{sw} are used, a decrease in the rate of convergence of y should be observed. In Fig. 6 it can be seen that this is indeed the case.

Taken together, the results of Figs. 5 and 6 give strong evidence that the controller behaves the way it has been predicted and that such a control strategy can indeed influence the transverse dynamics for real planar cable-driven systems.

5 Conclusion

Transverse vibrations are a real problem for planar cable-driven systems. While it is possible to improve manipulator stiffness within the motion plane with the use of internal antagonistic cable forces, the effect is minor in the planar normal direction and large undesired displacements remain a possibility. Such transverse vibrations are linearly uncontrollable; however, the results of this study demonstrate that it is possible to stabilize the system by using a nonlinear switching controller. With the developed controller, the closed-loop system is shown to be asymptotically stable using Lyapunov's stability theorem. The performance of the controller was then investigated and demonstrated in simulation and experimentally to be effective at eliminating out-of-plane translational vibrations.

For highly-stiff systems with high natural frequencies, such a control strategy may not be feasible due to the heavy impact on the cable drive system by the required high-frequency discontinuous switching control signal. Therefore, in order to effectively eliminate out-of-plane vibrations in systems with high-stiffness, it may be necessary to add an additional set of actuators to improve the controllability of the system in directions which are linearly uncontrollable. However, In systems that are less stiff, such that the cable drive system is able to keep up with the demands of the switching control signal, this may very well be an effective solution for eliminating out-of-plane platform vibrations.

Acknowledgements. The authors would like to acknowledge the financial support of the Natural Sciences and Engineering Research Council of Canada in this project.

References

1. Tang, X.: An overview of the development for cable-driven parallel manipulator. In: Advances in Mechanical Engineering (2014)
2. Kawamura, S., Choe, W., Tanaka, S., Pandian, S.R.: Development of an ultrahigh speed robot falcon using wire drive system. In: 1995 IEEE International Conference on Robotics and Automation, 1995, Proceedings, vol. 1, pp. 215–220. IEEE (1995)
3. Nan, R., Li, D., Jin, C., Wang, Q., Zhu, L., Zhu, W., Zhang, H., Yue, Y., Qian, L.: The five-hundred-meter aperture spherical radio telescope (fast) project. Int. J. Mod. Phys. D **20**(06), 989–1024 (2011)
4. Roberts, R.G., Graham, T., Lippitt, T.: On the inverse kinematics, statics, and fault tolerance of cable-suspended robots. J. Field Robot. **15**(10), 581–597 (1998)
5. Jamshidifar, H., Khajepour, A., Fidan, B., Rushton, M.: Kinematically-constrained redundant cable-driven parallel robots: modeling, redundancy analysis and stiffness optimization. IEEE/ASME Trans. Mechatron. **22**, 921–930 (2016)
6. Torres-Mendez, S., Khajepour, A.: Design optimization of a warehousing cable-based robot. In: ASME 2014 International Design Engineering Technical Conferences and Computers and Information in Engineering Conference, p. V05AT08A091. American Society of Mechanical Engineers (2014)
7. Williams, R.L., Gallina, P., Vadia, J.: Planar translational cable-direct-driven robots. J. Robot. Syst. **20**(3), 107–120 (2003)
8. Oh, S.-R., Agrawal, S.K.: Cable suspended planar robots with redundant cables: controllers with positive tensions. IEEE Trans. Robot. **21**(3), 457–465 (2005)
9. Khosravi, M.A., Taghirad, H.D.: Robust pid control of fully-constrained cable driven parallel robots. Mechatronics **24**(2), 87–97 (2014)
10. Rushton, M., Khajepour, A.: Optimal actuator placement for vibration control of a planar cable-driven robotic manipulator. In: 2016 American Control Conference (ACC), pp. 3020–3025. IEEE (2016)

11. Rushton, M.: Vibration control in cable robots using a multi-axis reaction system (2016)
12. Behzadipour, S., Khajepour, A.: Stiffness of cable-based parallel manipulators with application to stability analysis. J. Mech. Des. **128**(1), 303–310 (2006)
13. Torres-Mendez, S., Khajepour, A.: Analysis of a high stiffness warehousing cable-based robot. In: Proceedings of the ASME 2014 International Design Engineering Technical Conferences and Computers and Information in Engineering Conference, Buffalo, New York, USA, 17–20 August 2014 (2014)
14. Torres Mendez, S.J.: Low mobility cable robot with application to robotic warehousing (2014)

Application of a Differentiator-Based Adaptive Super-Twisting Controller for a Redundant Cable-Driven Parallel Robot

Christian Schenk[1(✉)], Carlo Masone[1], Andreas Pott[2], and Heinrich H. Bülthoff[1]

[1] Max Planck Institute for Biological Cybernetics, Spemannstraße 38, 72076 Tübingen, Germany
christian.schenk@tuebingen.mpg.de
[2] ISW Universität Stuttgart, 70174 Stuttgart, Germany
andreas.pott@isw.uni-stuttgart.de

Abstract. In this paper we present preliminary, experimental results of an Adaptive Super-Twisting Sliding-Mode Controller with time-varying gains for redundant Cable-Driven Parallel Robots. The sliding-mode controller is paired with a feed-forward action based on dynamics inversion. An exact sliding-mode differentiator is implemented to retrieve the velocity of the end-effector using only encoder measurements with the properties of finite-time convergence, robustness against perturbations and noise filtering. The platform used to validate the controller is a robot with eight cables and six degrees of freedom powered by 940 W compact servo drives. The proposed experiment demonstrates the performance of the controller, finite-time convergence and robustness in tracking a trajectory while subject to external disturbances up to approximately 400% the mass of the end-effector.

1 Introduction

Cable-Driven Parallel Robots (*CDPR*) are systems that use elastic cables to guide the so called end-effector through space. Each cable is connected on one side at the end-effector and guided by several pulleys to ground mounted motors on the other side. These motors coil and uncoil the cables, apply a wrench at the platform and by doing this, change the position and orientation of the end-effector. Compared to other manipulators such as Steward Platforms or serial manipulators, CDPR have some outstanding advantages such as large workspace, modularity, mobility and scalability. Furthermore, because of their parallel structure and light-weight construction, CDPR have the ability to exert high accelerations. Thanks to these properties, the potential applications of CDPR are numerous and diverse and include pick and place tasks [4], rehabilitation [19], entertainment [2], simulation of motion [11] and telescopes [3]. Regardless of the application, it is of paramount importance to ensure stability and adequate performance (according to some suitable metric) in spite of possible unmodelled effects, external disturbances (e.g. wind gusts for a radio telescope) and uncertain or varying parameters (e.g. varying mass in pick and place tasks).

Several strategies have been proposed in the literature of CDPR to deal with perturbations, and they are often based on the adaptation of kinematic/dynamic parameters ([1,4,6]). However, this approach only considers parametric uncertainties and assumes the knowledge of an upper bound of the perturbations to choose the control gains. Unfortunately, in practice such an upper bound is typically unknown and one must resort to overly conservative estimates of the perturbations, which means unnecessarily high control actions and noise amplification. Another approach to deal with perturbations is to use control laws that are implicitly robust, such as sliding mode controllers [4], yet the problem of tuning the gains according to a worst case scenario persists. Additionally, sliding-mode controllers suffer from chattering effects [17] which can reduce the performance of the system and even damage its components. To overcome these limitations we proposed a robust sliding-mode controller with adaptive gains ($ASTC$) [14], based on [18], that does not require the knowledge of an upper bound of the perturbations. In [14] the controller was successfully validated in a numerical simulation, demonstrating the capability of tracking desired trajectories in operational space (i.e. pose of the end-effector) in presence of parameter uncertainties and external disturbances. Moreover, the positive effect of the adaptive control gains was remarked by a comparison with a continuous sliding mode controller with fixed gains [4] showing that, for comparable tracking error, the adaptive strategy in the ASTC effectively reduces chattering.

In this paper we present the first tests of the ASTC with a real CDPR. Our goal is to assess whether the performance of the ASTC that was obtained in simulation still holds in practice and, if not, to get a better understanding of the limiting factors. Indeed, the results of this paper show that the resilience of the controller to perturbations and the property of finite time convergence are preserved but with worse performance in terms of tracking error. In comparison to [14], another novelty of this paper is the introduction of an adaptive sliding mode differentiator [17] to indirectly derive the velocity of the end-effector using only encoders at the winches and the Forward Kinematics (FK) model [12, 16], that is the barest minimum sensor information usually available in CDPR. The results demonstrate also the robustness and finite time convergence of the differentiator.

This paper is structured as follows: in Sect. 2 we present the model, in Sect. 3 we discuss the control strategy as well as a robust sliding-mode differentiator. Section 4 describes the experimental setup and the evaluation of the results.

2 Modeling

The control algorithm and the kinematic relations of the system considered in this paper are based on a simplified model of a redundant CDPR with $n = 6$ degrees of freedom (dof) and $m = 8$ massless and inextensible cables. The choice of neglecting the dynamics of the cables is an obvious abstraction of reality that is quite common in literature since this is an open topic of research. Additionally, we assume that each cable leaves the corresponding winch from a fixed point, i.e. the rotation of the winches is considered negligible. With this simplifications the generic i-th cable forms a straight line [15] that connects a point B_i on the end-effector (*onboard connection*) to the fixed point A_i (*offboard connection*) on the winch (Fig. 1). For the derivation of the main kinematic and dynamic relations of the system we define two frames of reference: a frame $\mathscr{F}_E = \{\mathcal{O}_E, \boldsymbol{X}_E, \boldsymbol{Y}_E, \boldsymbol{Z}_E\}$, fixed on the geometrical center of the end-effector and an inertial world frame $\mathscr{F}_W = \{\mathcal{O}_W, \boldsymbol{X}_W, \boldsymbol{Y}_W, \boldsymbol{Z}_W\}$. With this setting the pose of the end-effector w.r.t $\mathscr{F}_W$ will be denoted by the vector $\boldsymbol{x}_\nu = [\boldsymbol{p}^T \boldsymbol{\nu}^T]^T \in SE(3)$ where $\boldsymbol{p} = [x\,y\,z]^T \in \mathbb{R}^3$ is the position and $\boldsymbol{\nu} = [\phi\,\theta\,\psi]^T \in SO(3)$ is the orientation expressed as Euler angles[1].

2.1 Kinematics

The kinematics of CDPR describe the relation between the pose $\boldsymbol{x}_\nu$ of the end-effector and the vector $\boldsymbol{\rho} = [\rho_1, \rho_2, \ldots, \rho_n]^T \in \mathbb{R}^{n \times 1}$ of lengths of the cables. In particular the forward kinematics (*FK*) describes the mapping from $\boldsymbol{\rho}$ to $\boldsymbol{x}_\nu$, whereas the inverse kinematics (*IK*) describes the inverse mapping. Both the IK and FK are relevant for the implementation of the controller and are briefly discussed in the following.

Inverse Kinematics

Thanks to the simplifying assumptions made so far, the IK of a fully constrained CDPR has a closed form solution. Formally, the mapping from $\boldsymbol{x}_\nu$ to the length ρ_i of each cable i can be written in the form of a polygonal constraint (see Fig. 1), i.e.

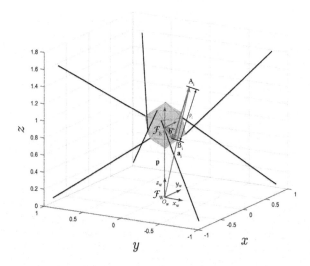

Fig. 1. Sketch of the cable robot.

---

[1] In the experiments we use the ZYX (roll-pitch-yaw) sequence.

$$\|^W l_i\| = \|\rho_i{}^W n_i\| = \rho_i = \|^W a_i - p - {}^W R_E{}^E b_i\|, \tag{1}$$

where $^W n_i$ is the support unit vector of the cable, $a_i \in \mathbb{R}^3$ is the position vector of A_i w.r.t. $\mathscr{F}_\mathscr{W}$, $^W R_E{}^E b_i = {}^W b_i \in \mathbb{R}^3$ is the vector $\overrightarrow{O_E B_i}$ expressed in $\mathscr{F}_\mathscr{W}$ and $^W R_E$ is the rotation matrix that expresses the orientation of $\mathscr{F}_\mathscr{E}$ w.r.t. $\mathscr{F}_\mathscr{W}$. Differentiating (1) with respect to time leads to the well-known differential equation

$$\dot{\rho} = \underbrace{\begin{bmatrix} -{}^W n_1^T & -({}^W R_E{}^E b_i \times {}^W n_i)^T \\ \vdots & \vdots \\ -{}^W n_n^T & -({}^W R_E{}^E b_n \times {}^W n_n)^T \end{bmatrix}}_{J} \dot{x}. \tag{2}$$

with the vector $\dot{x} = [p^T, {}^W \omega^T]^T$ describing the velocity of the rigid body end-effector by using angular velocities[2] and the Jacobian matrix J. It will be shown in Sect. 2.2 that the Jacobian J also plays a role in mapping forces exerted by the cables (cable tension) to wrenches applied to the end-effector.

Forward Kinematics

When considering the problem of tracking a trajectory of the end-effector using only the measurements given by the encoders at the winches (i.e. measurements of the lengths of the cables) we need to solve the FK problem to indirectly retrieve the current pose of the end-effector. Since the FK for a fully constrained CDPR in general does not have a unique solution [7] we address it as an optimization problem, i.e.

$$x_{FF}^\star = \min_{p^\star, \nu^\star} \Psi_{FF}(l, p, \nu) = \min_{p^\star, \nu^\star} \sum_{i=1}^n \{(\rho_i - \hat{\rho}_i(p, \nu))\} \tag{3}$$

in which the cost function $\Psi_{FF}(l, p, \nu)$ is the error between the measured cable lengths and the cable lengths from the kinematic relation (1), and $(\bullet)^\star$ denotes the optimal solution. It has been shown in [8,9] that problem 3 can be solved efficiently by using the method from levenberg-marquardt (lm), an iterative algorithm that updates the solution of the optimization problem with the following rule

$$x_{lm,k+1} = x_{lm,k} + h_{lm,k}, \tag{4}$$

where $h_{lm,k}$ is the update computed at the step k. The update step $h_{lm,k}$ in (4) is determined by solving the equation

$$\left(J(x_{lm,k})^T J(x_{lm,k}) + \lambda_{lm} I_n\right) h_{lm,k} = J(x_{lm,k})^T (\rho - \hat{\rho}(x_k)). \tag{5}$$

Note that in (5) the term λ_{lm} is a damping factor that modifies the singular values of the matrix $\left(J(x_{lm,k})^T J(x_{lm,k}) + I_n\right)$ to make it better conditioned close to singularities of J. The algorithm stops when the update step becomes small enough, i.e. when $\|x_{lm,k+1} - x_{lm,k}\| < \epsilon_{lm}$ with a predefined threshold

---

[2] Note that $\dot{x} \neq \frac{d}{dt} x_\nu$ since $^W \omega \neq \dot{\nu}$.

$\epsilon_{lm} \in \mathbb{R}$. In practice we found that an acceptable solution would be usually found in less than 20 iterations by using the previous solution as initial guess for the optimization [12].

2.2 Dynamics

The dynamics of the CDPR provide a relation between the tensions of the cables and their effect onto the motion of the end-effector. Given a nominal mass $m \in \mathbb{R}$, a nominal inertia $I_E \in \mathbb{R}^{3\times3}$ w.r.t. $\mathscr{F}_E$ and the position of the center of mass $^W\boldsymbol{c} = [c_x, c_y, c_z]^T$ in $\mathscr{F}_W$ and using either the Newton-Euler or Euler-Lagrange approach, the dynamics of the system are expressed by a second order ordinary differential equation

$$B(\boldsymbol{x}_\nu)\ddot{\boldsymbol{x}} + C(\boldsymbol{x}_\nu, \dot{\boldsymbol{x}})\dot{\boldsymbol{x}} - \boldsymbol{g}(\boldsymbol{x}_\nu) = \boldsymbol{u} = -J^T\boldsymbol{t} \tag{6}$$

$$B(\boldsymbol{x}_\nu) = \begin{bmatrix} mI_3 & m\,^W\boldsymbol{c}\times^T \\ m\,^W\boldsymbol{c}\times & H \end{bmatrix}, \tag{7}$$

$$C(\boldsymbol{x}_\nu, \dot{\boldsymbol{x}})\dot{\boldsymbol{x}} = \begin{bmatrix} m\,^W\boldsymbol{\omega}\times\,^W\boldsymbol{\omega}\times\,^W\boldsymbol{c} \\ ^W\boldsymbol{\omega}\times H\,^W\boldsymbol{\omega} \end{bmatrix}, \tag{8}$$

$$\boldsymbol{g}(\boldsymbol{x}_\nu) = \begin{bmatrix} 0 & 0 & -mg & -mc_yg & mc_xg & 0 \end{bmatrix}^T \tag{9}$$

$$H = {^WR_E}\,{^EI_E}\,{^ER_W} + m\,{^W}\boldsymbol{c}\times\,{^W}\boldsymbol{c}\times^T. \tag{10}$$

where $\boldsymbol{t} \in \mathbb{R}^{8\times1}$ is the vector of cable tensions, $\boldsymbol{u} \in \mathbb{R}^{6\times1}$ is the wrench applied to the end-effector and $(\bullet)\times$ is the well-known cross-product operator.

To formulate the control we need to rewrite the model (6) to (10) in regular form. For this purpose, let us first denote the full state of the system as the 12×1 vector $\bar{\boldsymbol{x}} = [\boldsymbol{x}_\nu^T\ \dot{\boldsymbol{x}}^T]^T = [\boldsymbol{p}^T\ \boldsymbol{\nu}^T\ \dot{\boldsymbol{p}}^T\ ^W\boldsymbol{\omega}^T]^T$ and the output as $\boldsymbol{x}_\nu = [\boldsymbol{p}^T\ \boldsymbol{\nu}^T]$. Now we define the diffeomorphism Ω that brings the system in regular form as

$$\boldsymbol{z} = \begin{bmatrix} \boldsymbol{z}_1 \\ \boldsymbol{z}_2 \end{bmatrix} = \Omega(\bar{\boldsymbol{x}}) = \begin{bmatrix} I_6 & 0_{6\times6} \\ 0_{6\times6} & \underbrace{\begin{bmatrix} I_3 & 0_{3\times3} \\ 0_{3\times3} & ^\nu E_\omega \end{bmatrix}}_{A(\boldsymbol{x}_\nu)} \end{bmatrix} \bar{\boldsymbol{x}}. \tag{11}$$

Finally, applying (11) and using the new state vector $\bar{\boldsymbol{x}}$ on (6) we obtain the regular form

$$\begin{cases} \dot{\boldsymbol{z}}_1 = \boldsymbol{z}_2 \\ \dot{\boldsymbol{z}}_2 = \boldsymbol{f}(\bar{\boldsymbol{x}}) + h(\bar{\boldsymbol{x}})\boldsymbol{u} \end{cases} \tag{12}$$

with

$$\boldsymbol{f}(\bar{\boldsymbol{x}}) = -A(\boldsymbol{x}_\nu)B^{-1}(\boldsymbol{x}_\nu)\left[C(\boldsymbol{x}_\nu, \dot{\boldsymbol{x}})\dot{\boldsymbol{x}} - \boldsymbol{g}(\boldsymbol{x}_\nu)\right] \tag{13}$$
$$+ \dot{A}(\boldsymbol{x}_\nu, \dot{\boldsymbol{x}})\dot{\boldsymbol{x}}$$
$$h(\bar{\boldsymbol{x}}) = A(\boldsymbol{x}_\nu)B^{-1}(\boldsymbol{x}_\nu) \tag{14}$$

Remark 1. *To ensure that $^{\nu}E_{\omega}$ is not singular and $\boldsymbol{h}(\tilde{\boldsymbol{x}})$ has rank m, we limit the pitch angle θ to $\left(-\frac{\pi}{2}, \frac{\pi}{2}\right)$.*

As mentioned earlier, this model is a simplified representation of reality and it may contain unmodelled effects such as slip-stick effects, breakaway torque, etc. The model (12) can be extended to incorporate model uncertainties and external wrench disturbances $\boldsymbol{\zeta}$ as follows

$$\begin{cases} \dot{z}_1 &= z_2 \\ \dot{z}_2 &= \boldsymbol{f}_n(\tilde{\boldsymbol{x}}) + \Delta \boldsymbol{f}(\tilde{\boldsymbol{x}}) + \boldsymbol{h}_n(\tilde{\boldsymbol{x}}) \cdot (\boldsymbol{u} + \boldsymbol{\zeta}) + \Delta \boldsymbol{h}(\tilde{\boldsymbol{x}}) \cdot (\boldsymbol{u} + \boldsymbol{\zeta}) \\ &= \boldsymbol{f}_n(\tilde{\boldsymbol{x}}) + \boldsymbol{h}_n(\tilde{\boldsymbol{x}})\boldsymbol{u} + \boldsymbol{\xi} \end{cases} \quad (15)$$

where

- $\boldsymbol{f}_n$ and $\boldsymbol{h}_n$ indicate the nominal model of the robot;
- $\Delta(\bullet)$ contains the unmodelled effects and parameter uncertainties;
- $\boldsymbol{\xi} = \boldsymbol{h}_n(\tilde{\boldsymbol{x}})\boldsymbol{\zeta} + \Delta \boldsymbol{f}(\tilde{\boldsymbol{x}}) + \Delta \boldsymbol{h}(\tilde{\boldsymbol{x}}) \cdot (\boldsymbol{u} + \boldsymbol{\zeta})$ is the vector of lumped perturbations.

As mentioned in Sect. 1, in nominal working conditions the lumped perturbations of a physical system reasonably have an upper bound but this upper bound is not known. Thus, we make the following assumption

Assumption 1. *$\boldsymbol{\xi}$ is bounded by an unknown upper bound ξ_{max} such that $0 \leq |\boldsymbol{\xi}|_2 \leq \xi_{max}$.*

3 Control and State Differentiation

Using the model developed in Sect. 2, we can now design the control input $\boldsymbol{u}$ in (15) to let the end-effector track a desired trajectory $\boldsymbol{z}_{1,d} = [\boldsymbol{p}_d^T \ \boldsymbol{\nu}_d^T]^T \in SE(3)$ that is provided with its derivatives $\boldsymbol{z}_{2,d} = [\dot{\boldsymbol{p}}_d^T \ \dot{\boldsymbol{\nu}}_d^T]^T$ and $\dot{\boldsymbol{z}}_{2,d} = [\ddot{\boldsymbol{p}}_d^T \ \ddot{\boldsymbol{\nu}}_d^T]^T$, assuming also the presence of perturbations. The solution that we propose combines a feed-forward control action $\boldsymbol{u}_{FF}$ based on the inversion of the nominal dynamic model, and a term based on a robust sliding-mode controller. The full control input then becomes:

$$\boldsymbol{u} = \boldsymbol{u}_{SM} + \boldsymbol{u}_{FF}. \quad (16)$$

where $\boldsymbol{u}_{FF}$ is a feed-forward input and $\boldsymbol{u}_{SM}$ is the control action of the sliding-mode controller.

Before venturing in the details of the terms forming the control wrench $\boldsymbol{u}$ we must point out that not all wrenches are feasible, due to the possible presence of singularities in the Jacobian $\boldsymbol{J}$ and to the fact that the tensions $\boldsymbol{t}$ that can be exerted by the cables are limited within a non-negative range[3], i.e. $0 \leq \boldsymbol{t}_{min} \leq \boldsymbol{t} \leq \boldsymbol{t}_{max}$. Hereinafter we will simply assume that $\boldsymbol{u}$ is within the

---

[3] The cables can only pull the end-effector, not push it. Hence the tension in a cable can never be negative.

admissible range of wrenches that defines the wrench-feasible-workspace of the robot. Nevertheless, in the next section it will be shown that the sliding-mode control u_{SM} embeds a saturation that can be tuned according to the feasible range of wrenches. Due to lack of space, we will not perform an analysis of the range of feasible wrenches for our robot, however the experiment presented in Sect. 4 will demonstrate that with suitable tuning of the controller and a reasonable desired trajectory the control wrench u is always feasible. Lastly, the cable tensions that need to be commanded by the motors to implement u can be safely obtained using one of the many tension distribution algorithms that are available in literature (e.g. [6,10,13]).

3.1 Control Design

ASTC: To steer the tracking error $e = z_{1,d} - z_1$ and its derivative $\dot{e} = z_{2,d} - z_2$ to zero we choose as sliding variable $\sigma = \dot{e} + \Lambda e$ where Λ is a positive definite matrix of proper size. To achieve $\sigma = \dot{\sigma} = 0$ with σ we implement u_{SM} as a super-twisting controller [17]. Moreover, to overcome the disadvantages of constant high chosen gains, we use the adaptation law proposed by Shtessel in [18]. The full expression of u_{SM} is given by:

$$u_{SM} = -\alpha |\sigma|^{\frac{1}{2}} \operatorname{sign}(\sigma) + v \tag{17}$$

$$\dot{v} = \begin{cases} -u_{SM} & \text{if } |u_{SM}| > \overline{u} \\ -\beta \operatorname{sign}(\sigma) & \text{if } |u_{SM}| \leq \overline{u} \end{cases} \tag{18}$$

$$\dot{\alpha} = \begin{cases} \omega_\alpha \sqrt{\frac{\gamma}{2}} \operatorname{sign}(|\sigma| - \mu), & \text{if } \alpha > \alpha_m \\ \eta, & \text{if } \alpha \leq \alpha_m \end{cases} \tag{19}$$

$$\beta = 2\epsilon\alpha \tag{20}$$

$$\mu(t) = 4\alpha(t) T_e, \tag{21}$$

where

- $\overline{u}, \underline{u}$ are the upper and lower bound for u_{SM};
- α, β are positive definite diagonal matrices of gains;
- $\omega_\alpha, \gamma, \eta, \alpha_m$ are positive constants that determine the update rate of the gains (we refer the reader to [18] for more details;
- T_e is the sampling period for the controller.

Feedforward: One important characteristic of the ASTC is that the input u_{SM} is continuous, since the discontinuity in (18) is passed through an integrator. While this helps to reduce numerical chattering, this effect is not completely removed. The feedforward input u_{FF} in (16) is meant to reduce the control effort of the ASTC and thus also reduce chattering and noise amplification. The term u_{FF} is based on dynamic inversion of the nominal model, using the reference trajectory $x_\nu d$, $\dot{x}_d$ and $\ddot{x}_d$. Formally, inverting the dynamics presented in Sect. 2.2 yields

$$u_{FF} = u = B(x_{\nu,d})\ddot{x}_d + C(x_{\nu,d}, \dot{x}_d)\dot{x}_d - g(x_{\nu,d}) \tag{22}$$

3.2 Sliding-Mode Differentiator

In order to implement the control input u_{SM} we need to measure both z_1 and z_2 or, equivalently, the pose x_ν and velocity $\dot{x}$ of the end-effector and use (11). For the the pose x_ν, as explained in Sect. 2.1, we can retrieve it from the encoders measurements by solving the FK problem. However, to obtain the velocity $\dot{x}$ from the positional encoders we must resort to a numerical differentiator. Hence, using the same principles of the ASTC, we implemented a first order sliding mode differentiator [17]. The advantage of this family of differentiators is that it achieves finite time convergence and exact stabilization, besides having the robustness properties of a sliding mode algorithm. For detailed information on sliding mode differentiators and their properties we refer the reader to [17]. Here we just show the final equation of the differentiator, i.e.

$$\dot{\tilde{x}}_0 = -\lambda_1 L^{\frac{1}{4}} \|\tilde{x}_0 - x_\nu\|^{\frac{1}{2}} \text{sign}(\tilde{x}_0 - x_\nu) + \tilde{x}_1$$
$$\dot{\tilde{x}}_1 = -\lambda_0 L \text{sign}(\tilde{x}_0 - x_\nu)$$
(23)

where

- $\tilde{x}_0$ and $\tilde{x}_1$ are the estimates of x_ν and $\dot{x}_\nu$, respectively[4].
- λ_1 and λ_2 are the gains of the differentiator. As suggested in [17], we tuned these gains as $\lambda_1 = 1.1$ and $\lambda_2 = 1.3$.
- L is the Lipschitz constant of the last derivative estimated by the differentiator, in this case $\dot{x}_\nu$. In practice this value determines the cutoff frequency of the differentiator. We used the value $L = 15$.

4 Experimental Validation

The robot used to validate the proposed control algorithm is a redundant CDPR with 6 dof that was custom built at the Max Planck Institute for Biological Cybernetics (see Fig. 2). The end-effector of the robot is an icosahedron platform made of aluminium. The nominal mass of the end-effector is $m = 2.6$ kg and its nominal inertia tensor with respect to the principal axes is $^E I_E = diag\{[0.04613, 0.04613, 0.04873]\}$ Nm². The system is powered by eight 940 W compact servo drives of type Beckhoff AM8033-0E21 with a drum diameter of 39.15 mm, that are located under the metal base of the frame. These actuators allow to exert a maximum cable force of 878 N.

On the software side, we used an off-the-shelves Beckhoff TwinCAT software (*Windows Control and Automation Technology*) to control the robot and interface with the sensors. TwinCAT offers the possibility to run simultaneously multiple Programmable Logic Controllers (*PLC*) in real-time and it provides both a runtime environment for real time NC-axis control and a programming environment for code development. Moreover, this software allows to implement

---

[4] Note that the differentiator estimates $\dot{x}_\nu$, however the transformation from $\dot{x}_\nu$ to $\dot{x}$ is straightforward.

Fig. 2. Mini CableRobot simulator.

the low-level control of the drives both giving a target cable length (position control) or a target torque (torque control). In our experiment we used this second mode of operation because the control input from our ASTC algorithm is the applied wrench, i.e. the tension of the cables.

Visible in Fig. 2 is also a set of near-infrared cameras (we have four in total) from VICON[5] with a sampling frequency of 250Hz that can be used to track the pose of the platform with a precision of 0.5mm thanks to a set of reflective markers mounted on the top of the end-effector. We want to stress that the VICON tracking system was not used in the experiments, during which we only used encoder measurements. This is due to the fact that an external vision based tracking system like the one from VICON is rarely used with large size CDPR like our CableMotion Simulator [11]. The VICON tracking system has only been used before the experiment to calibrate the kinematic parameters a_i and b_i in (1). To perform the calibration the end-effector was moved throught the operational workspace, taking care to excite as much as possible the full range of translations and rotations, meanwhile recording both the length measurements ρ from the encoders and the pose meaurement x_ν from the visual tracker. With these measurements the kinematic parameters a_i and b_i for the generic i-th cable are calibrated by solving a quadratic minimization problem over all N measurement samples, i.e.

$$\begin{bmatrix} a_i^\star \\ b_i^\star \end{bmatrix} = \min_{a_i, b_i} \left\{ \sum_{j=1}^{N} \rho_{i,j} - \|{}^W a_i - p - {}^W R_E {}^E b_i\|^2 \right\} \quad \text{s.t.} \quad a_i \in [\underline{a}, \overline{a}], \ b_i \in [\underline{b}, \overline{b}] \tag{24}$$

where $\underline{a}$, $\overline{a}$ and $\underline{b}$, $\overline{b}$ are the boundaries of the search region. The results are reported in Table 1.

---

[5] http://vicon.com/.

Table 1. Kinematic parameters

	a_1	a_2	a_3	a_4	a_5	a_6	a_7	a_8	b_1	b_2	b_3	b_4	b_5	b_6	b_7	b_8
x [m]	-0.907	-0.853	0.911	0.871	0.907	0.860	-0.900	-0.870	-0.127	-0.206	0.014	0.134	0.135	0.008	-0.143	-0.220
y [m]	0.871	0.965	0.820	0.918	-0.845	-0.928	-0.829	-0.935	0.177	0.065	0.239	0.175	-0.186	-0.241	-0.206	-0.082
z [m]	1.640	0.0675	1.590	0.079	1.610	0.079	1.600	0.063	0.115	-0.111	-0.102	0.122	-0.101	0.114	-0.092	0.107

4.1 Results

To validate the proposed ASTC algorithm we performed an experiment in which the end-effector is tasked to follow the predefined trajectory $\boldsymbol{p}_d = [0.15 + 0.15cos(\omega \cdot t - \frac{\pi}{2}), -0.145 + 0.15 + 0.15sin(\omega \cdot t - \frac{\pi}{2}), 0.85]$ m, $\boldsymbol{\nu}_d = [0, 0, 0]$ ° with $\omega = 0.5$. The parameters of the control algorithm (17) to (19) and (21) were set to $\epsilon = 0.3$, $\gamma = 1$, $\omega_\alpha = 21.213$.

In order to use the ASTC control law we first need to retrieve the measurement of the state of the system, i.e. the pose $\boldsymbol{x}_\nu$ and velocity $\dot{\boldsymbol{x}}$ of the end-effector. Recalling our previous discussion, the pose can be reconstructed from the encoder measurements ρ by solving the FK problem whereas the velocity, not being directly available, is obtained from the Sliding-Mode Differentiator (SMD) (23). The finite-time convergence of the differentiator is confirmed by Figs. 3a and b where it is visible that the pose from the SMD rapidly converges to the pose from the FK. For the velocity, lacking a measurement of $\dot{\boldsymbol{x}}$ to compare the output of the SMD with, we computed the numerical derivative of the encoder measurement ρ and then used the differential kinematics (2) to get $\dot{\boldsymbol{x}}$. Using this as a comparison, Figs. 3c and d show that, for our choice of parameters in Sect. 3.2, the velocity estimate from the SMD follows the mean of the numerical derivative but filters noise.

Having assessed the correct operation of the SMD we can now look at the performance of the ASTC, its robustness in particular. In this regard, the controller has to face three different kinds of perturbations. Firstly, parameter uncertainties, because the nominal physical parameters of the end-effector have been derived from an approximated CAD model and therefore it is reasonable to expect deviations from the real values. Secondly, unmodelled dynamics, since in our model we considered an extremely simplified power-train with ideal cables and without transients or other effects in the drives. Lastly, external disturbances, because to make the task more challenging we applied external wrenches to the platform while it was moving. More specifically:

1. From 55s to 61s a 5 kg disk ($\approx$ 192% the mass of the end-effector) was hooked under the platform, approximately equivalent to a disturbance wrench $w_{d,1} \approx [0, 0, -50N, 0, 0, 0]^T$.
2. From 116s to 127s the platform was pulled in approximately the $\mathbf{X}_W$ direction with a wrench (measured using a mass sensor) of $w_{d,2} \approx [-100N, 0, 0, 0, 0, 0]^T$ ($\approx$ 384% the mass of the end-effector).
3. From 143s to 155s the platform was pulled in approximately the $\mathbf{X}_W$ direction with a wrench (measured using a mass sensor) of $w_{d,3} = [100N, 0, 0, 0, 0, 0]^T$ ($\approx$ 384% the mass of the end-effector).

With this premise in mind, we can first look at the tracking error shown in Figs. 3e and f. The first thing worth noting in these plots is that despite the model uncertainties the end-effector converges extremely quickly (less than half a second) to the desired trajectory, thus providing a practical proof of the

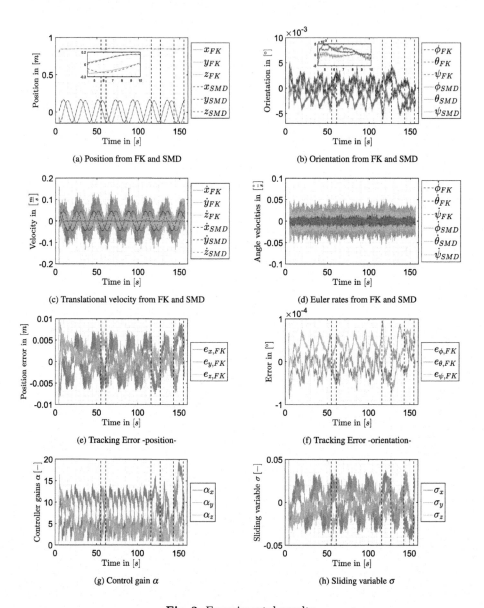

Fig. 3. Experimental results

finite-time convergence of the ASTC. Additionally, it is remarkable that throughout the whole experiment the tracking error remains in the same small range, even when the external disturbances are applied, confirming that the system is very robust and capable to quickly counteract even significant perturbations. Such a reactive behaviour is achieved thanks to the adaptation law (19). Indeed, the graph of the evolution of the controller gains α (Fig. 3g) reveals that in correspondence to the external disturbances the gains along the affected direction of motion increase very quickly and decrease as quickly once the disturbance is removed.

Another reason for including the adaptive gains, as explained earlier, was to reduce the effect of chattering that is common to sliding-mode controllers based on the sign() function. However, the evolution of the sliding variable σ (see Fig. 3h) shows that the chattering is still significant. The reason for this might be partially due to the fact that the parameters of the controller, in particular the saturation thresholds in the adaptation law, were not well tuned. While we do not have yet a decisive explanation to this issue, we can speculate that the dominant cause of the chattering be unmodelled dynamics of the drive train (e.g. friction, slip-stick effects, hysteresis, creeping, inhomogeneous materials, elasticity, thermal expansion and flattening of cables) or changes in the cable configuration [8] that inject vibrations in the system. For example, we think that static friction had a significant effect because we needed to apply torques much larger than expected to start moving the robot. In conclusion, this experiment suggests that further investigation is needed to quantify and understand the cause of the chattering, i.e. whether it is due to the numerical implementation of the algorithm or to the aforementioned unmodelled effects.

5 Conclusion

In this paper we presented preliminary experimental validation of a Super-Twisting sliding mode controller with adaptive gains applied to a redundant CDPR with eight cables and six degrees of freedom. The experiment showed promising characteristics of the ASTC, in terms of tracking accuracy and robustness to significant perturbations, particularly a remarkable reactivity to sudden disturbances. At the same time, the experiment also raised questions about the presence of chattering. From the preliminary results in this paper we think that unmodelled dynamics of the power train (e.g. friction [5]) or changes in the cable configuration ([8]) might be connected to this effect, but further investigation is required. Additionally, we plan to explore possible improvements of the ASTC by combining it with interconnection-damping-assignment.

References

1. Babaghasabha, R., Khosravi, M.A., Taghirad, H.D.: Adaptive control of KNTU planar cable-driven parallel robot with uncertainties in dynamic and kinematic parameters. In: Pott, A., Bruckmann, T. (eds.) Cable-Driven Parallel Robots. Mechanisms and Machine Science, vol. 32, pp. 145–159. Springer (2015)
2. Cone, L.L.: Skycam-an aerial robotic camera system. Byte **10**(10), 122 (1985)
3. Duan, B., Qiu, Y., Zhang, F., Zi, B.: Analysis and experiment of the feed cable-suspended structure for super antenna. In: 2008 IEEE/ASME International Conference on Advanced Intelligent Mechatronics, pp. 329–334 (2008)
4. El-Ghazaly, G., Gouttefarde, M., Creuze, V.: Adaptive terminal sliding mode control of a redundantly-actuated cable-driven parallel manipulator: CoGiRo. In: Pott, A., Bruckmann, T. (eds.) Cable-Driven Parallel Robots. Mechanisms and Machine Science, vol. 32, pp. 179–200. Springer (2015)
5. Kraus, W., Kessler, M., Pott, A.: Pulley friction compensation for winch-integrated cable force measurement and verification on a cable-driven parallel robot. In: 2015 IEEE International Conference on Robotics and Automation (ICRA), pp. 1627–1632. IEEE (2015)
6. Lamaury, J., Gouttefarde, M., Chemori, A., Herve, P.E.: Dual-space adaptive control of redundantly actuated cable-driven parallel robots. In: 2013 IEEE/RSJ International Conference on Intelligent Robots and Systems, pp. 4879–4886 (2013)
7. Merlet, J.P.: Solving the forward kinematics of a gough-type parallel manipulator with interval analysis. Int. J. Robot. Res. **23**(3), 221–235 (2004)
8. Merlet, J.P.: Checking the cable configuration of cable-driven parallel robots on a trajectory. In: 2014 IEEE International Conference on Robotics and Automation (ICRA), pp. 1586–1591. IEEE (2014)
9. Miermeister, P., Kraus, W., Pott, A.: Differential kinematics for calibration, system investigation, and force based forward kinematics of cable-driven parallel robots. In: Bruckmann, T., Pott, A. (eds.) Cable-Driven Parallel Robots. Mechanisms and Machine Science, vol. 12, pp. 319–333. Springer, Heidelberg (2013)
10. Mikelsons, L., Bruckmann, T., Hiller, M., Schramm, D.: A real-time capable force calculation algorithm for redundant tendon-based parallel manipulators. In: 2008 IEEE International Conference on Robotics and Automation, pp. 3869–3874 (2008)
11. Philipp, M., Masone, C., Heinrich, H., Joachim, T.: The cablerobot simulator - large scale motion platform based on cable robot technology. In: 2016 IEEE/RSJ International Conference on Intelligent Robots and Systems (2016)
12. Pott, A.: An algorithm for real-time forward kinematics of cable-driven parallel robots. In: Pott, A. (ed.) Advances in Robot Kinematics: Motion in Man and Machine, pp. 529–538. Springer (2010)
13. Pott, A., Bruckmann, T., Mikelsons, L.: Closed-form force distribution for parallel wire robots. In: Computational Kinematics, pp. 25–34. Springer (2009)
14. Schenk, C., Bülthoff, H., Masone, C.: Robust adaptive sliding mode control of a redundant cable driven parallel robot. In: International Conference on System Theory, Control and Computing, pp. 427–434 (2015)
15. Schenk, C., Miermeister, P., Masone, C., Bülthoff, H.: Modeling and analysis of cable vibrations on a cable-driven parallel robot. In: ICIAIEEE International Conference on Information and Automation, pp. 427–434 (2017)
16. Schmidt, V., Kraus, W., Pott, A.: Presentation of experimental results on stability of a 3 DOF 4-cable-driven parallel robot without constraints. In: Bruckmann, T., Pott, A. (eds.) Cable-Driven Parallel Robots. Mechanisms and Machine Science, vol. 12, pp. 87–99. Springer, Heidelberg (2013)

17. Shtessel, Y., Edwards, C., Fridman, L., Levant, A.: Sliding Mode Control and Observation. Springer, New York (2014)
18. Shtessel, Y.B., Taleb, M., Plestan, F.: A novel adaptive-gain supertwisting sliding mode controller: Methodology and application. Automatica **48**(5), 759–769 (2012)
19. Surdilovic, D., Bernhardt, R.: String-man: a new wire robot for gait rehabilitation. In: IEEE International Conference on Robotics and Automation, New Orleans, Louisiana, pp. 2031–2036 (2004)

Tension Distribution Algorithm for Planar Mobile Cable-Driven Parallel Robots

Tahir Rasheed[1], Philip Long[2], David Marquez-Gamez[3], and Stéphane Caro[4(✉)]

[1] École Centrale de Nantes, Laboratoire des Sciences du Numérique de Nantes,
UMR CNRS 6004, 1, Rue de la Noë, 44321 Nantes, France
Tahir.Rasheed@ls2n.fr, Tahir.Rasheed@irccyn.ec-nantes.fr

[2] RIVeR Lab, Department of Electrical and Computing Engineering,
Northeastern University, Boston, USA
philip.long@northeastern.edu

[3] IRT Jules Verne, Chemin du Chaffault, 44340 Bouguenais, France
david.marquez-gamez@irt-jules-verne.fr

[4] CNRS, Laboratoire des Sciences du Numérique de Nantes,
UMR CNRS 6004, 1, Rue de la Noë, 44321 Nantes, France
stephane.caro@ls2n.fr, stephane.caro@irccyn.ec-nantes.fr

Abstract. Cable-Driven Parallel Robots (CDPRs) contain numerous advantages over conventional manipulators mainly due to their large workspace. Reconfigurable Cable-Driven Parallel Robots (RCDPRs) can increase the workspace of classical CDPRs by modifying the geometric architecture based on the task feasibility. This paper introduces a novel concept of RCDPR, which is a Mobile CDPR (MCDPR) mounted on multiple mobile bases allowing the system to autonomously reconfigure the CDPR. A MCDPR composed of two mobile bases and a planar CDPR with four cables and a point mass is studied as an illustrative example. As the mobile bases containing the exit points of the CDPR are not fixed to the ground, the static and dynamic equilibrium of the mobile bases and the moving-platform of the MCDPR are firstly studied. Then, a real time Tentensions onto the mobilesion Distribution Algorithm (TDA) that computes feasible and continuous cable tension distribution while guaranteeing the static stability of mobile bases and the equilibrium of the moving-platform of a $n = 2$ Degree of Freedom (DoF) CDPR driven by n+2 cables is presented.

Keywords: Cable-Driven Parallel Robot · Mobile robot · Reconfigurability · Tension Distribution Algorithm · Equilibrium

1 Introduction

A Cable-Driven Parallel Robot (CDPR) is a type of parallel robot whose moving-platform is connected to the base with cables. The lightweight properties of the CDPR makes them suitable for multiple applications such as constructions [1,10], industrial operations [3], rehabilitation [11] and haptic devices [4].

A general CDPR has a fixed cable layout, *i.e.* fixed exit points and cable configuration. This fixed geometric structure may limit the workspace size of the manipulator due to cable collisions and some extrernal wrenches that cannot be accepted due to the robot configuration. As there can be several configurations for the robot to perform the prescribed task, an optimized cable layout is required for each task considering an appropriate criterion. Cable robots with movable exit and/or anchor points are known as Reconfigurable Cable-Driven Parallel Robots (RCDPRs). By appropriately modifying the geometric architecture, the robot performance can be improved e.g. lower cable tensions, larger workspace and higher stiffness. The recent work on RCDPR [2,3,9,12,15] proposed different design strategies and algorithms to compute optimized cable layout for the required task, while minimizing appropriate criteria such as the robot energy consumption, the robot workspace size and the robot stiffness. However, for most existing RCDPRs, the reconfigurability is performed either discrete and manually or continuously, but with bulky reconfigurable systems.

This paper deals with the concept of Mobile Cable-Driven Parallel Robots (MCDPRs). The idea for introducing MCDPRs is to overcome the manual and discrete reconfigurability of RCDPRs such that an autonomous reconfiguration can be achieved. A MCDPR is composed of a classical CDPR with m cables and a n degree-of-freedom (DoF) moving-platform mounted on p mobile bases. Mobile bases are four-wheeled planar robots with two-DoF translational motions and one-DoF rotational motion. A concept idea of a MCDPR is illustrated in Fig. 1 with $m = 8$, $n = 6$ and $p = 4$. The goal of such system is to provide a low cost and versatile robotic solution for logistics using a combination of mobile bases and CDPR. This system addresses an industrial need for fast pick and place operations while being easy to install, keeping existing infrastructures and covering large areas. The exit points for the cable robot is associated with the position of its respective mobile bases. Each mobile base can navigate in the environment thus allowing the system to alter the geometry of the CDPR. Contrary to classical CDPR, equilibrium for both the moving-platform and the mobile bases should be considered while analyzing the behaviour of the MCDPR.

A Planar Mobile Cable-Driven Parallel Robot with four cables ($m = 4$), a point mass ($n = 2$) and two mobile bases ($p = 2$), shown in Fig. 2, is considered throughout this paper as an illustrative example. This paper is organized as follows. Section 2 presents the static equilibrium conditions for mobile bases using the free body diagram method. Section 3 introduces a modified real time Tension Distribution Algorithm (TDA), which takes into account the dynamic equilibrium of the moving-platform and the static equilibrium of the mobile bases. Section 4 presents the comparison between the existing and modified TDA on the equilibrium of the MCDPR under study. Finally, conclusions are drawn and future work is presented in Sect. 5.

2 Static Equilibrium of Mobile Bases

This section aims at analyzing the static equilibrium of the mobile bases of MCDPRs. As both the mobile bases should be in equilibrium during the motion

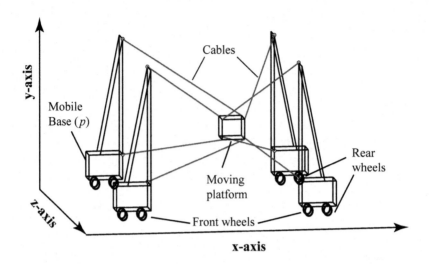

Fig. 1. Concept idea for Mobile Cable-Driven Parallel Robot (MCDPR) with eight cables ($m = 8$), a six degree-of-freedom moving-platform ($n = 6$) and four mobile bases ($p = 4$)

of the end-effector, we need to compute the reaction forces generated between the ground and the wheels of the mobile bases. Figure 2 illustrates the free body diagram for the jth mobile base. $\mathbf{u}_{ij}$ denotes the unit vector of the ith cable attached to the jth mobile base, $i, j = 1, 2$. $\mathbf{u}_{ij}$ is defined from the point mass P of the MCDPR to the exit point A_{ij}. Using classical equilibrium conditions for the jth mobile base p_j, we can write:

$$\sum \mathbf{f} = 0 \Rightarrow m_j \mathbf{g} + \mathbf{f}_{1j} + \mathbf{f}_{2j} + \mathbf{f}_{r1j} + \mathbf{f}_{r2j} = 0 \quad (1)$$

All the vectors in Eq. (1) are associated with the superscript x and y for respective horizontal and vertical axes. Gravity vector is denoted as $\mathbf{g} = [0 \ -g]^T$ where g = 9.8 m.s^{-2}, $\mathbf{f}_{1j} = [f_{1j}^x \ f_{1j}^y]^T$ and $\mathbf{f}_{2j} = [f_{2j}^x \ f_{2j}^y]^T$ are the reaction forces due to cable tensions onto the mobile base p_j, C_{1j} and C_{2j} are the front and rear wheels contact points having ground reaction forces $\mathbf{f}_{r1j} = [f_{r1j}^x \ f_{r1j}^y]^T$ and $\mathbf{f}_{r2j} = [f_{r2j}^x \ f_{r2j}^y]^T$, respectively. In this paper, wheels are assumed to be simple support points and the friction between those points and the ground is supposed to be high enough to prevent the mobile bases from sliding. The moment at a point O about z-axis for the mobile base to be in equilibrium is expressed as:

$$M_O^z = 0 \Rightarrow \mathbf{g}_j^T \mathbf{E}^T m_j \mathbf{g} + \mathbf{a}_{1j}^T \mathbf{E}^T \mathbf{f}_{1j} + \mathbf{a}_{2j}^T \mathbf{E}^T \mathbf{f}_{2j} + \mathbf{c}_{1j}^T \mathbf{E}^T \mathbf{f}_{r1j} + \mathbf{c}_{2j}^T \mathbf{E}^T \mathbf{f}_{r2j} = 0 \quad (2)$$

with

$$\mathbf{E} = \begin{bmatrix} 0 & -1 \\ 1 & 0 \end{bmatrix} \quad (3)$$

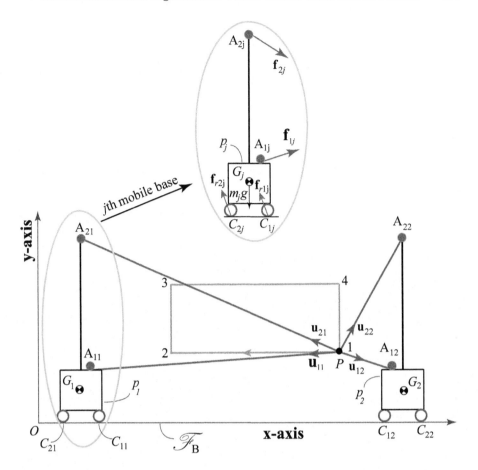

Fig. 2. Point mass Mobile Cable-Driven Parallel Robot with $p = 2$, $n = 2$ and $m = 4$ (Color figure online)

$\mathbf{a}_{1j} = [a_{1j}^x \ a_{1j}^y]^T$ and $\mathbf{a}_{2j} = [a_{2j}^x \ a_{2j}^y]^T$ denote the Cartesian coordinate vectors of the exit points A_{1j} and A_{2j}, $\mathbf{c}_{1j} = [c_{1j}^x \ c_{1j}^y]^T$ and $\mathbf{c}_{2j} = [c_{2j}^x \ c_{2j}^y]^T$ denote the Cartesian coordinate vectors of the contact points C_{1j} and C_{2j}. $\mathbf{g}_j = [g_j^x \ g_j^y]^T$ is the Cartesian coordinate vector for the center of gravity G_j of the mobile base p_j. The previous mentioned vector are all expressed in the base frame $\mathcal{F}_B$. Solving simultaneously Eqs. (1) and (2), the vertical components of the ground reaction forces take the form:

$$f_{r1j}^y = \frac{m_j g(c_{2j}^x - g_j^x) + f_{1j}^y(a_{1j}^x - c_{2j}^x) + f_{2j}^y(a_{2j}^x - c_{2j}^x) - f_{1j}^x a_{1j}^y - f_{2j}^x a_{2j}^y}{c_{2j}^x - c_{1j}^x} \quad (4)$$

$$f_{r2j}^y = m_j g - f_{1j}^y - f_{2j}^y - f_{r1j}^y \quad (5)$$

Equations (4) and (5) illustrate the effect of increasing the external forces (cable tensions) onto the mobile base. Indeed, the external forces exerted onto the mobile base may push the latter towards frontal tipping. It is apparent that the higher the cable tensions, the higher the vertical ground reaction force $f^y_{r_{1j}}$ and the lower the ground reaction force $f^y_{r_{2j}}$. There exists a combination of cable tensions such that $f^y_{r_{2j}} = 0$. At this instant, the rear wheel of the jth mobile base will lose contact with the ground at point C_{2j}, while generating a moment M_{C1j} about z-axis at point C_{1j}:

$$M^z_{C1j} = (\mathbf{g}_j - \mathbf{c}_{1j})^T \mathbf{E}^T m_j \mathbf{g} + (\mathbf{a}_{1j} - \mathbf{c}_{1j})^T \mathbf{E}^T \mathbf{f}_{1j} + (\mathbf{a}_{2j} - \mathbf{c}_{1j})^T \mathbf{E}^T \mathbf{f}_{2j} \quad (6)$$

Similarly for the rear tipping $f^y_{r_{1j}} = 0$, the jth mobile base will lose the contact with the ground at C_{1j} and will generate a moment M_{c2j} about z-axis at point C_{2j}:

$$M^z_{C2j} = (\mathbf{g}_j - \mathbf{c}_{2j})^T \mathbf{E}^T m_j \mathbf{g} + (\mathbf{a}_{1j} - \mathbf{c}_{2j})^T \mathbf{E}^T \mathbf{f}_{1j} + (\mathbf{a}_{2j} - \mathbf{c}_{2j})^T \mathbf{E}^T \mathbf{f}_{2j} \quad (7)$$

As a consequence, for the first mobile base p_1 to be always stable, the moments generated by the external forces should be counter clockwise at point C_{11} while it should be clockwise at point C_{21}. Therefore, the stability conditions for mobile base p_1 can be expressed as:

$$M^z_{C11} \geq 0 \quad (8)$$
$$M^z_{C21} \leq 0 \quad (9)$$

Similarly, the stability constraint conditions for the second mobile base p_2 are expressed as:

$$M^z_{C12} \leq 0 \quad (10)$$
$$M^z_{C22} \geq 0 \quad (11)$$

where M^z_{C12} and M^z_{C22} are the moments of the mobile base p_2 about z-axis at the contact points C_{12} and C_{22}, respectively.

3 Real-Time Tension Distribution Algorithm

In this section an existing Tension Distribution Algorithm (TDA) defined for classical CDPRs is adopted to Mobile Cable-driven Parallel Robots (MCDPRs). The existing algorithm, known as barycenter/centroid algorithm is presented in [7,8]. Due to its geometric nature, the algorithm is efficient and appropriate for real time applications [5]. First, the classical Feasible Cable Tension Domain (FCTD) is defined for CDPRs based on the cable tension limits. Then, the stability (static equilibrium) conditions for the mobile bases are considered in order to define a modified FCTD for MCDPRs. Finally, a new TDA aiming at obtaining the centroid/barycenter of the modified FCTD is presented.

3.1 FCTD Based on Cable Tension Limits

The dynamic equilibrium equation of a point mass platform is expressed as:

$$\mathbf{W}\mathbf{t}_p + \mathbf{w}_e = 0 \Longrightarrow \mathbf{t}_p = -\mathbf{W}^+\mathbf{w}_e \qquad (12)$$

where $\mathbf{W} = [\mathbf{u}_{11}\ \mathbf{u}_{21}\ \mathbf{u}_{12}\ \mathbf{u}_{22}]$ is $n \times m$ wrench matrix mapping the cable tension space defined in $\mathbb{R}^m$ onto the available wrench space defined in $\mathbb{R}^{(m-n)}$. $\mathbf{w}_e$ denotes the external wrench exerted onto the moving-platform. $\mathbf{W}^+$ is the Moore Penrose pseudo inverse of the wrench matrix $\mathbf{W}$. $\mathbf{t}_p = [t_{p11}\ t_{p21}\ t_{p12}\ t_{p22}]^T$ is a particular solution (Minimum Norm Solution) of Eq. (12). Having redundancy $r = m - n = 2$, a homogeneous solution $\mathbf{t}_n$ can be added to the particular solution $\mathbf{t}_p$ such that:

$$\mathbf{t} = \mathbf{t}_p + \mathbf{t}_n \Longrightarrow \mathbf{t} = -\mathbf{W}^+\mathbf{w}_e + \mathbf{N}\boldsymbol{\lambda} \qquad (13)$$

where $\mathbf{N}$ is the $m \times (m-n)$ null space of the wrench matrix $\mathbf{W}$ and $\boldsymbol{\lambda} = [\lambda_1\ \lambda_2]^T$ is a $(m-n)$ dimensional arbitrary vector that moves the particular solution into the feasible range of cable tensions. Note that the cable tension t_{ij} associated with the ith cable mounted onto the jth mobile base should be bounded between a minimum tension $\underline{t}$ and a maximum tension $\overline{t}$ depending on the motor capacity and the transmission system at hand. According to [5,7], there exists a 2-D affine space Σ defined by the solution of Eq. (12) and another m-dimensional hypercube Ω defined by the feasible cable tensions:

$$\Sigma = \{\mathbf{t}\ |\ \mathbf{W}\mathbf{t} = \mathbf{w}_e\} \qquad (14)$$

$$\Omega = \{\mathbf{t}\ |\ \underline{t} \leq \mathbf{t} \leq \overline{t}\} \qquad (15)$$

The intersection between these two spaces amounts to a 2-D convex polygon also known as feasible polygon. Such a polygon exists if and only if the tension distribution admits a solution at least that satisfies the cable tension limits as well as the equilibrium of the moving-platform defined by Eq. (12). Therefore, the feasible polygon is defined in the $\boldsymbol{\lambda}$-space by the following linear inequalities:

$$\underline{t} - \mathbf{t}_p \leq \mathbf{N}\boldsymbol{\lambda} \leq \overline{t} - \mathbf{t}_p \qquad (16)$$

The terms of the $m \times (m-n)$ null space matrix $\mathbf{N}$ are defined as follows:

$$\mathbf{N} = \begin{bmatrix} \mathbf{n}_{11} \\ \mathbf{n}_{21} \\ \mathbf{n}_{12} \\ \mathbf{n}_{22} \end{bmatrix} \qquad (17)$$

where each component $\mathbf{n}_{ij}$ of the null space $\mathbf{N}$ in Eq. (17) is a (1×2) row vector.

3.2 FCTD Based on the Stability of the Mobile Bases

This section aims at defining the FCTD while considering the cable tension limits and the stability conditions of the mobile bases. In order to consider the stability of the mobile bases, Eqs. (8–11) must be expressed into the λ-space. The stability constraint at point C_{11} from Eq. (8) can be expressed as:

$$0 \leq (\mathbf{g}_1 - \mathbf{c}_{11})^T \mathbf{E}^T m_1 \mathbf{g} + (\mathbf{a}_{11} - \mathbf{c}_{11})^T \mathbf{E}^T \mathbf{f}_{11} + (\mathbf{a}_{21} - \mathbf{c}_{11})^T \mathbf{E}^T \mathbf{f}_{21} \quad (18)$$

$\mathbf{f}_{ij}$ is the force applied by the ith cable attached onto the jth mobile base. As $\mathbf{f}_{ij}$ is opposite to $\mathbf{u}_{ij}$ (see Fig. 2), from Eq. (13) $\mathbf{f}_{ij}$ can be expressed as:

$$\mathbf{f}_{ij} = -[t_{pij} + \mathbf{n}_{ij}\boldsymbol{\lambda}] \, \mathbf{u}_{ij} \quad (19)$$

Substituting Eq. (19) in Eq. (18) yields:

$$(\mathbf{c}_{11} - \mathbf{g}_1)^T \mathbf{E}^T m_1 \mathbf{g} \leq (\mathbf{c}_{11} - \mathbf{a}_{11})^T \mathbf{E}^T [t_{p11} + \mathbf{n}_{11}\boldsymbol{\lambda}]\mathbf{u}_{11} + (\mathbf{c}_{11} - \mathbf{a}_{21})^T \mathbf{E}^T [t_{p21} + \mathbf{n}_{21}\boldsymbol{\lambda}]\mathbf{u}_{21} \quad (20)$$

$$\underline{M}_{C11} \leq (\mathbf{c}_{11} - \mathbf{a}_{11})^T \mathbf{E}^T [\mathbf{n}_{11}\boldsymbol{\lambda}]\mathbf{u}_{11} + (\mathbf{c}_{11} - \mathbf{a}_{21})^T \mathbf{E}^T [\mathbf{n}_{21}\boldsymbol{\lambda}]\mathbf{u}_{21} \quad (21)$$

Term $[\mathbf{n}_{ij}\boldsymbol{\lambda}]\mathbf{u}_{ij}$ is the mapping of homogeneous solution $\mathbf{t}_{nij}$ for the ith cable carried by the jth mobile base into the Cartesian space. $\underline{M}_{C11}$ represents the lower bound for the constraint (8) in the λ-space:

$$\underline{M}_{C11} = (\mathbf{c}_{11} - \mathbf{g}_1)^T \mathbf{E}^T m_1 \mathbf{g} + (\mathbf{a}_{11} - \mathbf{c}_{11})^T \mathbf{E}^T \mathbf{t}_{p11} + (\mathbf{a}_{21} - \mathbf{c}_{11})^T \mathbf{E}^T \mathbf{t}_{p21} \quad (22)$$

Simplifying Eq. (21) yields:

$$\underline{M}_{C11} \leq \begin{bmatrix} (\mathbf{c}_{11} - \mathbf{a}_{11})^T \mathbf{E}^T \mathbf{u}_{11} & (\mathbf{c}_{11} - \mathbf{a}_{21})^T \mathbf{E}^T \mathbf{u}_{21} \end{bmatrix} \begin{bmatrix} \mathbf{n}_{11} \\ \mathbf{n}_{21} \end{bmatrix} \begin{bmatrix} \lambda_1 \\ \lambda_2 \end{bmatrix} \quad (23)$$

Equation (23) can be written as:

$$\underline{M}_{C11} \leq \mathbf{n}_{C11} \boldsymbol{\lambda} \quad (24)$$

where $\mathbf{n}_{C11}$ is a 1×2 row vector. Similarly the stability constraint at point C_{21} from Eq. (9) can be expressed as:

$$\mathbf{n}_{C21} \boldsymbol{\lambda} \leq \overline{M}_{C21} \quad (25)$$

where:

$$\overline{M}_{C21} = (\mathbf{c}_{21} - \mathbf{g}_1)^T \mathbf{E}^T m_1 \mathbf{g} + (\mathbf{a}_{11} - \mathbf{c}_{21})^T \mathbf{E}^T \mathbf{t}_{p11} + (\mathbf{a}_{21} - \mathbf{c}_{21})^T \mathbf{E}^T \mathbf{t}_{p21} \quad (26)$$

$$\mathbf{n}_{C21} = \begin{bmatrix} (\mathbf{c}_{21} - \mathbf{a}_{11})^T \mathbf{E}^T \mathbf{u}_{11} & (\mathbf{c}_{21} - \mathbf{a}_{21})^T \mathbf{E}^T \mathbf{u}_{21} \end{bmatrix} \begin{bmatrix} \mathbf{n}_{11} \\ \mathbf{n}_{21} \end{bmatrix} \quad (27)$$

Equations (24) and (25) define the stability constraints of the mobile base p_1 in the λ- space for the static equilibrium about frontal and rear wheels. Similarly, the above procedure can be repeated to compute the stability constraints in the λ-space for mobile base p_2. Constraint Eqs. (10) and (11) for point C_{12} and C_{22} can be expressed in the λ-space as:

$$\mathbf{n}_{C12}\lambda \leq \overline{M}_{C12} \tag{28}$$

$$\underline{M}_{C22} \leq \mathbf{n}_{C22}\lambda \tag{29}$$

Considering the stability constraints related to each contact point (Eqs. (24), (25), (28) and (29)) with the cable tension limit constraints (Eq. (16)), the complete system of constraints to calculate the feasible tensions for MCDPR can be expressed as:

$$\begin{bmatrix} \underline{\mathbf{t}} - \mathbf{t}_p \\ \underline{\mathbf{M}} \end{bmatrix} \leq \begin{bmatrix} \mathbf{N} \\ \mathbf{N_c} \end{bmatrix} \begin{bmatrix} \lambda_1 \\ \lambda_2 \end{bmatrix} \leq \begin{bmatrix} \overline{\mathbf{t}} - \mathbf{t}_p \\ \overline{\mathbf{M}} \end{bmatrix} \tag{30}$$

where:

$$\mathbf{N_c} = \begin{bmatrix} \mathbf{n}_{C11} \\ \mathbf{n}_{C21} \\ \mathbf{n}_{C12} \\ \mathbf{n}_{C22} \end{bmatrix}, \quad \underline{\mathbf{M}} = \begin{bmatrix} \underline{M}_{C11} \\ -\infty \\ -\infty \\ \underline{M}_{C22} \end{bmatrix}, \quad \overline{\mathbf{M}} = \begin{bmatrix} \infty \\ \overline{M}_{C21} \\ \overline{M}_{C12} \\ \infty \end{bmatrix}, \tag{31}$$

The terms $-\infty$ and ∞ are added for the sake of algorithm [5] as the latter requires bounds from both ends. The upper part of Eq. (30) defines the tension limit constraints while the lower part represents the stability constraints for both mobile bases.

3.3 Tracing FCTD into the λ-space

The inequality constraints from Eq. (30) are used to compute the feasible tension distribution among the cables using the algorithm in [5] for tracing the feasible polygon P_I. Each constraint defines a line in the λ-space where the coefficients of λ define the slope of the corresponding lines. The intersections between these lines form a feasible polygon. The algorithm aims to find the feasible combination for λ_1 and λ_2 (if it exists), that satisfies all the inequality constraints. The algorithm can start with the intersection point $\mathbf{v}_{ij}$ between any two lines L_i and L_j where each intersection point $\mathbf{v}$ corresponds to a specific value for λ. After reaching the intersection point $\mathbf{v}_{ij}$, the algorithm leaves the current line L_j and follows the next line L_i in order to find the next intersection point $\mathbf{v}_{ki}$ between lines L_k and L_i.

The feasible polygon P_I is associated with the feasible index set I, which contains the row indices in Eq. (30). At each intersection point, the feasible index set is unchanged or modified by adding the corresponding row index of Eq. (30). It means that for each intersection point, the number of rows from Eq. (30) satisfied at current intersection point should be greater than or equal to the number of rows satisfied at previous visited points. Accordingly, the algorithm makes sure to converge toward the solution. The algorithm keeps track of the intersection points and updates the first vertex $\mathbf{v}_f$ of the feasible polygon, which depends on the update of feasible index set I. If the feasible index set is updated

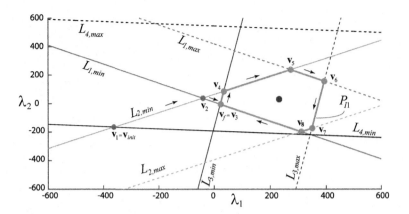

Fig. 3. Feasible Polygon considering only tension limit constraints

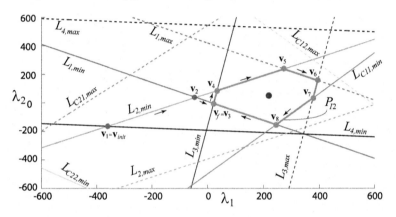

Fig. 4. Feasible Polygon considering both tension limit and stability constraints

at intersection point $\mathbf{v}$, the first vertex of the polygon is updated as $\mathbf{v}_f = \mathbf{v}$. Let's consider that the algorithm has reached a point $\mathbf{v}_{ki}$ by first following line L_j, then following L_i intersecting with line L_k. The feasible index set I_{ki} at $\mathbf{v}_{ki}$ should be such that $I_{ij} \subseteq I_{ki}$. If index k is not available in I_{ij}, then $I_{ki} = I_{ij} \cup k$ as the row k is now satisfied. At each update of the feasible index set I, a new feasible polygon is achieved and the first vertex $\mathbf{v}_f$ of the polygon is replaced by the current intersection point. This procedure is repeated until a feasible polygon (if it exists) is found, which is determined by visiting $\mathbf{v}_f$ more than once. After computing the feasible polygon, its centroid, namely the solution furthest away from all the constraints is calculated. The $\boldsymbol{\lambda}$ coordinates of the centroid is used to calculate the feasible tension distribution using Eq. (13).

For the given end-effector position in static equilibrium (see Fig. 2), the feasible polygon P_{I1} based only on the cable tension limits is illustrated in Fig. 3 while the feasible polygon P_{I2} based on the cable tension limits and the stability of the mobile bases is illustrated in Fig. 4. It can be observed that P_{I2} is smaller than P_{I1} and, as a consequence, their centroids are different.

4 Case Study

The stability of the mobile bases is defined by the position of their Zero Moment Point (ZMP). This index is commonly used to determine the dynamic stability of the humanoid and wheeled robots [6,13,14]. It is the point where the moment of contact forces is reduced to the pivoting moment of friction forces about an axis normal to the ground. Here the ZMP amounts to the point where the sum of the moments due to frontal and rear ground reaction forces is null. Once the feasible cable tensions are computed using the constraints of the modified TDA, the ZMP d_j of the mobile base p_j is expressed by the equation:

$$M_{dj}^z = \widetilde{M_O^z} - f_{rj}^y d_j \qquad (32)$$

where f_{rj}^y is the sum of all the vertical ground reaction forces computed using Eqs. (4) and (5), M_{dj} is the moment generated at ZMP for the jth mobile base such that $M_{dj}^z = 0$. $\widetilde{M_O}$ is the moment due to external forces, i.e., weight and cable tensions, except the ground reaction forces at O given by the Eq. (2). As a result from Eq. (32), ZMP d_j will take the form:

$$d_j = \frac{\widetilde{M_O^z}}{f_{rj}^y} = \frac{\mathbf{g}_j^T \mathbf{E}^T m_j \mathbf{g} + \mathbf{a}_{1j}^T \mathbf{E}^T \mathbf{f}_{1j} + \mathbf{a}_{2j}^T \mathbf{E}^T \mathbf{f}_{2j}}{f_{rj}^y} \qquad (33)$$

For the mobile base p_j to be in static equilibrium, ZMP d_j must lie within the contact points of the wheels, namely,

$$c_{21}^x \leq d_1 \leq c_{11}^x \qquad (34)$$

$$c_{12}^x \leq d_2 \leq c_{22}^x \qquad (35)$$

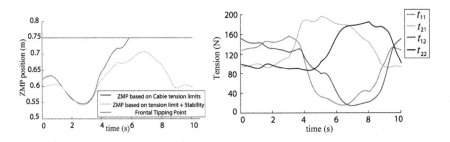

Fig. 5. (a) Evolution of ZMP for mobile base p_1 (b) Cable tension profile (Color figure online)

Modified Algorithm for MCDPRs is validated through simulation on a rectangular test trajectory (green path in Fig. 2) where each corner of the rectangle is a zero velocity point. A 8 kg point mass is used. Total trajectory time is 10 s having 3 s for 1–2 and 3–4 paths while 2 s for 2–3 and 4–1 paths. The size of each mobile base is 0.75 m × 0.64 m × 0.7 m. The distance between the two mobile bases is 5 m with exit points A_{2j} located at the height of 3 m. The evolution of ZMP for mobile base p_1 is illustrated in Fig. 5a. ZMP must lie between 0 and 0.75, which corresponds to the normalized distance between the two contact points of the wheels, for the first mobile base to be stable. By considering only cable tension limit constraints in the TDA, the first mobile base will tip over the front wheels along the path 3–4 as ZMP goes out of the limit (blue in Fig. 5a). While considering both cable tension limits and stability constraints, the MCDPR will complete the required trajectory with the ZMP satisfying Eqs. (34) and (35). Figure 5b depicts positive cable tensions computed using modified FCTD for MCDPRs.

A video showing the evolution of the feasible polygon as a function of time considering only tension limit constraints and both tension limits and stability constraints can be downloaded at[1]. This video also shows the location the mobile base ZMP as well as some tipping configurations of the mobile cable-driven parallel robot under study.

5 Conclusion

This paper has introduced a new concept of Mobile Cable-Driven Parallel Robots (MCDPR). The idea is to autonomously navigate and reconfigure the geometric architecture of CDPR without any human interaction. A new real time Tension Distribution algorithm is introduced for MCDPRs that takes into account the stability of the mobile bases during the computation of feasible cable tensions. The proposed algorithm ensures the stability of the mobile bases while guaranteeing a feasible cable tension distribution. Future work will deal with the extension of the algorithm to a 6-DoF MCDPR by taking into account frontal as well as sagittal tipping of the mobile bases and experimental validation thanks to a MCDPR prototype under construction in the framework of the European ECHORD++ "FASTKIT" project.

Acknowledgements. This research work is part of the European Project ECHORD++ "FASTKIT" dealing with the development of collaborative and mobile cable-driven parallel robots for logistics.

References

1. Albus, J., Bostelman, R., Dagalakis, N.: The NIST SPIDER, a robot crane. J. Res. Natl. Inst. Stan. Technol. **97**(3), 373–385 (1992)
2. Gagliardini, L., Caro, S., Gouttefarde, M., Girin, A.: A reconfiguration strategy for reconfigurable cable-driven parallel robots. In: 2015 IEEE International Conference on Robotics and Automation (ICRA), pp. 1613–1620. IEEE (2015)

---

[1] https://www.youtube.com/watch?v=XMBdLRZZ5jQ.

3. Gagliardini, L., Caro, S., Gouttefarde, M., Girin, A.: Discrete reconfiguration planning for cable-driven parallel robots. Mech. Mach. Theor. **100**, 313–337 (2016)
4. Gallina, P., Rosati, G., Rossi, A.: 3-dof wire driven planar haptic interface. J. Intell. Robot. Syst. **32**(1), 23–36 (2001)
5. Gouttefarde, M., Lamaury, J., Reichert, C., Bruckmann, T.: A versatile tension distribution algorithm for-dof parallel robots driven by cables. IEEE Trans. Robot. **31**(6), 1444–1457 (2015)
6. Lafaye, J., Gouaillier, D., Wieber, P.-B.: Linear model predictive control of the locomotion of pepper, a humanoid robot with omnidirectional wheels. In: 2014 IEEE-RAS International Conference on Humanoid Robots, pp. 336–341. IEEE (2014)
7. Lamaury, J., Gouttefarde, M.: A tension distribution method with improved computational efficiency. In: Cable-Driven Parallel Robots, Springer. pp. 71–85 (2013)
8. Mikelsons, L., Bruckmann, T., Hiller, M., Schramm, D.: A real-time capable force calculation algorithm for redundant tendon-based parallel manipulators. In: Proceedings of the 2008 IEEE International Conference on Robotics and Automation (ICRA 2008), pp. 3869–3874. IEEE (2008)
9. Nguyen, D.Q., Gouttefarde, M., Company, O., Pierrot, F.: On the analysis of large-dimension reconfigurable suspended cable-driven parallel robots. In: 2014 IEEE International Conference on Robotics and Automation (ICRA), pp. 5728–5735. IEEE (2014)
10. Pott, A., Meyer, C., Verl, A.: Large-scale assembly of solar power plants with parallel cable robots. In: Robotics (ISR), 2010 41st International Symposium on and 2010 6th German Conference on Robotics (ROBOTIK), pp. 1–6. VDE (2010)
11. Rosati, G., Andreolli, M., Biondi, A., Gallina, P.: Performance of cable suspended robots for upper limb rehabilitation. In: 2007 IEEE 10th International Conference on Rehabilitation Robotics, pp. 385–392. IEEE (2007)
12. Rosati, G., Zanotto, D., Agrawal, S.K.: On the design of adaptive cable-driven systems. J. Mech. Robot. **3**(2), 021004 (2011)
13. Sardain, P., Bessonnet, G.: Forces acting on a biped robot. center of pressure-zero moment point. IEEE Trans. Syst. Man Cybern. Part A Syst. Hum. **34**(5), 630–637 (2004)
14. Vukobratović, M., Borovac, B.: Zero-moment point-thirty five years of its life. Int. J. Hum. Robot. **1**(01), 157–173 (2004)
15. Zhou, X., Tang, C.P., Krovi, V.: Analysis framework for cooperating mobile cable robots. In: Proceedings of the 2012 IEEE International Conference on Robotics and Automation (ICRA 2012), pp. 3128–3133. IEEE (2012)

Improvement of Cable Tension Observability Through a New Cable Driving Unit Design

Mathieu Rognant[1(✉)] and Eric Courteille[2]

[1] Onera, DTIS, 2 Avenue Edouard Belin, 31000 Toulouse, France
mathieu.rognant@onera.fr
[2] Université Bretagne-Loire, INSA-LGCGM-EA 3913,
20, Avenue des Buttes de Cöesmes, 35043 Rennes, France
eric.courteille@insa-rennes.fr

Abstract. The context of this paper is a research project on a Cable-Driven Parallel Manipulator uses to reproduce the free flight condition of aircraft models in wind tunnels (SACSO Project). The force control of the CDPM enables to simulate the thrust of the aircraft engine, and to modify the scale model mass and inertia. A very important point for an efficient force control is an accurate estimation of the cables tension. In this paper, an original Cable Driving Unit with an integrated 3D force sensor is developped to improve the cable tension observability. An associated Extended Kalman Filter implementation is then proposed to estimate the cables tension.

1 Introduction

This study on cable-driven parallel manipulators (CDPMs) takes place within the SACSO project [3, 10]. This project deals with the study of aircraft behavior and aims to set up an active suspension which would enable free flight simulations in wind tunnels. This suspension sustains a scale model in a wind tunnel. The device has to reproduce propulsion forces as well as creating virtual mass and inertia in order to respect the similitude coefficients. The suspension must have displacement abilities for the scale model installation and for standard tests purposes. It must also have a high bandwidth force control to simulate the propulsion effects and to confer an artificial inertia to the scale model. These two control capabilities have to be ensured along 6 Degrees-Of-Freedom (DOF). During those free flight simulations, damping coefficients and dynamic derivatives of aircraft models can be identified. The suspension should not disturb the streamline flow and must be implemented in existing wind tunnels with the slightest modifications. Series type robot structures cannot fulfill all theses constraints. A cable driven suspension manipulator has been retained.

A CDPM is formed by a base, a mobile platform connected to the base through flexible cables. The motion control of the platform is performed by Cable Driving Units (CDUs) which modify cables lengths. As cables can be winded over great lengths, CDPM are very relevant for large workspace application [4, 16], and their light weight allows fast motion [8, 14]. A key challenge of the control of CDPM is that cables can only apply tensile forces. Cables must be kept in tension in the whole workspace of the robot to ensure the controllability of the end-effector. In order to fulfill this requirement, control loop needs sensors and associated fusion algorithm which guarantee an accurate estimation of cables tension.

Besides, accurate identification of damping coefficients and dynamic derivatives of the aircraft model depends on the accuracy of the measurements of the aerodynamic wrench applied to the scale model. A reliable evaluation of the wrench due to the cables tension is then needed. In the case of fully constrained CDPMs, which are characterized by more cables than DOF, there is an infinity of solutions for one resultant wrench [5]. Consequently, the resultant wrench measurement on the end-effector is not sufficient to ensure the observability of each cable tension. Cable tension Estimation on the joint space is also required.

This paper focuses on the improvement of the cable tension observability through an original CDU design. First, the mechanical design and associated sensors chosen during SACSO's former development will be presented. Although these solutions could guarantee the force observability required by the wind-tunnel application, they cause some restrictions on implementation that will also be detailed. Some modifications of the extant device are suggested here. Then, a new CDU design with an integrated 3D force sensor is presented. Finally, the approach based on an Extended Kalman Filter (EKF) is then proposed to increase the accuracy in the estimation of the cable tension. Preliminary results obtained on a mono-axial prototype are presented.

2 SACSO

Designed and created within the framework of an ONERA internal research project, SACSO is a CDPM which has been installed and tested in a low speed vertical wind tunnel at ONERA-Lille [3]. This set-up allows 6 DOF dynamic motions.

2.1 Wind Tunnel Application

The functionalities provided by this CDPM allow more complete tests for analysis and modeling of the flight dynamics, such as:

- generating movements according to the 6 DOF: that way, the parameters of the aerodynamic model can be isolated and identified, which is often impossible to achieve on the existing dynamic setups;
- simulating free, or semi-free, flight (according to chosen DOF), which means that motions with important aerodynamics couplings (wing-rock on combat aircraft for example) and time-dependent behavior can be studied.

Limits on the size and the mass of the tested scale models are 1 m long and 5 kg. These scale models are held by a carbon fiber beam from a 6-components aerodynamic balance, which is suspended in the flow by 9 cables (Fig. 1b). The workspace of this fully constrained CDPM is a cube of 1 m side, and the angular limits are ±15° in heading, ±45° in roll, and −20° to +45° in pitch. The maximum speeds and accelerations follow the similarity conditions for the motions of a military aircraft: $3\,\mathrm{m.s}^{-1}$ and $15\,\mathrm{m.s}^{-2}$ vertically and laterally, $80°.s^{-1}$ and $1400°.s^{-2}$ in yaw, $400°.s^{-1}$ and $7000°.s^{-2}$ in pitch and $440°.s^{-1}$ and $8500°.s^{-2}$ in roll.

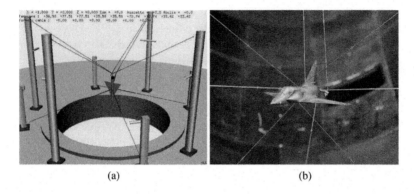

Fig. 1. SACSO: (a) SACSO layout on wind tunnel (b) Scale Model on SACSO

2.2 Control Design

The control algorithm of this CDPM is a position-force hybrid coordinated control structure, with two distinct controllers: a task space controller and a joint space controller. The task space controller calculates tension distribution. A quadratic programming algorithm is used to ensure that results stay in the operational range. Three control schemes are implemented:

- To follow 6-DOF trajectories imposed on the scale model: This is a position control where cable tensions are computed from the end-effector tracking error;
- To leave the scale model in 6-DOF free flight: This is a force control where cable tensions are calculated to simulate the thrust of the engines, and to modify the scale model mass and inertia by generating the corresponding inertial forces;
- To impose motion along some of the 6 DOF: This is an hybrid position-force control.

Outputs from the task space controller are used by the joint space controller to calculate the torque to apply by each winch. These control loops also need the estimation of tension, length and elongation of each cable. In practice, the two controllers are synchronous and activated with the same sample frequency of 2 kHz.

2.3 Data Processing and Limitation

The states of the CDPM are estimated by fusion of sensors data. The scale model includes three gyrometers, seven accelerometers and a 6-components aerodynamic balance. The length and tension of each driving cable are respectively measured with an incremental encoder implemented on the winch and a force sensor fixed on the last pulley of the CDU. An external location device composed of two cameras provides the 6D-pose of the scale model.

The fitting of the aerodynamic balance inside the scale model is a complex and an expensive task. Moreover, scale model of aircraft without fuselage, like rotary wings UAV, can not be equipped with such a sensor. In order to extend the range of compatible aircraft and to reduce the cost of the scale model, the aerodynamic balance must be removed. The development of a new CDU design embedding a force sensor would improve the cable tension observability, and hence the resultant wrench estimation.

3 Proposed Cable Driving Unit

3.1 Principle

CDU is an essential component of a CDPM. Different design concepts for modifying the length of cables have been proposed in the literature. The MARIONET-REHAB robot uses a linear actuator with a pulley system which allows to reach very high speed cable length modification with accuracy [13]. It use remains limited to low cable tension [12]. Donohoe *et al.* [2] propose an original winch system that actuates the cable over a rotatable pulley arm. Winch systems with a rotary motor that turns a drum remain the most frequent actuation scheme.

The design phase of winch system with drum is guided by two main constraints.

The first one is to keep fixed the position of the point at which the cable is drawn out from the drum. The accurate estimation of the unwound cable length is essential for an effective position control. In order to fulfill this requirement, the drum can translate while rotating about a fixed screw shaft [1,7]. The pitch of the drum and the screw must be equal. Another concept is the winch design proposed by Pott *et al.* for the IPANEMA [14]. A pulley translates in parallel to the axis of the drum and another one at the end of the CDU fixes the output point of the cable.

The second constraint on the design of the winches is the cable tension measurement. In fact, as mentioned in the previous sections, the cable tension measurement was highlighted as a critical issue for which different solutions have been applied:

- Cable force sensors are directly integrated on the end-effector [11]. The accessibility and the data dispatch make the practical implementation difficult.
- The cable tension is estimated by using the motor current measurement [2]. The bandwidth is relatively limited and the measurements can be noisy due to the gear ratio of the transmission and the pulleys.

- The sensors can also be integrated as measurement units using pulleys [9]. The pulleys disturb the cable force measurement accuracy due to friction. The modeling of the friction is essential for its compensation.
- The force sensors can be directly integrated in the winch [7,17].

3.2 Proposed Design

The adopted strategy hinges the two previous constraints around which a new CDU design was devised to improve the cable tension observability (Fig. 2). As shown in Fig. 2a, the kinematic solution uses a screw/nut joint. The synchronization of the translational and rotational movements allows the cable to be kept in a fixed vertical position. The CAD cross-sectional view of the actuator system is given in Fig. 2b. The whole actuator system motor-drum both sits on a linear guide composed of two slides and the nut of the ball-screw. The motor shaft is supported at its free end by a ball bearing. The originality of this design lies in aligning the rotation axis of the pulley block with the cable at the exit of the drum. The diameter of the pulley is adjusted so that the cable will tangent

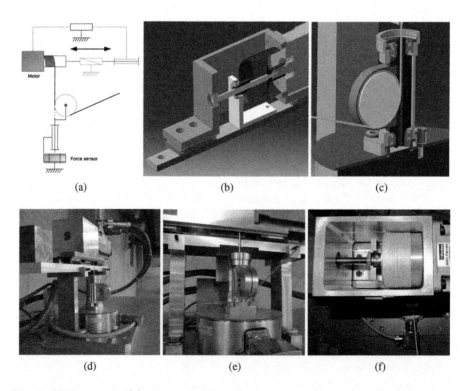

Fig. 2. CDU concept: (a) Mechanical kinematic scheme (b) CAD cross-sectional view of Actuator system (c) CAD cross-sectional view of the pulley block (d) Cable Driving Unit (e) Pulley block and 3D force sensor (f) Winch details

it to the entry point. The rotational guidance of the pulley block around the vertical axis is ensured by a ball bearing (Fig. 2c). The pulley also sits on two ball bearings. This concept allows to integrate under the CDU a 3D force sensor in the vertical axis of the cable with a high geometric accuracy.

The prototype of the CDU is shown in Figs. 2d, e and f. The winch drum is powered by a NX310EAP Parvex motor. The motor should run up to 4000 rpm with a maximal torque of 2 N.m. The cable used is made of Dyneema (diameter: 1.4 mm). The diameter and the pitch of the drum are respectively 60 mm and 1.5 mm. Therefore, the screw has a pitch of 1.5 mm.

3.3 Integrated Force Measurement

Two solutions have been studied to integrate the 3D force sensor (Fig. 3). The used sensor is a CMC 301 distributed by TME and with a range of 25 daN.

3.3.1 Cable Driving Unit Measured

The first solution is to have the whole of the CDU carried directly by the force sensor (Fig. 3a). Thus reaction forces exerted by the cable are directly measured by the 3D force sensor. This solution has been tested. Analyses and evaluations of the first prototype reveal that the inertia forces inherent to the translation of the drum are negligible even for high winding or unwinding speed. The measurement noise due to the rotation of the geared motor remains limited. Nevertheless, in this configuration, two critical aspects emerged:

- The translation of the drum leads to a shift of the center of gravity of the CDU which must be compensated in the cable tension estimation.
- The mass of the whole CDU suspended on the stiffness of the force sensor lowers the first eigenmode of the assembly.

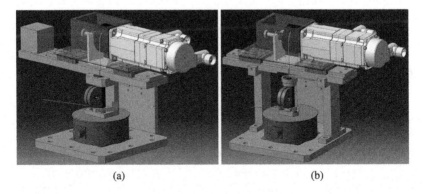

Fig. 3. Integrated force measurement: (a) Cable Driving Unit measured (b) Pulley block measured

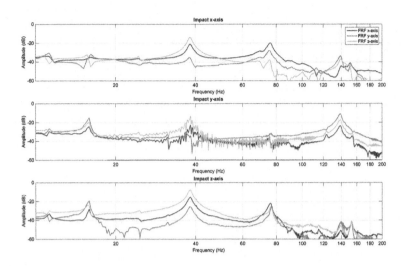

Fig. 4. Frequency Response Functions obtained from an impact testing

Among these two points, the last one is the most troublesome in the estimation of the cable tension. Impact testing has been done on this CDU configuration for measuring Frequency Response Functions (FRFs). From Fig. 4, the first eigenmode of the assembly is around 14 Hz, in the CDPM bandwidth.

3.3.2 Pulley Measured

For the second solution, only the pulley block is mounted on the force sensor (Fig. 3b). The low mass of the pulley block rejects the first eigenmode measured by sensor outside the CDPM bandwidth. In closed loop, the force control allows in this configuration to guarantee a bandwidth of 500 Hz. It remains that the pulley friction may disturb the cable tension estimation from the force measurement. This point can be averted by developing a suitable sensor fusion framework as described in the following section.

4 Cable Tension Estimation

As mentionned in Sect. 2.2, our application requires an estimation of tension of cables, with a sample frequency of 2 kHz. Because each sensor measures at its own frequency (location device at 50 Hz; accelerometers and gyrometers at 400 Hz and incremental encoders, motor resolvers and force sensor at 2 kHz), an EKF has to be used to proceed to the sensor data fusion and to provide an accurate estimation of the device state to the controller. In this section, prediction and measurement models implemented by this EKF are presented.

4.1 CDPM Dynamic Model

The dynamic model of a CDPM end-effector is derived from the Newton Euler's equations:

$$m_p \ddot{\mathbf{p}} = m_p \mathbf{g} + \sum_i^m \mathbf{t}_i + \mathbf{f}_{ext}, \tag{1}$$

$$\mathbf{I}_p \dot{\boldsymbol{\omega}} = -\boldsymbol{\omega} \times \mathbf{I}_p \boldsymbol{\omega} + \sum_i^m \mathbf{r}_i \times \mathbf{t}_i + \boldsymbol{\tau}_{ext}, \tag{2}$$

where m is the number of cables. m_p and $\mathbf{I}_p$ are mass and inertial tensor of the scale model. g is the gravitational acceleration. $\mathbf{p}$ and $\boldsymbol{\omega}$ are the position and angular velocity of the scale model. $\mathbf{f}_{ext}$ and $\boldsymbol{\tau}_{ext}$ are the external forces and torques (in our case aerodynamic wrench). $\mathbf{r}_i$ is the relative position of the ith cable anchor point to the gravity center of the scale model. $\mathbf{t}_i$ is the cable tension vector of the ith cable.

4.2 CDU Model

The parametrization of the ith CDU is summarized in Fig. 5. At each CDU a reference frame $\Re_i$ ($\mathbf{o}_i, x_i, y_i, z_i$) is associated. Vectors $\mathbf{a}_i$ and $\mathbf{c}_i$ are respectively the position of the ith cable anchor point A_i, and the center of the pulley axis C_i. Angles ϕ_i, α_i and β_i are cable angle values relative to the $z_i x_i$ plane, the $x_i y_i$ plane, and the vector ($\mathbf{c}_i - \mathbf{a}_i$) respectively.

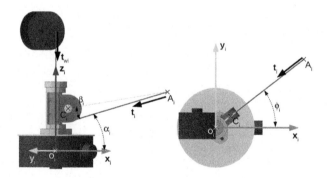

Fig. 5. CDU modelling

Forces applied by the cable on both the winch and the end-effector are described by the vectors $\mathbf{t}_{w_i}$ and $\mathbf{t}_i$. The direction $^i\mathbf{u}_i$ of the tension vector $\mathbf{t}_i$ within the frame $\Re_i$ is a function of angles ϕ_i and α_i:

$$^i\mathbf{t}_i = t_i {}^i\mathbf{u}_i = t_i \begin{bmatrix} \cos\alpha_i \cos\phi_i;\ \cos\alpha_i \sin\phi_i;\ \sin\alpha_i \end{bmatrix}^T. \tag{3}$$

These angle values are deduced from $\mathbf{a}_i$, $\mathbf{o}_i$ and $\mathbf{c}_i$ coordinates and the pulley radius r_{p_i} as follows:

$$\phi_i = \arctan\left(\frac{(\mathbf{a}_i-\mathbf{o}_i)|_{y_i}}{(\mathbf{a}_i-\mathbf{o}_i)|_{x_i}}\right) ; (\alpha_i - \beta_i) = \arctan\left(\frac{(\mathbf{a}_i-\mathbf{c}_i)|_{z_i}}{(\mathbf{a}_i-\mathbf{c}_i)|_{x_iy_i}}\right) ; \beta_i = \arcsin\frac{\|\mathbf{c}_i-\mathbf{a}_i\|}{r_{p_i}} \quad (4)$$

where $|_{x_i}$, $|_{y_i}$, $|_{z_i}$ and $|_{x_iy_i}$ are projection operators onto x_i, y_i, z_i axis and the x_iy_i plane, respectively. Regarding the cable tension t_i, it depends on the cable elongation ϵ_i and the linear stiffness ratio of the cable k_c. ϵ_i can be deduced to the end-effector position and the drum angle as follows:

$$t_i = \frac{k_c}{l_i}\epsilon_i = \frac{k_c}{\|\mathbf{c}_i-\mathbf{a}_i\|}(r_{w_i}\theta_i - \|\mathbf{c}_i - \mathbf{a}_i\|) \quad (5)$$

where r_{w_i} is the drum radius. Under the assumption of perfect joints (ball-screw and drum), one can derive the dynamic model of the winch:

$$j_{w_i}\ddot{\theta}_i + r_{w_i}t_{w_i} = \tau_i \quad (6)$$

where j_{w_i} is the inertia driven by the motor. Finally, inertial and friction effects of the pulley are modelled by a coefficient η_i which sign depends on the cable velocity:

$$t_{w_i} = (1 + \eta_i)t_i. \quad (7)$$

More advanced pulley model could be involved as presented in [9]. However, this new CDU design associated with a location device provides an estimation of t_i insensitive to the pulley model, as it will be demonstrated in Sect. 4.4. Hence, only a simple pulley description is considered in this paper.

4.3 Prediction Model

The overall system dynamic model can be obtained by combining Eqs. (1) and (6) and by deriving Eqs. (4), (5) and (7). The state and control of the CDPM are then described by vectors $\mathbf{x}$ and $\mathbf{u}$, respectively:

$$\mathbf{u} = \boldsymbol{\tau} \; ; \; \mathbf{x} = \left[\underbrace{\mathbf{p}^T \; \boldsymbol{\vartheta}^T \; \dot{\mathbf{p}}^T \; \boldsymbol{\omega}^T}_{\text{scale model}} \; \underbrace{\boldsymbol{\theta}^T \; \dot{\boldsymbol{\theta}}^T}_{\text{winches}} \; \underbrace{\mathbf{t}^T \; \boldsymbol{\alpha}^T \; \boldsymbol{\phi}^T}_{\text{cables}} \; \underbrace{\boldsymbol{\eta}^T}_{\text{pulleys}} \right]^T \quad (8)$$

with $\boldsymbol{\tau}$ the vector of the motor torques; $\boldsymbol{\vartheta}$ the orientation of the scale model (roll, pitch, yaw Euler angles representation); $\boldsymbol{\theta}$ and $\dot{\boldsymbol{\theta}}$ vectors of angles θ_i and angular velocities $\dot{\theta}_i$ of m winches; $\mathbf{t}$ the vector of n cable tensions t_i; and $\boldsymbol{\alpha}$ and $\boldsymbol{\phi}$ vectors of orientation angles α_i and ϕ_i of m cables. The prediction model can be written:

$$\dot{\mathbf{x}} = \mathbf{f}(\mathbf{x}, \mathbf{u}) + \mathbf{w} \quad (9)$$

where $\mathbf{w}$ is a vector of independent, zero mean, Gaussian noise processes of the covariance matrices $\mathbf{Q}$ [15]. The first 6 equations of Eq. (9) are simple integrators, and the following 6 equations related to $\ddot{\mathbf{p}}$ and $\dot{\boldsymbol{\omega}}$ are deduced from Eq. (1).

Associated **Q** coefficients fit to the expected range of $\mathbf{f}_{ext}$ and $\boldsymbol{\tau}_{ext}$ contributions. 2 m equations related to winches are deduced from Eq. (6). Associated **Q** coefficients fit to the perfect joints assumption. 3 m equations of $\dot{\mathbf{t}}$, $\dot{\boldsymbol{\alpha}}$ and $\dot{\boldsymbol{\phi}}$ related to cables dynamics can be expressed by derivation of Eqs. (4) and (5) as a function of $\dot{\mathbf{p}}$ and $\dot{\boldsymbol{\theta}}$. Last m equations related to the pulleys efficiency variation are set to 0 and associated **Q** coefficients fit to the expected range of η_i.

4.4 Measurement Model

States observability related to the scale model and winches is already insure by extant sensors: accelerometers and gyrometers embedded on the end-effector provide direct measurements of $\ddot{\mathbf{p}}$ and $\boldsymbol{\omega}$, the location device provides an estimation of $\mathbf{p}$ and $\boldsymbol{\vartheta}$ and incremental encoders and motor resolvers provide $\boldsymbol{\theta}$ and $\dot{\boldsymbol{\theta}}$. Components measured by the force sensor can be modeled by:

$$\mathbf{f}_{m_i} = \begin{bmatrix} f_{mx_i}; f_{my_i}; f_{mz_i} \end{bmatrix}^T = \begin{bmatrix} t_i \cos\alpha_i \cos\phi_i; t_i \cos\alpha_i \sin\phi_i; t_i \sin\alpha_i + t_{w_i} \end{bmatrix}^T. \tag{10}$$

From the two first components, new pseudo-measurements can be defined as follows:

$$f_{mxy_i} = (\sqrt{f_{mx_i}^2 + f_{my_i}^2}) = t_i \cos\alpha_i \; ; \; \frac{f_{my_i}}{f_{mx_i}} = \tan\phi_i. \tag{11}$$

From Eq. (11) the estimation of t_i can be calculated independently of the pulley friction by $t_i = f_{my_i}/\cos\alpha_i$. The Jacobian of this relation allows to evaluate the upper bound of σ_{t_i}, the standard deviation of the estimated t_i:

$$\sigma_{t_i} = \begin{bmatrix} \frac{1}{\cos\alpha_i} & \frac{t_i \sin\alpha_i}{\cos\alpha_i} \end{bmatrix} \begin{bmatrix} \sigma_{f_{mxy_i}} \\ \sigma_{\alpha_i} \end{bmatrix} \tag{12}$$

where $\sigma_{f_{mxy_i}}$, the standard deviation of f_{mxy_i}, is 1% of the sensor range. The external location device provides an estimation of $\mathbf{a}_i$ with a standard deviation $\sigma_{\mathbf{a}_i}$ equal to 5 mm. The standard deviation of α_i can be approximated by $\sigma_{\alpha_i} \approx \sigma_{\mathbf{a}_i}/\|\mathbf{a}_i\|$. Then one can derive the relative standard deviation of f_i due to the α_i uncertainty:

$$\sigma_{t_i}^{\alpha_i} = (\sin\alpha_i \sigma_{\mathbf{a}_i}) / (\cos\alpha_i \|\mathbf{a}_i\|). \tag{13}$$

From the $\sigma_{\mathbf{a}_i}$ value, mean and max values of $\sigma_{t_i}^{\alpha_i}$ are equal to 0.2% and 1.3% of t_i, respectively. The operating range considered is $\alpha \in [0°; 70°]$ and $\|\mathbf{a}_i\| \in [1\,m; 4\,m]$.

In the case of the use of force measurement only, the modeling and compensation of the friction of the unique pulley are necessary to keep an acceptable accuracy in the cable tension estimation for CDPM force control [6,9].

5 Conclusion

The estimation of cable tensions of fully-constrained CDPMs is the major study of this paper. The original CDU design presented improves cable tension observability while being relatively easy to manufacture and suitable to different CDPM dynamical requirements. A 3D force sensor is integrated into the CDU which guides the cable from the drum to a pulley with a fixed direction. A screw/nut joint is used to synchronize translational and rotational movement of the drum from the rotation of the geared motor unit. The pulley block is directly mounted on the axis of the force sensor. An Extended Kalman Filter is derived from the dynamic model of the new CDU to estimate the cable tension. The method based on data fusion of force sensor measurement and absolute end-effector pose estimation coming from external cameras prevents taking into account the effect of the pulley friction. The new CDU design with the external location device and its involvement in the different control loops are tested on a mono-axial prototype. The experimental validation of the cable tension estimation approach will be the focus of future research.

References

1. Borgstrom, P.H., Jordan, B.L., Borgstrom, B.J., Stealey, M.J., Sukhatme, G.S., Batalin, M.A., Kaiser, W.J.: Nims-pl: a cable-driven robot with self-calibration capabilities. IEEE Trans. Robot. **25**(5), 1005–1015 (2009)
2. Donohoe, S.P., Velinsky, S.A., Lasky, T.A.: Mechatronic implementation of a force optimal underconstrained planar cable robot. IEEE/ASME Trans. Mechatron. **21**(1), 69–78 (2016)
3. Farcy, D., Llibre, M., Carton, P., Lambert, C.: Sacso: wire-driven parallel set-up for dynamic tests in wind tunnel-review of principles and advantages for identification of aerodynamic models for flight mechanics. In: 8th ONERA-DLR Aerospace Symposium, Göttingen (2007)
4. Gagliardini, L., Caro, S., Gouttefarde, M., Girin, A.: A reconfiguration strategy for reconfigurable cable-driven parallel robots. In: 2015 IEEE International Conference on Robotics and Automation (ICRA), pp. 1613–1620. IEEE (2015)
5. Gosselin, C., Grenier, M.: On the determination of the force distribution in overconstrained cable-driven parallel mechanisms. Meccanica **46**(1), 3–15 (2011)
6. Heo, J.M., Choi, S.H., Park, K.S.: Workspace analysis of a 6-DOF cable-driven parallel robot considering pulley bearing friction under ultra-high acceleration. Microsyst. Technol., 1–13 (2016)
7. Izard, J.B., Gouttefarde, M., Michelin, M., Tempier, O., Baradat, C.: A reconfigurable robot for cable-driven parallel robotic research and industrial scenario proofing. In: Cable-Driven Parallel Robots, pp. 135–148. Springer (2013)
8. Kawamura, S., Kino, H., Won, C.: High-speed manipulation by using parallel wire-driven robots. Robotica **18**(1), 13–21 (2000)
9. Kraus, W., Kessler, M., Pott, A.: Pulley friction compensation for winch-integrated cable force measurement and verification on a cable-driven parallel robot. In: 2015 IEEE International Conference on Robotics and Automation (ICRA), pp. 1627–1632. IEEE (2015)
10. Lafourcade, P., Llibre, M., Reboulet, C.: Design of a parallel wire-driven manipulator for wind tunnels. In: Proceedings of the Workshop on Fundamental Issues and Future Research Directions for Parallel Mechanisms and Manipulators, pp. 187–194 (2002)
11. Mayhew, D., Bachrach, B., Rymer, W.Z., Beer, R.F.: Development of the macarm-a novel cable robot for upper limb neurorehabilitation. In: 9th International Conference on Rehabilitation Robotics, ICORR 2005, pp. 299–302. IEEE (2005)

12. Merlet, J.P.: Comparison of actuation schemes for wire-driven parallel robots. In: New Trends in Mechanism and Machine Science, pp. 245–254. Springer (2013)
13. Merlet, J.P., Daney, D.: A new design for wire-driven parallel robot. In: Proceedings 2nd International Congress, Design and Modelling of Mechanical Systems (2007)
14. Pott, A., Mütherich, H., Kraus, W., Schmidt, V., Miermeister, P., Verl, A.: Ipanema: a family of cable-driven parallel robots for industrial applications. In: Cable-Driven Parallel Robots, pp. 119–134. Springer (2013)
15. Song, Y., Grizzle, J.W.: The extended kalman filter as a local asymptotic observer for nonlinear discrete-time systems. In: American Control Conference, pp. 3365–3369. IEEE (1992)
16. Zi, B., Duan, B., Du, J., Bao, H.: Dynamic modeling and active control of a cable-suspended parallel robot. Mechatronics **18**(1), 1–12 (2008)
17. Zitzewitz, J.v., Rauter, G., Steiner, R., Brunschweiler, A., Riener, R.: A versatile wire robot concept as a haptic interface for sport simulation. In: IEEE International Conference on Robotics and Automation, ICRA 2009, pp. 313–318. IEEE (2009)

A Fast Algorithm for Wrench Exertion Capability Computation

Giovanni Boschetti[1(✉)], Chiara Passarini[1], Alberto Trevisani[1], and Damiano Zanotto[2]

[1] DTG - University of Padova, Stradella San Nicola 3, 36100 Vicenza, Italy
giovanni.boschetti@unipd.it
[2] Stevens Institute of Technology, 1 Castle Point Terrace, Hoboken, NJ 07030, USA
dzanotto@stevens.edu

Abstract. Computational efficiency is a critical issue in most real time applications. In general, non-iterative computational algorithms are less demanding than iterative ones. This paper proposes a new geometry-based algorithm to determine the maximum force or moment a cable driven parallel manipulator can exert in a given direction. A method to find a feasible set of cable tensions corresponding to the desired maximum wrench is also presented, and the proposed approach is validated using three illustrative examples. These algorithms show promise for applications requiring real-time planning.

1 Introduction

A cable driven parallel robot, or simply cable robot, mainly consists of a moving platform connected to a fixed base through cables wound around actuated pulleys. The length of the cables can be adjusted to control the degrees of freedom of the moving platform.

The performance of a parallel manipulator can be studied globally, in terms of workspace volume, or locally, in terms of available wrench set for a given pose. In particular, the wrench capability of a parallel manipulator is defined as the maximum force/moment that can be exerted on its moving platform. Such capability is highly pose-dependent and varies significantly with the direction of motion. In [14] the authors define the maximum force that an n-DoF manipulator can apply on the moving platform as the boundary of a polytope. Such polytope describes the available force set in the n-dimensional space. The force polytope was generated by computing the maximum force in any given direction with an optimization-based algorithm. Thereafter, many authors proposed several methods to define the available wrench set (i.e., the wrench polytope) for parallel manipulators. In [18] Zibil et al. proposed an explicit method based on scaling-factors to define the wrench polytope with improved accuracy and efficiency. Other studies that make use of the wrench polytope to describe the robot capabilities can be found in [8,9].

When dealing with cable robots, an additional constraint has to be considered: cables can only exert tensile forces and hence the minimum acceptable tension in each cable must be strictly positive. Local performance indexes based on the wrench polytope have been used in [15] in the context of adaptive cable-driven robots. The proposed method, which was formalized in [16] for a generic local index and experimentally validated in [17], allows designers to derive the optimal trajectories of the cable attachment points that guarantee a target level of the performance index within a given workspace.

In [6] a new method to define the wrench polytope of cable driven parallel manipulators was presented. Such method takes into account the unilateral force exertion capability of cables and proposes a non-iterative algorithm to obtain the *H-representation* of the polytope. In [7] the authors proposed a new method to identify vertexes, edges, and faces of the wrench polytope with the aim of exploring the relationship between the polytope and the cable robot structure. Moreover, convex analysis is exploited in [12] to perform the force-closure and workspace analysis of cable driven manipulators.

To investigate the capabilities of a cable driven robot, a new approach for performance evaluation was presented in [4]. Based on the computation of the maximum exertable force in a given direction, this approach can be applied to any redundant cable robot. Following that study, a novel index for cable robots called WEC (Wrench Exertion Capability) was presented in [5], and extended to underactuated cable robots. This index describes the maximum wrench that can be exerted along a given direction d while keeping null all the other wrench components. Furthermore, as proposed in [3], such index can be exploited to find the force, and hence acceleration, limits when facing the motion planning problem.

The objective of this paper is to introduce a new geometry-based algorithm aimed at finding the Wrench Exertion Capability of a cable driven robot. Compared to iterative algorithms, geometric algorithms can generate the wrench polytope in a more efficient way. Indeed, optimization-based methods are time expensive in terms of computation and hard to implement in real-time control because of their iterative nature [11]. The importance of having efficient algorithms that can be implemented in real-time applications is also underlined in [2,13].

The new geometry-based algorithm proposed here is more suitable than the recursive linear programming method proposed initially in [4], enabling the use of WEC index in real time applications.

This paper is organized as follows: Sect. 2 provides an overview on polytopes as representations of the linear transformation between the tension space and the force/moment space of the moving platform. Section 3 proposes a new methodology to find the maximum exertable wrench in a given direction, while a procedure to find the corresponding set of feasible tensions is presented in Sect. 4. These algorithms are validated in Sect. 5 using illustrative examples and numerical simulations. Finally, conclusions are stated in Sect. 6.

2 Definition of the Wrench Polytope

Let us consider a cable driven robot with m cables controlling n degrees of freedom. Define $\mathbf{w} = [\mathbf{f}^T, \mathbf{m}^T]^T$ as the wrench vector consisting of forces $\mathbf{f}$ and moments $\mathbf{m}$ exerted by cables on the moving platform. The relation between the wrench vector $\mathbf{w} \in \mathbb{R}^n$ and the cable tension vector $\boldsymbol{\tau} \in \mathbb{R}^m$ can be described by the following static equilibrium equation:

$$\mathbf{w} = \mathbf{S}\boldsymbol{\tau} \qquad (1)$$

subjected to the constraints

$$\boldsymbol{\tau}_{min} \preccurlyeq \boldsymbol{\tau} \preccurlyeq \boldsymbol{\tau}_{max} \qquad (2)$$

The symbol $\preccurlyeq$ stands for the componentwise inequality; it means that not only the cable tensions must be kept under a maximum limit $\boldsymbol{\tau}_{max}$ to avoid cable breakages, but they must also be greater than a lower positive limit $\boldsymbol{\tau}_{min}$, to avoid slack cables and ensure stiffness to the mechanism. The matrix $\mathbf{S} \in \mathbb{R}^{n \times m}$ (i.e., the structure matrix) is defined as:

$$\mathbf{S} = \begin{bmatrix} \mathbf{u}_1 & \mathbf{u}_2 & \cdots & \mathbf{u}_m \\ \mathbf{r}_1 \times \mathbf{u}_1 & \mathbf{r}_2 \times \mathbf{u}_2 & \cdots & \mathbf{r}_m \times \mathbf{u}_m \end{bmatrix} \qquad (3)$$

where:

- $\mathbf{u}_i$ is the unit vector directed along the i-th cable (starting from the attachment point of the mobile platform).
- $\mathbf{r}_i$ is the vector connecting the centre of mass of the moving platform to the attachment point of the i-th cable to the moving platform.

The first three rows of the structure matrix ($\mathbf{S}_f$) describe the relation between the cable tensions and the forces exerted on the moving platform, while the second three rows ($\mathbf{S}_t$) refer to the torques.

The wrench $\mathbf{w}$ can be rewritten in a more convenient form by defining a new reference frame having the x-axis aligned with a direction of interest d. A rotational matrix $\mathbf{R}$ can be computed to describe the relation between the fixed reference frame and the new one. Hence, it is possible to define the wrench $\mathbf{w}_d$ by means of the following expressions:

$$\mathbf{w}_d = \begin{bmatrix} \mathbf{R}^T & 0 \\ 0 & \mathbf{R}^T \end{bmatrix} \begin{bmatrix} \mathbf{S}_f \\ \mathbf{S}_t \end{bmatrix} \boldsymbol{\tau} = \mathbf{S}_d \boldsymbol{\tau} \qquad (4)$$

From a geometric point of view, the structure matrix $\mathbf{S}$ describes an affine transformation $\varGamma$ from the m-dimensional tension space onto the n-dimensional wrench space. In the m-dimensional space, the bounded region of acceptable cable tensions can be defined as an orthotope $\mathbf{T}$

$$\mathbf{T} = \left\{ \boldsymbol{\tau} | \boldsymbol{\tau} = [\tau_1 ... \tau_m]^T \in \mathbb{R}^m \ s.t. \ \tau_i \in [\tau_{min}, \tau_{max}] \right\} \qquad (5)$$

It can be noticed that, if τ_{min} and τ_{max} are fixed for all cables, then the orthotope is actually a hypercube. The hypercube $\mathbf{T}$ has 2^m vertexes $\boldsymbol{V}_j$, each corresponding to a particular tension configuration $\boldsymbol{\tau}_j = [\tau_{j_1}...\tau_{j_m}]^T$, where all the components τ_i take either their maximum or their minimum value (i.e., $\tau_{j_i} \in \{\tau_{min}; \tau_{max}\}$).

All the vertexes $\boldsymbol{V}_j$ are projected onto the n-dimensional wrench space by the affine transformation Γ to obtain 2^m characteristic points $\boldsymbol{U}_j$ such that $\boldsymbol{U}_j = \Gamma(\boldsymbol{V}_j)$. The convex hull defined by the points $\boldsymbol{W}_h$ that encloses all the characteristics points $\boldsymbol{U}_j$ is the available wrench set Ω for the given pose:

$$\Omega = \{\mathbf{w}_d \in \mathbb{R}^n | \mathbf{w}_d = \mathbf{S}_d \boldsymbol{\tau}, s.t. \boldsymbol{\tau} \in \mathbf{T}\} \tag{6}$$

It follows that $\boldsymbol{W}_h$ are the vertexes of the wrench polytope Ω. If $n=m$, all the characteristic points $\boldsymbol{U}_j$ are vertexes of Ω; however, in the more general case $n<m$, only some $\boldsymbol{U}_j$ are vertexes while others lie inside Ω or on its surface [7]. An example is given in Fig. 1 for a simple planar point-mass robot with three cables controlling two degrees of freedom. The example describes the available force set in terms of F_d and F_o defined as $[F_d \ F_o]^T = \mathbf{S}_d [\tau_1 \ \tau_2 \ \tau_3]^T$. Matrix $\mathbf{S}_d$ is defined in Eq. (4) for $d = \pi/6$ (depicted with an orange arrow in Fig. 1). The manipulator layout is depicted on the left: the point-mass end effector is connected to the fixed frame through three cables attached to the points $\mathbf{A}(-1, -1)$ m, $\mathbf{B}(0, 1)$ m and $\mathbf{C}(1, -1)$ m. On the right, the cube represents $\mathbf{T}$ in the 3-dimensional tension space (axes τ_1, τ_2 and τ_3); tension limits for this example are set to $\tau_{min} = 10$ N and $\tau_{max} = 70$ N. The eight vertexes of the cube are projected onto the 2-dimensional force space (axes F_d and F_o) to form the force polygon Ω. The projections of two of these vertexes lie inside the polygon. It can be proved that $\mathbf{T}$ and Ω are both convex polytopes [6]. Hence, it is possible to describe Ω by vertexes (*V-representation*) or by hyperplanes supporting its faces (*H-representation*) [10]. The *V-representation* requires an iterative algorithm such as quickhull [1] to identify which characteristic points are actually vertexes. Even if quickhull is usually fast, a non-iterative algorithm is more desirable for real-time applications. Hence, in this work we exploit the hyperplane-shifting method proposed in [6] to get the *H-representation* of the wrench polytope.

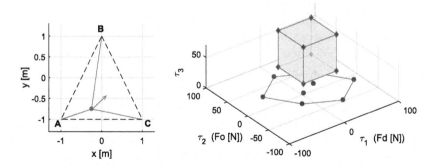

Fig. 1. Tension and wrench polytopes for a m = 3 n = 2 cable robot

In its *H-representation*, the available wrench set Ω is defined as a bounded intersection of closed half-spaces

$$\Omega = \{\mathbf{w}_d \in \mathbb{R}^n | \mathbf{N}\mathbf{w}_d \preccurlyeq \boldsymbol{\delta}\} \tag{7}$$

where:

- $\mathbf{N}$ is a matrix having in each row the transpose of a unit vector $\mathbf{n}_i$ which is normal to the hyperplane supporting a face, directed away from the polytope (outer unit normal vector)
- $\boldsymbol{\delta}$ is a vector whose element δ_i can be expressed using a known point $\mathbf{w}_{i0}$ belonging to the hyperplane, such that $\delta_i = \mathbf{n}_i^T \mathbf{w}_{i0}$.

In [6] it is shown that the wrench set is a zonotope in which every face has at least one other face parallel to it. Consequently, for each vector $\mathbf{n}_i$ orthogonal to one face, there is another vector $-\mathbf{n}_i$ orthogonal to the corresponding parallel face.

To identify $\mathbf{n}_i$, the hyperplane shifting method [6] consists in taking a set of $n-1$ linearly independent unit wrenches $\mathbf{s}_i$ from the matrix $\mathbf{S}_d$. A normal vector $\mathbf{n}_i$ is then obtained by the normalization of the generalized cross product among these $n-1$ vectors. This step must be repeated for all the feasible permutations of $n-1$ unit wrenches. From combinatorics, the total number of permutations is $\frac{m!}{k!(n-1)!}$, where $k=m-n+1$. However, only the permutations involving $n-1$ linearly independent vectors generate hyperplanes. Hence, if n_p indicates the number of hyperplanes, this process leads to a total of $2n_p$ normal vectors, the first n_p are the vectors $\mathbf{n}_i$, followed by their opposite $-\mathbf{n}_i$.

The matrix $\mathbf{N} \in \mathbb{R}^{(2n_p \times n)}$ can be written as:

$$\mathbf{N} = \begin{bmatrix} \mathbf{n}_1 \ \mathbf{n}_2 \ \cdots \ \mathbf{n}_{n_p} \ -\mathbf{n}_1 \ -\mathbf{n}_2 \ \cdots \ -\mathbf{n}_{n_p} \end{bmatrix}^T \tag{8}$$

A detailed description on how to determine the points $\mathbf{w}_{i0}$ can be found in [6]. The normal vector $\mathbf{n}_i$ and the corresponding point $\mathbf{w}_{i0}$ define an hyperplane $\Pi_i \in \mathbb{R}^{n-1}$ supporting a face of the wrench polytope.

3 Determination of the Maximum Exertable Wrench

Once all the hyperplanes have been identified, it is possible to analyse the boundaries of the polytope to find the maximum exertable force in the desired direction.

Two characteristics can be exploited to find the extreme wrench $\mathbf{w}_{max}$ (the same reasoning is valid for $\mathbf{w}_{min}$).

1. $\mathbf{w}_{max}$ lies in the hull of the wrench polytope, i.e., it belongs to at least one of its faces. Hence, for at least one of the $2n_p$ inequalities $\mathbf{N}\mathbf{w}_d \preccurlyeq \boldsymbol{\delta}$, the expression $\mathbf{n}_j^T \mathbf{w}_{max}$ will reach its maximum value, such that $\mathbf{n}_j^T \mathbf{w}_{max} = \delta_j$. In other words, $\mathbf{w}_{max} \in \Pi_j$, where Π_j is the hyperplane identified by $\mathbf{n}_j$.
2. Since $\mathbf{w}_{max}$ has all null wrench components except for the one in the direction of interest, it belongs to the straight line defined as $r = \lambda \mathbf{e}_1$, where $\lambda \in \mathbb{R}$ and $\mathbf{e}_1$ is the first column of the identity matrix $\mathbf{I}_n$: $\mathbf{w}_{max} = \lambda_j \mathbf{e}_1$.

For a generic wrench $\mathbf{w}_i \in \mathbb{R}^n$ that satisfies the above conditions, it is true that
$$\begin{cases} n_{i_1} * \lambda_i \leq \delta_i, & \forall i \in [1, 2n_p] \\ \exists j \in [1, 2n_p] \text{ s.t. } n_{j_1} * \lambda_j = \delta_j \end{cases} \quad (9)$$
where n_{i_1} is the projection of $\mathbf{n}_i$ along $\mathbf{e}_1$.

Let us consider the set $\mathbf{I} = \{i \in [1, 2n_p]\}$; such a set can be divided into three subsets: $\mathbf{P} = \{i \in [i, 2n_p] \mid n_{i_1} > 0\}$, $\mathbf{Q} = \{i \in [i, 2n_p] \mid n_{i_1} < 0\}$ and $\mathbf{S} = \{i \in [i, 2n_p] \mid n_{i_1} = 0\}$.

For the following analysis to be valid, it is necessary to remove from $\mathbf{N}$ the faces whose normal vector is orthogonal to r (that is, those identified by the set of indexes $\mathbf{S}$); indeed, such faces cannot intersect the straight-line r and hence can be excluded *a priori*.

Equation (9) can be therefore rewritten as:
$$\begin{cases} \lambda_i \leq \dfrac{\delta_i}{n_{i_1}}, & \forall i \in \mathbf{P} \\ \lambda_i \geq \dfrac{\delta_i}{n_{i_1}}, & \forall i \in \mathbf{Q} \\ \exists j \in [1, 2n_p] \setminus \mathbf{S} \text{ s.t. } n_{j_1} * \lambda_j = \delta_j \end{cases} \quad (10)$$

where the first two inequalities are equivalent to: $\max\limits_{i \in \mathbf{Q}} \left(\dfrac{\delta_i}{n_{i_1}}\right) \leq \lambda_i \leq \min\limits_{i \in \mathbf{P}} \left(\dfrac{\delta_i}{n_{i_1}}\right)$.

By adding the third condition, the previous inequalities can be further rewritten as: $\max\limits_{i \in \mathbf{Q}} \left(\dfrac{\delta_i}{n_{i_1}}\right) \leq \dfrac{\delta_j}{n_{j_1}} \leq \min\limits_{i \in \mathbf{P}} \left(\dfrac{\delta_i}{n_{i_1}}\right)$ from which one can infer that

$$\lambda_{j1,2} = \max\limits_{i \in \mathbf{Q}} \left\{\dfrac{\delta_i}{n_{i_1}}\right\}, \min\limits_{i \in \mathbf{P}} \left\{\dfrac{\delta_i}{n_{i_1}}\right\} \quad (11)$$

are two solutions of (10) and therefore must be the two desired intersection points (i.e., the maximum and minimum exertable wrench in the desired direction). In other words, $\mathbf{R} = \{\mathbf{w}_i \mid \mathbf{w}_i = \lambda_i * \mathbf{e}_1, \forall i \in \mathbf{P} \cup \mathbf{Q}\}$ is the set of intersection points $\mathbf{w}_i$ between the $2n_p$ supporting hyperplanes $\boldsymbol{\Pi}_i$; only two of these points represent feasible wrenches (i.e., $\mathbf{w}_{max}$ and $\mathbf{w}_{min}$) while the others are points outside Ω.

The example in Fig. 2 shows the intersection points $\mathbf{w}_{max}$ and $\mathbf{w}_{min}$ corresponding to the cable robot and the desired direction shown in Fig. 1 referring to the planar cable robot introduced in Sect. 2. The wrench space is actually a 2-dimensional force space, and the polytope Ω is a polygon with six edges corresponding to six straight-lines (dashed-lines). The straight-line r coincides with the x-axis, and intersects the polygon Ω in the two points $\mathbf{w}_{max}$ and $\mathbf{w}_{min}$. The black line between these two points represents the feasible range for a force directed along $d = \frac{\pi}{6}$. The points $\mathbf{w}_i$, i.e., the intersections between the dashed lines and the x-axis, are depicted with blue circles.

The method can be applied even in presence of external loading such as, for example, the gravity force. Due to this external wrench, the zonotope is shifted in the wrench space. The direction of each supporting hyperplane does

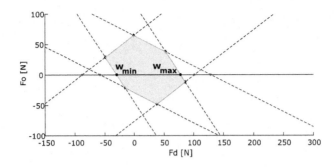

Fig. 2. Tension and wrench polytopes for a m = 3 n = 2 cable robot

not change, while the point identifying the hyperplane position is shifted such that $\mathbf{w}'_{i0} = \mathbf{w}_{i0} + \mathbf{w}_e$.

4 Determination of the Tension Configuration

In some applications it might be useful to find a feasible set of cable tensions corresponding to the local WEC, in addition to the WEC index.

Because all the points on the hull of the zonotope are images of points that belong to the hull of the hypercube, the tension configuration corresponding to the extreme wrenches $\mathbf{w}_{max}$ and $\mathbf{w}_{min}$ has to be sought in the hull of the tension hypercube $\mathbf{T}$.

Indeed, each vertex $\mathbf{W}_h \in \Omega$ is the projection of at least one vertex $\mathbf{V}_j \in \mathbf{T}$. Each $\mathbf{V}_j$ is directly connected to other m vertexes to form m edges. Moving along one edge of $\mathbf{T}$, the value of one tension force changes linearly from τ_{min} to τ_{max}. Similarly, the edges of the wrench zonotope are the projections of some of the edges of the tension hypercube. Hence, the wrenches of one edge are the result of the same tension configuration, except for one tension value: this value is τ_{min} in one vertex and τ_{max} in the other. It should be noted that this is not true when two or more vertexes of the wrench zonotope overlap and hence more than two $\mathbf{W}_h$ belongs to one edge of Ω. Therefore, in the following we exclude these special cases and assume that each vertex of the wrench polytope is the image of a single vertex of the tension hypercube.

The pre-images $\mathbf{V}_j$ of the vertexes $\mathbf{W}_h$ that delimit one face of the wrench polytope, have $m-n+1$ common tension values; for all wrenches lying on the face, these values remain constant while the other $n-1$ values change.

The normal vector that identifies each supporting hyperplane describes which tension values are fixed and common to the whole face (i.e., the ones related to the $m-n+1$ unit wrenches that define the position of the face, in terms of distance along $\mathbf{n}_i$ between such face and the initial parallel hyperplane passing through the origin). Consequently, the values of the remaining $n-1$ tensions corresponding to the desired wrench have to be determined.

Let us indicate the WEC index value as w_{max} (i.e., $\mathbf{w}_{max} = w_{max} * \mathbf{e}_1$). This wrench can be obtained following the procedure described in Sect. 3. Moreover, let $\mathbf{n}_{max}$ be the vector normal to the hyperplane Π_{max}, i.e., the one supporting the face to which $\mathbf{w}_{max}$ belongs. The contribution of each unit wrench to $\mathbf{w}_{max}$ can be highlighted by rewriting (1) as:

$$\mathbf{w}_{max} = \mathbf{s}_1 \tau_1 + \mathbf{s}_2 \tau_2 + \cdots + \mathbf{s}_m \tau_m \qquad (12)$$

where $\mathbf{s}_i$ is the i-th column of the structure matrix $\mathbf{S}$. Starting from $\mathbf{n}_{max}$, it is possible to trace back which are the unknown tensions that have to be calculated (i.e., $\boldsymbol{\tau}_u \in \mathbb{R}^{n-1}$) by looking at the specific permutation of $n-1$ linearly independent unit wrenches $\mathbf{s}_i$ used to generate $\mathbf{n}_{max}$. The remaining $m-n+1$ tensions represent the known vector $\boldsymbol{\tau}_k \in \mathbb{R}^{m-n+1}$.

Hence, the contributions to the exerted wrench are divided into two parts:

$$\mathbf{w}_{max} = \mathbf{S}_k \boldsymbol{\tau}_k + \mathbf{S}_u \boldsymbol{\tau}_u \qquad (13)$$

where $\mathbf{S}_k \in \mathbb{R}^{n \times m-n+1}$ is the matrix obtained from $\mathbf{S}_d$ by selecting only the columns related to $\boldsymbol{\tau}_k$, and similarly $\mathbf{S}_u \in \mathbb{R}^{n \times n-1}$ is the matrix obtained by $\mathbf{S}_d$ by selecting only the columns related to $\boldsymbol{\tau}_u$. The objective is then to find $\boldsymbol{\tau}_u$, such that:

$$\mathbf{S}_u \boldsymbol{\tau}_u = \mathbf{w}_{max} - \mathbf{S}_k \boldsymbol{\tau}_k \qquad (14)$$

Equation (14) represents an overdetermined linear system of n equations with $n-1$ unknowns. Such a linear system has a feasible solution because $\mathbf{w}_{max}$ lies inside the available wrench set. The overdetermined system can be easily solved by applying numerical methods such as Gaussian Elimination.

5 Simulation and Results

In this section, the proposed methodology is applied to three different cable robots. The chosen topologies have $n \leq 3$ degrees of freedom to allow the visualization of the wrench zonotope. All the examples aim at finding the maximum and minimum exertable force in the direction of interest d. Such direction is depicted with an orange arrow in Figs. 3, 5 and 6.

The forces $\mathbf{F}$ and torques $\mathbf{M}$ exerted on the moving platform are rewritten in a more suitable reference frame as follows

$$\mathbf{w}_d = \begin{bmatrix} F_d & F_{o1} & F_{o2} & M_d & M_{o1} & M_{o2} \end{bmatrix}^T = \mathbf{R}^T \mathbf{S}_d \boldsymbol{\tau} \qquad (15)$$

where $\mathbf{R}$, $\mathbf{S}_d$ and $\boldsymbol{\tau}$ have been defined in Sect. 2.

1. *Cable suspended configuration*

    Figure 3 shows the cable robot configuration on the left and the corresponding wrench zonotope for the given pose on the right.

    The three degrees of freedom of the point-mass end effector are controlled by three cables attached to the points $\mathbf{A}$ $(-1;1;0)$m, $\mathbf{B}$ $(0;1;0)$m and $\mathbf{C}$

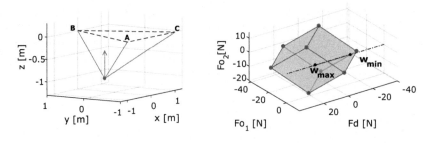

Fig. 3. m = 3 n = 3 robot configuration (left) and available wrench set (right)

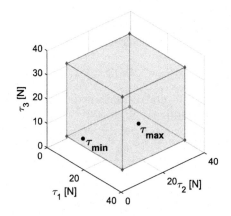

Fig. 4. Tension cube

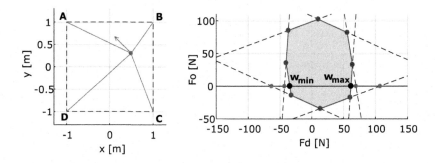

Fig. 5. m = 4 n = 2 point mass cable robot (Color figure online)

(1;−1;0)m. The direction of interest d is aligned with the z-axis. The green parallelepiped identifies the available wrench set in the 3-dimensional space. The dash-dotted line is the straight-line r; the intersection between r and Ω, depicted as a black segment, is the feasible range for a force exerted along d. Hence, it is possible to identify the two points $\mathbf{w}_{max} = [16\text{N}, 0, 0]^T$ and $\mathbf{w}_{min} = [-12\text{N}, 0, 0]^T$ on the surface of Ω.

As for the tension configurations associated with the extreme wrenches $\mathbf{w}_{min}$ and $\mathbf{w}_{max}$, they belong to the tension hypercube hull, and have $m-n+1$ tensions at their maximum or minimum values, i.e., τ_k.

For the cable suspended configuration showed in Fig. 3, τ_k is actually a scalar value. In particular, the tension configuration $\boldsymbol{\tau}_{max}$ exerting the wrench $\mathbf{w}_{max}$ belongs to the face $\tau_1 = \tau_{max}$, while the tension configuration $\boldsymbol{\tau}_{min}$ exerting the wrench $\mathbf{w}_{min}$ belongs to the face $\tau_2 = \tau_{min}$.

Figure 4 shows the cube $\mathbf{T}$ in the 3-dimensional tension space; in this simulation the tension limits are chosen as: $\tau_{min} = 5\text{N}$ and $\tau_{max} = 35\text{N}$. The points $\boldsymbol{\tau}_{max} = [\tau_{max}, 13\text{N}, 20.5\text{N}]^T$ and $\boldsymbol{\tau}_{min} = [13.5\text{N}, \tau_{min}, 7.9\text{N}]^T$ can be easily identified on the cube surface.

2. *Over-constrained configuration*

The second example considers a planar point-mass manipulator controlled by four cables. Figure 5 shows a schematic representation of the robot layout on the left and the available wrench set for the given pose on the right.

The four cables are attached to the vertexes of a square with side length of 2m; the origin of the fixed reference frame is located in the centroid of the square. The available wrench set is computed with reference to a direction of interest $\theta = \dfrac{3}{4}\pi$ (depicted with an orange arrow). In this case, Ω is the octagon depicted with a green area in Fig. 5.

Again, the intersection between the x-axis (i.e., the straight line r) and the green area is a segment whose extreme points are $\mathbf{w}_{max} = [61.7\text{N}, 0]^T$ and $\mathbf{w}_{min} = [-34.5\text{N}, 0]^T$. Similarly to the example presented in Sect. 4, the straight-lines supporting the edges of the polygon are depicted with dashed lines intersecting r in the points $\mathbf{w}_i$.

Looking at the corresponding tension configurations, the hypercube $\mathbf{T}$ belongs to the 4-dimensional tension space. For the point-mass cable robot, the known tensions vector τ_k belongs to $\mathbb{R}^3$; specifically, for the given pose $P = [0.5, 0.3]^T m$, the tension configurations exerting the maximum and the minimum force in the direction of interest are respectively $\boldsymbol{\tau}_{max} = [\tau_{max}, \tau_{max}, \tau_{min}, 37\text{N}]^T$ and the $\boldsymbol{\tau}_{min} = [\tau_{min}, 56\text{N}, \tau_{max}, \tau_{min}]^T$, where $\tau_{min} = 20\text{N}$ and $\tau_{max} = 70\text{N}$.

3. *Fully-constrained configuration*

The third example refers to a planar cable robot having a moving platform controlled by four cables. The cable output points are located at the vertexes of a square whose size is the same as in Fig. 5. The moving platform is rectangular, with height 0.2 m and width 0.35m. The orange arrow indicates the direction of interest (i.e., $\theta = \pi/4$) along which the platform should exert the maximum force while keeping constant its orientation. The available wrench set is a 3-dimensional figure, describing two translational and one rotational degrees of freedom. It is possible to identify the two intersection points $\mathbf{w}_{max} = [28.8\text{N}, 0, 0]^T$ and $\mathbf{w}_{min} = [-48.4\text{N}, 0, 0]^T$ between the straight line r and Ω.

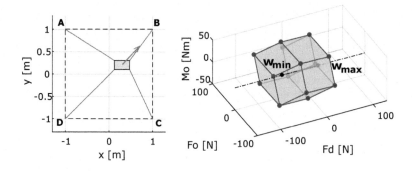

Fig. 6. m = 4 n = 2 point mass cable robot

In this case, the vector of known tensions has $m - n + 1 = 2$ elements. Specifically, for the given pose $P = [0.3, 0.2]^T m$, the results are $\boldsymbol{\tau}_{max} = [60.9N, \tau_{max}, 66.5N, \tau_{min}]^T$ and the $\boldsymbol{\tau}_{min} = [\tau_{min}, 27.7N, 14.9N, \tau_{max}]^T$, with $\tau_{min} = 10N$ and $\tau_{max} = 70N$.

6 Conclusion and Future Works

This paper introduced an efficient geometry-based algorithm to determine the maximum exertable wrench in a given direction. The proposed algorithm takes advantage of the *H-representation* of the wrench polytope and is not iterative. A method to obtain the set of cable tensions yielding the desired maximum wrench was also presented. Future studies will extend these methods to include special cases (i.e., poses of the end effector) which were not investigated in this work. The effectiveness of the proposed algorithms for real-time motion planning and control of cable-driven robots will then be verified through testing on a real prototype.

References

1. Barber, C.B., Dobkin, D.P., Huhdanpaa, H.: The quickhull algorithm for convex hulls. ACM Trans. Mathe. Softw. (TOMS) **22**(4), 469–483 (1996)
2. Borgstrom, P.H., Jordan, B.L., Sukhatme, G.S., Batalin, M.A., Kaiser, W.J.: Rapid computation of optimally safe tension distributions for parallel cable-driven robots. IEEE Trans. Robot. **25**(6), 1271–1281 (2009)
3. Boschetti, G., Passarini, C., Trevisani, A.: A strategy for moving cable driven robots safely in case of cable failure. In: Advances in Italian Mechanism Science, pp. 203–211. Springer (2017)
4. Boschetti, G., Trevisani, A.: Performance evaluation for cable direct driven robot. In: ASME 2014 12th Biennial Conference on Engineering Systems Design and Analysis. American Society of Mechanical Engineers (2014)
5. Boschetti, G., Trevisani, A.: On the use of the wrench exertion capability as a performance index for cable driver robot. In: Proceedings of ECCOMAS Thematic Conference on Multibody Dynamics, pp. 375–384 (2015)

6. Bouchard, S., Gosselin, C., Moore, B.: On the ability of a cable-driven robot to generate a prescribed set of wrenches. J. Mech. Robot. **2**, 011010 (2010)
7. Dai, X., Zhang, Y., Wang, D., Song, J.: Structural characteristics of force/moment polytopes of cable driven parallel mechanisms. In: Advances in Reconfigurable Mechanisms and Robots II, pp. 1482–1493. Springer (2016)
8. Firmani, F., Zibil, A., Nokleby, S.B., Podhorodeski, R.P.: Wrench capabilities of planar parallel manipulators. Part i: wrench polytopes and performance indices. Robotica **26**(06), 791–802 (2008)
9. Firmani, F., Zibil, A., Nokleby, S.B., Podhorodeski, R.P.: Wrench capabilities of planar parallel manipulators. Part ii: redundancy and wrench workspace analysis. Robotica **26**(06), 803–815 (2008)
10. Grünbaum, B.: Convex Polytopes, volume 221 of Graduate Texts in Mathematics (2003)
11. Lamaury, J., Gouttefarde, M.: A tension distribution method with improved computational efficiency. In: Cable-driven parallel robots, pp. 71–85. Springer (2013)
12. Lim, W.B., Yang, G., Yeo, S.H., Mustafa, S.K.: A generic force-closure analysis algorithm for cable-driven parallel manipulators. Mech. Mach. Theory **46**(9), 1265–1275 (2011)
13. Mikelsons, L., Bruckmann, T., Hiller, M., Schramm, D.: A real-time capable force calculation algorithm for redundant tendon-based parallel manipulators. In: IEEE International Conference on Robotics and Automation, ICRA 2008, pp. 3869–3874. IEEE (2008)
14. Nokleby, S.B., Fisher, R., Podhorodeski, R.P., Firmani, F.: Force capabilities of redundantly-actuated parallel manipulators. Mech. Mach. Theory **40**(5), 578–599 (2005)
15. Rosati, G., Zanotto, D.: A novel perspective in the design of cable-driven systems. In: ASME 2008 International Mechanical Engineering Congress and Exposition, pp. 617–625. American Society of Mechanical Engineers (2008)
16. Rosati, G., Zanotto, D., Agrawal, S.K.: On the design of adaptive cable-driven systems. J. Mech. Robot. **3**(2), 021004 (2011)
17. Zanotto, D., Rosati, G., Minto, S., Rossi, A.: Sophia-3: a semiadaptive cable-driven rehabilitation device with a tilting working plane. IEEE Trans. Robot. **30**(4), 974–979 (2014)
18. Zibil, A., Firmani, F., Nokleby, S.B., Podhorodeski, R.P.: An explicit method for determining the force-moment capabilities of redundantly actuated planar parallel manipulators. J. Mech. Des. **129**(10), 1046–1055 (2007)

Design and Applications

Design and Analysis of a Novel Cable-Driven Haptic Master Device for Planar Grasping

Kashmira S. Jadhao[1], Patrice Lambert[1(✉)], Tobias Bruckmann[2], and Just L. Herder[1]

[1] Delft University of Technology, Mekelweg 2, 2628 CD Delft, The Netherlands
jadhao.kashmira@gmail.com, {p.lambert,j.l.herder}@tudelft.nl
[2] University of Duisburg-Essen, Forsthauswweg 2, 47057 Duisburg, Germany
tobias.bruckmann@uni-due.de

Abstract. This paper introduces a novel cable-driven planar haptic device with 4 DOF, six actuators, and two end-effectors, which can be used to provide planar motion and grasping capabilities. In this design, the rigid end-effector that is found in regular cable-driven robots is replaced by a configurable platform, i.e. a closed-loop that possess some internal DOF, which in the present case is made of cables in tension. Both the position and the configuration of the platform can be fully controlled through motors located on the frame of the device, offering a novel solution to provide grasping capabilities in parallel cable-driven mechanisms. After establishing the governing kinematics and statics relations, a workspace analysis of the novel mechanism is presented. Then, a proof-of-concept prototype has been developed in order to validate the kinematics. Finally, an optimization of the design parameters for maximal compactness of the system is presented. This design is expected to find applications in haptics technology due to its unique gripping mechanism and high structural stiffness architecture.

1 Introduction

In order to render realistic force feedbacks in impedance controlled haptic device, a device must possess a high mechanical bandwidth, such that the high-frequency content of the forces occurring during contact with stiff environment can be rendered properly. Parallel mechanisms [1] are often used now a days as haptic devices since they offer higher stiffness and lower inertia than comparable serial devices [2]. This is mainly due to the fact that all their motors are generally located on or near the base.

In some haptic applications, it is interesting to allow the operator to interact with the remote or virtual environment via multiple contact points, allowing the user to feel the shape and stiffness of the manipulated object. When grasping capabilities are needed, a conventional approach is to mount an additional grasper with a dedicated motor at the top of the parallel manipulator [3]. This however results in additional inertia and lower mechanical bandwidth, since mass is added at the point that is the furthest from the base.

Cable-driven parallel manipulators have also been used as haptic devices [4–6] because, aside from the advantages of rigid-link parallel manipulators, they can also offer a practically unlimited translational workspace and even lower inertia. However, adding an actuated haptic gripper to a cable-driven mechanisms would be more complicated due to the absence of rigid links to attach the power and communication wires, and would proportionally have a worst effect on the total inertia of the system.

In this paper, we propose a new type of cable-driven parallel mechanism, which offers planar motion and grasping capabilities with force feedback, while all the motors are located at the base. The innovative structure is based on the use of a multi end-effectors configurable platform [7], also formed by cables in tension, for which both the position and configuration can be fully controlled from the actuator at the base.

After a general presentation of the system in Sect. 2, the notation used and the geometric, kinematic and static analysis of the system are presented in Sect. 3 and their governing equations are summarized. Section 4 attains to the workspace definitions and analysis to describe the reach of the robot. Section 5 presents a demonstrator developed to validate the design and summarizes the obtained experimental results. Finally, Sect. 6 presents a way to optimize the system for maximum workspace to robot frame ratio.

2 System Description

This section introduces the novel cable-driven architecture on which the haptic device is based. The innovation in this architecture is in the use of a configurable platform made of two cables in tension which allows the interface to interact with the operator via two end-effectors (EEs), allowing grasping capabilities.

Figure 1 shows the schematic of the 4 DOF (two EEs, each having x and y translation capabilities) six actuators, cable-robot with the novel grasping mechanism. The red dots represent the location of the actuators on the robot frame, with cables connecting to the blue colored EEs and the rings (indicated as circles). The rings are allowed to slide along the two platform cables in order for the system to achieve equilibrium. While the cable lengths attached to the motors can vary with different positions of the EEs, the summarized lengths of the two cables between the two EEs are kept constant. The cable lengths between the EEs are based on the desired gripping distances, i.e. the desired minimum and maximum distance between the EEs. The gripping distances considered are 0.04 m to 0.10 m for ergonomic purposes. To enable these gripping distances, the cable lengths between the EEs are considered to be 0.105 m each which is the minimum requirement for a 0.10 m gripping distance and collisions avoidance. Although the minimum number of cable needed to fully control a 4 DOF mechanism is 5, an over-redundant design with six cables was preferred in order to achieve symmetry in the structure and more symmetry in the robot workspace and performance. Compared to the use of a dedicated grasping motor on the end-effector, enabling grasping capabilities from the coupled action of the motors located on the base reduces the inertia of the haptic device, improving the mechanical bandwidth of the mechanism.

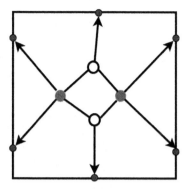

Fig. 1. Schematic of the 4 DOF, two end-effectors cable mechanism.

3 Geometric, Kinematic and Static Analysis

In this section, the kinematic relations between the end-effectors positions and the actuated cable lengths will be established based on the geometry of the architecture. The relations between the tensions in the actuated cables and the forces created at the end-effectors will also be introduced based on static equilibrium. We first introduce the notation and analysis of classical cable-driven robots and show how these concepts can be extended to a cable-driven robot with a configurable platform.

3.1 Modeling and Notation in Regular Cable-Driven Mechanisms

In case of planar robots, the orientation of each cable can be described with a unit vector $\mathbf{d}_i$ along the cable i as

$$\mathbf{d}_i = \begin{bmatrix} d_{ix} \\ d_{iy} \end{bmatrix} \quad (1)$$

When tension f_i is applied, the cable i exerts a pure force $f_i \mathbf{d}_i$ on the end-effector. As a cable robot can function only when the cables are in tension, f_i always has to be positive. For a mechanism with a point-shaped end-effector, the total force applied on end-effector must be zero under static equilibrium. These conditions can be represented with the following system of equations:

$$\mathbf{A}^T \mathbf{f} + \mathbf{w} = \mathbf{0}, \quad f_i > 0 \quad (2)$$

where the structure matrix is represented by $\mathbf{A}^T = [\mathbf{d}_1 \ \mathbf{d}_2 \ \mathbf{d}_3 \ \mathbf{d}_4 \ .... \ \mathbf{d}_m]$, the forces on the cables are summarized in $\mathbf{f} = [f_1 \ f_2 \ f_3 \ f_4 \ .... \ f_m]^T$ and $\mathbf{w}$ represents other external forces on the EE, like the gravity or other user defined forces.

The cable lengths (l_i) at each pose can be calculated by inverse kinematics. As explained by Tobias et al. in [12], it is possible to get satisfying force values, for low velocities with applied low minimum force limits. However, with increasing velocities and accelerations the gripper starts to wobble due to slack cables. To avoid this, it is suggested to set appropriate upper (f_{max}) and lower (f_{min}) bounds on the system forces. For our system, forces were bounded by $f_{min} = 5$ N and $f_{max} = 100$ N. Since the mechanism exhibits force redundancy, multiple values of feasible forces for each robot pose are possible. In order to get at least one dimensional solution set for force distribution for a specific robot pose, we select the minimum Euclidean norm of vector **f** as the applied forces. This optimal force solution set is computed for the desired system by using the MATLAB inbuilt optimization algorithm *fmincon*.

3.2 System Realization

Using the cable robot generic system equation as given in Eq. (2), the cable robot force equilibrium can be modeled.

$$\begin{bmatrix} d_{1x} & d_{2x} & d_{3x} \\ d_{1y} & d_{2y} & d_{3y} \end{bmatrix} \begin{bmatrix} f_1 \\ f_2 \\ f_3 \end{bmatrix} + \begin{bmatrix} 0 \\ 0 \end{bmatrix} = \begin{bmatrix} 0 \\ 0 \end{bmatrix} \tag{3}$$

Equation (3) represents the system equation for a three cables robot shown in Fig. 2(a). To extend this representation to our mechanism, we consider the complete architecture of the 4 DOF mechanism, as a collection of regular cable-driven mechanisms that are sharing some cables. For example, the system shown in Fig. 2(b) can be considered as two 3-cables mechanisms, sharing a common cable. This common cable introduces force constraints in the system. Forces f_3 and f_4 are equal in magnitude and opposite in directions as they represent the same cable. By concatenating the structure matrices for the two robots and adding an extra row to model the identical tension magnitudes in cables 3 & 4, the structure matrix for this arrangement is given by:

$$\begin{bmatrix} \mathbf{d_1} & \mathbf{d_2} & \mathbf{d_3} & 0 & 0 & 0 \\ 0 & 0 & 0 & \mathbf{d_4} & \mathbf{d_5} & \mathbf{d_6} \\ 0 & 0 & 1 & -1 & 0 & 0 \end{bmatrix} \begin{bmatrix} f1 \\ f2 \\ f3 \\ f4 \\ f5 \\ f6 \end{bmatrix} + \begin{bmatrix} 0 \\ 0 \\ 0 \\ 0 \\ 0 \end{bmatrix} = \begin{bmatrix} 0 \\ 0 \\ 0 \\ 0 \\ 0 \end{bmatrix} \tag{4}$$

Following this principle, the 4 DOF mechanism can be modeled as four mechanisms (2 nos. each of 3-cables & 4 cables mechanisms) sharing cables, as shown in Fig. 3(a). Besides the force constraints introduced by the shared cables on the system, the rings introduce additional constraints. The rings are not fixed entities like the *EEs* but are allowed to slide along the platform cables to achieve equilibrium when the *EEs* attain a particular configuration. The cables on the two sides of the rings are indeed the same cable. This results in the force equalities $f_5 = f_7$ and $f_{12} = f_{14}$. Concatenating the 4 mechanisms and adding these constraints due to the shared cables, the system equation for a 6 cable robot with 2 *EEs* is obtained.

A Novel Cable-Driven Haptic Master Device for Planar Grasping 311

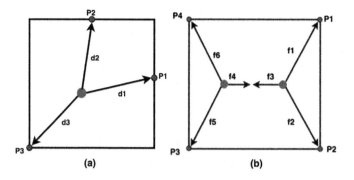

Fig. 2. Geometric modeling.

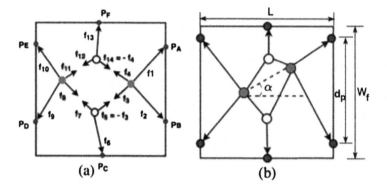

Fig. 3. (a) 4 DOF six cable robot with force equality constraints, the shared cables have equal and opposite forces for respective robots. (b) system configuration with end-effectors at an angle.

$$\begin{bmatrix} d_1 & d_2 & d_3 & d_4 & 0 & 0 & 0 & 0 & 0 & 0 & 0 & 0 & 0 & 0 \\ 0 & 0 & 0 & 0 & d_5 & d_6 & d_7 & 0 & 0 & 0 & 0 & 0 & 0 & 0 \\ 0 & 0 & 0 & 0 & 0 & 0 & 0 & d_8 & d_9 & d_{10} & d_{11} & 0 & 0 & 0 \\ 0 & 0 & 0 & 0 & 0 & 0 & 0 & 0 & 0 & 0 & 0 & d_{12} & d_{13} & d_{14} \\ 0 & 0 & 0 & 1 & -1 & 0 & 0 & 0 & 0 & 0 & 0 & 0 & 0 & 0 \\ 0 & 0 & 0 & 0 & 0 & 0 & 1 & -1 & 0 & 0 & 0 & 0 & 0 & 0 \\ 0 & 0 & 0 & 0 & 0 & 0 & 0 & 0 & 0 & 1 & -1 & 0 & 0 & 0 \\ 0 & 0 & 1 & 0 & 0 & 0 & 0 & 0 & 0 & 0 & 0 & 0 & 0 & -1 \\ 0 & 0 & 0 & 0 & 1 & 0 & -1 & 0 & 0 & 0 & 0 & 0 & 0 & 0 \\ 0 & 0 & 0 & 0 & 0 & 0 & 0 & 0 & 0 & 0 & 1 & 0 & -1 \end{bmatrix} \begin{bmatrix} f1 \\ f2 \\ f3 \\ f4 \\ f5 \\ f6 \\ f7 \\ f8 \\ f9 \\ f10 \\ f11 \\ f12 \\ f13 \\ f14 \end{bmatrix} + \begin{bmatrix} 0 \\ 0 \\ 0 \\ 0 \\ 0 \\ 0 \\ 0 \\ 0 \\ 0 \\ 0 \\ 0 \\ 0 \end{bmatrix} = \begin{bmatrix} 0 \\ 0 \\ 0 \\ 0 \\ 0 \\ 0 \\ 0 \\ 0 \\ 0 \\ 0 \\ 0 \\ 0 \end{bmatrix} \quad (5)$$

Equation (5) represents the system equation of the 6 cable robot. In order to simulate this system of equations, it is necessary to know the positions of the rings and end-effectors a priori. The positions of the EEs are the input to inverse kinematics. If a ring can slide freely on a platfrom cable while keeping the cable in tension, it will trace the trajectory of an ellipse with the two EEs at the focii as shown in Fig. 4(a). In order to compute the positions of the rings, two hypothesis were laid down.

(1) The rings will be positioned on the ellipse at a point which is at shortest distance from the pulley connected to that ring.
(2) The line joining the pulley and the ring on the ellipse, aligns with the angle bisector of the angle formed by joining the two focii to the ring as in Fig. 4(b).

The second hypothesis comes from the fact that, at each junction (of rings/end-effectors), the forces are in equilibrium. Which means force f_6 in Fig. 3(a) is a resultant of the forces f_5 and f_7. In theory, the resultant should be equal and opposite to force f_6. The second hypothesis in fact validates the first hypothesis. Simulating the two conditions proved that the angle bisector described in hypothesis (2) is indeed the shortest distance discussed in hypothesis (1).

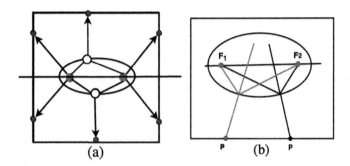

Fig. 4. System description (a) the ellipse concept with EEs at the foci and rings tracing the ellipse trajectory. (b) the shortest distance and angle bisector hypothesis, F are the ellipse foci.

4 Workspace Analysis

This section deals with the determination of workspace for the desired system [8,9,11]. The position of the mechanism is defined as the point at the mid distance between the two end-effectors. A point is considered a part of the workspace, if for its defined pose, all the cables have positive forces within the stated bounds [10]. To determine the workspace of the desired system, each point in the workspace is scanned to check and Eq. if all the required conditions are satisfied. Figure 5(a) and (b) represent the robot workspace with an end-effector angle of 0° & 10° respectively. The grasping distance is 0.10 m and the design parameters as shown in Fig. 3(b) are $d_p = 0.15$ m and $L = W_f = 0.30$ m.

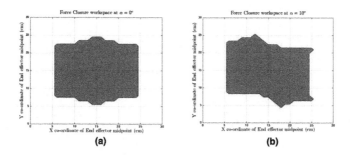

Fig. 5. Workspace with (a) EEs at $0°$ (b) EEs at $10°$.

This being a haptic device, it is important for the operator that the workspace is easily conceivable. To deal with this problem, regular shaped workspace within the actual workspace was determined which involved computing the largest conceivable shape possible in the obtained contour. The new retrieved workspace will henceforth be referred as *Useful Workspace*, as shown in Fig. 6(a).

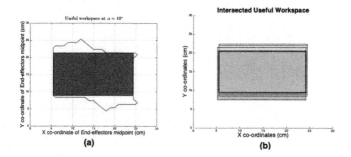

Fig. 6. (a) Useful workspace with EEs at $10°$. (b) intersected Useful workspace area.

A point is a part of the translational workspace if all postures of robot are possible at that point i.e. at all defined gripper distances (distance between two EEs, g) and EE angles (represented by α in Fig. 3(b)) within certain ranges. It is observed that workspace size reduces with decreasing gripper distance and increasing magnitude of angle between the EEs. For the purpose of this paper, a gripper distance of 0.04 m to 0.10 m and an angular range for α of $\pm$ 10° is considered. By varying geometric and design parameters d_p, W_f, L, α (as shown in Fig. 3(b)) and the gripping distance, different translational workspaces can be obtained.

An useful translation workspace for all gripper distances and different EE angles is computed as the intersection of useful workspaces for each gripping distance and EE angle values. As expected, this final workspace is much smaller than the initially computed useful workspace for individual gripping distances and EE angles. This whole process is depicted in Fig. 6(b).

5 Proof of Concept and Kinematic Validation

In order to validate the kinematic and static model of the novel architecture, a proof-of-concept demonstrator of the system was built and experiments were conducted for various tensions in the cables. This section illustrates the technical design of the robot architecture based on the proposed kinematic model. Task performance of the system is judged with positioning accuracy. The design is done keeping in mind haptics interface, which is considered as the final application of the system, and attention has been paid to ergonomics. These factors will govern the values of design parameters of the robot.

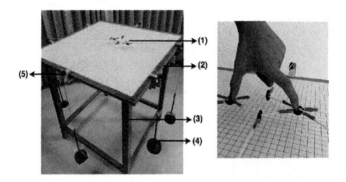

Fig. 7. Experimental set up (1) gripper (2) wooden plank, robot frame (3) table frame (4) weights and sandbags (5) pulley. Also shown: Gripping mechanism with two EEs and rings.

Figure 7(a) shows the experimental set up. It consists of a metallic table frame to hold the robot frame. A wooden plank of 0.75 m × 0.75 m served as the robot frame. Six pulleys mounted on adjustable C-clamps were screwed to the frame. The distance between the two pulleys on the same edge of the frame was fixed at 0.19 m each from the center, while the single pulleys were positioned in the middle of the edge. Nylon wires with low elasticity were used as cables. For the gripper, two cups and two rings were used as shown in Fig. 7.

The 3D printed gripper cups have a dimension of 1.5×10^{-2} m diameter for a human finger to fit in. Gripper distances can be varied from 0.04 – 0.10 m, these figures are again ergonomically influenced, considering the minimum and maximum grasping possible for a human hand. For providing tensions in the cables, discrete weights with sand bags for weight flexibility were hanged from the cables. A grid paper of 1×10^{-2} m x 1×10^{-2} m grids was stuck on the wooden plank for measurements.

While carrying out the experiment, it was important to take into account that the mechanical forces operating at the grippers, must be within a range that a human can exert. The considered force bounds for the system are 5–100 N which correspond to weights of 0.5 – 10 kg, operators can comfortably exert forces of any value in this range. Friction and other external forces on the system have not been accounted for in this design, however considering these forces will only alter the vector **w** of the system Eq. (2), which will not be a difficult task.

The main objective for kinematic validation is to test if the proposed designed kinematic model works in a real set-up. Seven different positions of EEs, with different angles and gripping distances were given to the MATLAB code as inputs and the force values were noted. These positions were randomly selected and are presented in Table 1. Weights equivalent to these forces were hanged from the cables and the gripper was allowed to attain equilibrium. Once the gripper achieved stability, the positions of the EEs with angle and gripping distances were noted and compared with the simulated values. At every test point, the end-effector positions were marked with a pencil on the grid paper and the required measurements were taken from these markings after removing the EEs from the position. This helped in avoiding parallax to some extent. Table 1 shows a comparison between the simulated positions and the experimental positions, for the same cable tensions.

Table 1. Simulation & experimental data analysis

Simulated values			Experimental values			Error		
Angle (α) (deg)	Gripping dist. (g) (m)	Position (x,y) (m)	Angle (α) (deg)	Gripping dist.(g) (m)	Position (x,y) (m)	$\Delta \alpha$	Δg	($\Delta x, \Delta y$)
−10	0.06	(0.25, 0.26)	−11	0.064	(0.253, 0.27)	−1	0.4	(0.3, 1)
10	0.08	(0.58, 0.29)	9.2	0.83	(0.587, 0.293)	−0.8	0.3	(0.7, 0.3)
−5	0.07	(0.40, 0.40)	−5.4	0.072	(0.383, 0.39)	−0.4	0.2	(−1.7, −1)
−5	0.06	(0.29, 0.54)	−4.6	0.061	(0.292, 0.537)	0.4	0.1	(0.2, −0.3)
0	0.09	(0.20, 0.40)	0	0.096	(0.205, 40.05)	0	0.6	(0.5, 0.5)
5	0.08	(0.34, 0.34)	4.8	0.083	(0.33, 0.344)	−0.2	0.3	(−1, 0.4)
0	0.10	(0.47, 0.47)	0	0.095	(0.476, 0.47)	0	−0.5	(0.6, 0)

A number of factors affect the positioning accuracy of a system such as wear of parts, dimensional drifts, tolerances, assembly errors and limitations, friction, component manufacturing errors, measurement errors etc. These factors can explain the small deviations between the actual kinematic parameters and their nominal/experimentally obtained values.

6 Optimization

The workspace varies with size of the robot frame (length L and width W_f) and also with the distance between the two pulleys d_p as shown in Fig. 3(b). These three parameters namely L, W_f and d_p formed the design variables used for optimization of the useful workspace. Since the whole device can be scaled up or down, the optimization objective is defined as the ratio between area of the frame and area of the intersected useful workspace.

$$Ratio = \frac{Area\ of\ frame}{Area\ of\ useful\ workspace} \tag{6}$$

Table 2 briefly presents the optimization parameters, their roles and considered bounds.

Table 2. Optimization parameters

Parameter	Role	Notation	Bounds
Ratio	Objective function	Ratio	NA
Frame width	Design variable	W_f	10 to 20
Frame length	Design variable	L	10 to 20
Dist. between pulleys	Design variable	d_p	2.5 to W

Pattern Search and Grid Search optimization algorithms were used for this purpose and the optimizer was made to compute the workspace for angle values of $-5°$, $5°$ and $0°$ and gripping distances of 0.04 m – 0.08 m. These ranges are different than those considered for experimentation to aid optimization and reduce the simulation time. Grid Search algorithm performed a discrete crude search while a Pattern Search algorithm was used to refine the search. The entire workspace is scanned for feasible robot positions with a resolution of 0.01 m. The values for step size and resolution are considered to be discrete for the purpose of optimization as the complexity of the program and hence the computation time increases exponentially with increasing parameter and resolution precision.

Table 3. Optimization results for grid & pattern search algorithm

Parameter	Non-optimized value	Optimized value (grid search)	Optimized value (pattern search)
d_p (m)	0.10	0.14	0.16
L (m)	0.20	0.20	0.197
W_f (m)	0.20	0.20	0.197
Frame area (m^2)	0.04	0.04	0.039
Useful workspace (m^2)	93.5×10^{-4}	137.49×10^{-4}	137.49×10^{-4}
Ratio	4.28	2.91	2.83

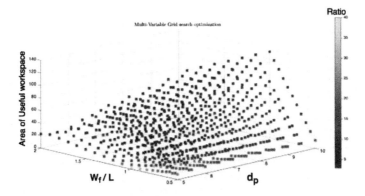

Fig. 8. Grid search plot.

Table 3 shows the optimization results for both the algorithms. Figure 8 show the variation of area of useful workspace with d and W_f/L.

Pattern Search algorithm was used to refine the search to cover the data skipped by Grid Search. However, there is not a significant difference between the results obtained in both the cases. This step to optimize the structure was done to get an idea about how the effective useful workspace changes with geometric parameters. Figure 8 shows how the workspace area varies with the chosen design parameters.

7 Conclusions

This paper presented a novel cable-driven master device for planar grasping for haptics interface. The need for a traditional gripper at the robot EE is eliminated by the unique gripper design with the two EEs and rings. Fixing the positions of the two EE, the position of the rings was geometrically computed for each pose of the robot. The wrench feasibility of the cable robot was investigated by $fmincon$ algorithm, such that all cables have positive tensions. The optimum values of cable forces were obtained which were set to fluctuate within pre-set minimum and maximum values of cable tension. The kinematic theory was validated through a working demonstrator. A comparative study of the results obtained experimentally and through simulations was done to support the design validation. Finally, a design optimization was carried out to maximize the workspace of the device for a given frame dimension. This design is expected to find applications in haptics technology due to its unique gripping mechanism and light and sturdy architecture.

References

1. Jean-Pierre Merlet. Parallel Robots, vol. 53 (2013)
2. Pandilov, Z., Dukovski, V.: Comparison of the characteristics between serial and parallel robots. Fascicule **1**, 2067–3809 (2014)
3. Tobergte, A., Helmer, et al.: The sigma. 7 haptic interface for MiroSurge: a new bi-manual surgical console. In: 2011 IEEE/RSJ International Conference on Intelligent Robots and Systems (IROS), pp. 3023–3030. IEEE (2014)
4. Robert, L., Williams, I.I.: Cable-suspended haptic interface. Int. J. Virtual Reality **3**(3), 13–21 (1998)
5. Gosselin, C.: Cable-driven parallel mechanisms: state of the art and perspectives. Mech. Eng. Rev. **1**(1), DSM0004 (2014)
6. Tobias, B., Lars, M., Thorsten, B., Manfred, H., Dieter, S.: Design approaches for wire robots. ASME Conference Proceedings (49040), pp. 25–34 (2009)
7. Lambert, P.: Parallel robots with configurable platforms: fundamental aspects of a new class of robotic architectures. Proc. Inst. Mech. Engi. Part C J. Mech. Eng. Sci. **230**(3), 463–472 (2016)
8. Ghasemi, A., Eghtesad, M., Farid, M.: Workspace analysis of planar and spatial redundant cable robots. In: Proceedings of the American Control Conference, pp. 2389–2394 (2008)
9. Berti, A., Merlet, J.P., Carricato, M.: Workspace analysis of redundant cable-suspended parallel robots. Mech. Mach. Sci. **32**, 41–53 (2015)
10. Bosscher, P., Riechel, A.T., Ebert-Uphoff, I.: Wrench-feasible workspace generation for cable-driven robots. IEEE Trans. Robot. **22**(5), 890–902 (2006)
11. Loloei, A.Z. Aref, M.M., Taghirad, H.D.: Wrench feasible workspace analysis of cable-driven parallel manipulators using LMI approach. In: IEEE/ASME International Conference on Advanced Intelligent Mechatronics, AIM, pp. 1034–1039 (2009)
12. Bruckmann, T., Pott, A., Franitza, D., Hiller, M.: A modular controller for redundantly actuated tendon-based stewart platforms. In: EuCoMes, Obergurgl, Austria (2006)

On the Design of a Three-DOF Cable-Suspended Parallel Robot Based on a Parallelogram Arrangement of the Cables

Dinh-Son Vu, Eric Barnett, Anne-Marie Zaccarin, and Clément Gosselin[✉]

Robotics Laboratory, Department of Mechanical Engineering,
Pavillon Adrien-Pouliot, Université Laval, Québec, QC G1V 0A6, Canada
{dinh-son.vu.1,eric.barnett.1,anne-marie.zaccarin.1}@ulaval.ca,
clement.gosselin@gmc.ulaval.ca

Abstract. An original design for a cable-suspended mechanism based on six cables, but actuated with three motors, is proposed in this paper. Each pair of cables is wound by a single actuator and their attachment points on the mobile platform and on the fixed base form a parallelogram, so that the orientation of the mobile platform remains constant while performing translational movements. First, the paper presents the architecture of the three-degree-of-freedom (three-DOF) manipulator and its corresponding kinematic equations. Then, the static workspace of the mechanism is determined analytically based on the simplification of the Jacobian matrix for a constant orientation of the mobile platform. Finally, the static workspaces of several cable arrangements are compared in order to assess the capabilities of the presented mechanism. In particular, one configuration of the three-DOF system with crossing cables is studied in more detail.

1 Introduction

Cable-driven parallel mechanisms offer numerous advantages compared to rigid-link robots, including large workspaces, high dynamic movement capabilities, effective payload-to-mass ratios, and ease of implementation. However, the inherent drawback of cable-driven robots is that cables cannot push on the moving platform; they can only exert pulling forces. Fully constrained end-effectors require $(n+1)$ cables in order to control n degrees of freedom. However, cable-suspended parallel robots use gravity in order to maintain cable tension, which acts similarly to a cable pulling downward. Therefore, the number of physical cables required to drive the effector is equal to the number of actuated degrees of freedom, assuming that some limitations on the platform accelerations are satisfied.

Workspace assessment is a critical step during cable mechanism design. Different approaches have been explored to determine the achievable positions of the mobile platform for different robot configurations. The static equilibrium workspace of a point-mass effector [1] gives all the positions for which the tension of all cables remains positive. Similarly, the constant orientation workspace of a mobile platform, which has been studied for the Robocrane [2] and for a six-DOF cable-suspended robot [3], provides its possible positions for a given constant orientation, assuming positive cable tensions. The calculation of the interference-free workspace [4] examines the static workspace of cable robots while taking into account the possible contacts between the cables. The volume obtained by setting a maximum and minimum tension in the cables and a particular wrench load applied at the platform is defined as the wrench-feasible workspace [5–7] and its shape is a zonotope whose boundaries depend on the tension limits.

Examples of six-DOF cable mechanisms that control the movement of the end-effector in both translation and orientation can be found in [2,3,8]. However, in many tasks for which cable-suspended parallel mechanisms are potential candidates, only translational motions are required. For example, in [9], a six-DOF cable-driven robot comprising eight actuators is proposed for pick-and-place operations and in [10], a six-DOF cable-driven robot that uses six actuators is proposed for the 3D printing of large objects, an application that only requires translational motion at constant orientation. In such cases, it is advantageous to reduce the number of actuators to the number of degrees of freedom required at the platform. Therefore, similarly to what was done in [11–13], this paper introduces the design of a three-DOF cable-suspended mechanism for the three spatial translations with constant orientation of the end-effector. First, the possible geometric arrangements of the three-DOF mechanism are described. Then, the general kinematic modelling of a six-DOF cable-suspended mechanism and the analytical calculation of the equations defining its static workspace for constant platform orientation are derived. This general result is then applied to the particular architecture of the proposed three-DOF mechanism with the cables crossing over the platform. Finally, the static equilibrium workspace obtained with the three-DOF mechanism for a few different configurations is compared with the workspace of a six-DOF cable-suspended robot used for large-scale 3D printing [10].

2 Geometry of the Three-DOF Cable-Suspended Mechanism with Parallelogram Architecture

Six individually actuated cables are generally used to constrain the suspended mobile platform in the six spatial DOFs. However, as shown in [11–13], if only translations are needed with a constant orientation, cables can be arranged as parallelograms and driven using only three independent motors. However, the geometric arrangement of the parallelograms proposed in [11–13] is based on a "convergent" design, which significantly limits the static workspace, as it will be shown in this paper.

The architecture proposed in this paper is based on parallelograms but the geometry is inspired from that proposed in [8]. The cables are crossing over the moving platform, thereby increasing the workspace without inducing cable interferences. It should also be pointed out that cable-driven mechanisms using

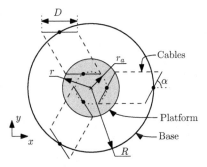

Fig. 1. Arrangement of the parallelograms based on the parallelogram width D, the orientation of the platform attachment points α and the distance to the attachment points relative to the centre of the moving platform r_a. The platform and the base are parallel horizontal planes at different elevations.

one motor to drive two parallel cables have also been proposed in the literature [14] in order to constrain *planar* cable-driven mechanisms to the plane of motion.

2.1 Architecture of the Three-DOF Mechanism

Figure 1 presents the parameters that are used to describe the geometry of the attachment points, and thus the arrangement of the parallelogram of each pair of cables driven by the same actuator. The parallelogram width (distance between two cables in a pair) and orientation are described by parameters D and α, respectively. Parameter r_a is the radius of the circle on the moving platform on which the midpoint of the line segment joining each pair of attachment points is placed. The size of the moving platform and the base are directly linked to the radii r and R respectively, which are primary design parameters that affect the footprint of the mechanism. Radius R corresponds to the circle on the base on which the centre of each pair of attachment points is located, while r is the radius of the smallest disk on the platform that includes all attachment points.

(a) Trivial architecture designed to avoid interference between cables.

(b) Architecture inspired by the geometry proposed in [8] for platforms with six actuators and crossing cables.

Fig. 2. Possible architectures of the three-DOF cable-suspended robot based on a parallelogram arrangement of the cables.

2.2 Example Architectures

The geometric parameters defined above strongly influence the achievable workspace and can be used to describe a broad variety of architectures [15]. Figure 2 shows two examples of possible architectures for a three-DOF cable-suspended mechanism based on a parallelogram arrangement of the cables, with the design parameters listed in Table 1. The trivial architecture shown in Fig. 2a avoids mechanical interference among the cables. This architecture corresponds to the design proposed in [11–13]. Crossing the cables over the mobile platform is known to increase the workspace of the mechanism, at the cost of potential interference among the cables [4,16]. However, by crossing only one cable of each pair that forms a parallelogram over the mobile platform, the static workspace of the robot can potentially be increased without generating this undesired interference. To maintain the parallelogram architecture, the fixed-frame cable attachment points must also be moved, as shown in Fig. 2b. This architecture, proposed here and inspired by that disclosed in [8], guarantees that there will be no cable-cable interference, a claim which can be proven by performing an analysis similar to that shown in [8] for a six-DOF robot.

Table 1. Parameters for the two architectures shown in Fig. 2. Position vector $\mathbf{b}_i$ goes from the centre of the platform to cable attachment point B_i, for the architecture with crossing cables, as shown in Fig. 3.

	r_a	D	α
Trivial architecture	$\frac{\sqrt{3}}{2}r$	r	$-\frac{\pi}{2}$
Architecture with crossing cables	0	$2r$	$-\frac{\pi}{3}$

$\mathbf{b}_1$	$\mathbf{b}_2$	$\mathbf{b}_3$	$\mathbf{b}_4$	$\mathbf{b}_5$	$\mathbf{b}_6$
$r[\frac{1}{2}, \frac{-\sqrt{3}}{2}, 0]^T$	$r[\frac{-1}{2}, \frac{\sqrt{3}}{2}, 0]^T$	$r[\frac{1}{2}, \frac{\sqrt{3}}{2}, 0]^T$	$r[\frac{-1}{2}, \frac{-\sqrt{3}}{2}, 0]^T$	$r[-1, 0, 0]^T$	$r[1, 0, 0]^T$

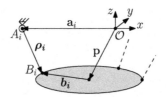

Fig. 3. Kinematic modelling of the cable platform.

3 Analytical Determination of the Static Workspace

The static equilibrium workspace corresponds to the set of positions of the moving platform for which the cables are in tension. Its analytical determination for constant orientation is based on the analysis of the Jacobian matrix of the manipulator.

3.1 Kinematic Modelling of a Six-DOF Cable-Suspended Robot

Even though the mechanism is actuated with only three motors, six cables transmit forces to the platform so the equations that describe the motion resemble those of a six-DOF cable-suspended mechanism. Figure 3 shows the notation used to establish the kinematic equations of the mechanism. For cable i, position vector $\mathbf{a}_i$ goes from the origin of the base frame to the base attachment point A_i, while position vector $\mathbf{b}_i$ goes from the centre of the platform (reference point) to the platform attachment point B_i. The position of the reference point of the moving platform with respect to the origin of the base frame is referred to as $\mathbf{p} = [x, y, z]^T$. The vector along cable i, connecting point A_i to point B_i is denoted $\boldsymbol{\rho}_i$. The relationship between vectors $\mathbf{a}_i$, $\mathbf{b}_i$, and $\boldsymbol{\rho}_i$ is as follows

$$\boldsymbol{\rho}_i = \mathbf{p} + \mathbf{b}_i - \mathbf{a}_i. \tag{1}$$

Since the orientation of the platform remains constant with respect to the base, the expression of the attachment points $\mathbf{a}_i$ and $\mathbf{b}_i$ can be written as $\mathbf{a}_i = [a_{ix}, a_{iy}, 0]^T$ and $\mathbf{b}_i = [b_{ix}, b_{iy}, 0]^T$, where a_{ix}, a_{iy}, b_{ix} and b_{iy} are design parameters related to the radii r and R of the moving platform and the base. The above equations assume that points A_i lie in one horizontal plane, and points B_i lie in a lower horizontal plane. Assuming massless straight cables and using the Newton-Euler approach to determine the static model of the mechanism yields

$$\mathbf{Mt} = \mathbf{g} \tag{2}$$

where $\mathbf{M}$ is the Jacobian matrix of the mechanism, whose expression is written as

$$\mathbf{M} = \begin{bmatrix} \boldsymbol{\rho}_1 & \boldsymbol{\rho}_2 & \boldsymbol{\rho}_3 & \boldsymbol{\rho}_4 & \boldsymbol{\rho}_5 & \boldsymbol{\rho}_6 \\ \boldsymbol{\rho}_1 \times \mathbf{b}_1 & \boldsymbol{\rho}_2 \times \mathbf{b}_2 & \boldsymbol{\rho}_3 \times \mathbf{b}_3 & \boldsymbol{\rho}_4 \times \mathbf{b}_4 & \boldsymbol{\rho}_5 \times \mathbf{b}_5 & \boldsymbol{\rho}_6 \times \mathbf{b}_6 \end{bmatrix}. \tag{3}$$

Parameter $\mathbf{t}$ is the vector containing the tensions in the cables per unit cable length and per unit platform mass and $\mathbf{g}$ the load wrench per unit platform mass, which is the external force applied on the effector. Their expressions are given as

$$\mathbf{t} = [t_1, t_2, t_3, t_4, t_5, t_6]^T, \qquad \mathbf{g} = [0, 0, g, 0, 0, 0]^T \tag{4}$$

where g is the gravitational acceleration and where it is assumed that the centre of mass of the platform is located at point $\mathbf{p}$, i.e., the centre of the platform.

3.2 Determination of the Static Workspace

The static workspace under constant orientation is now determined analytically based on Eq. 2. The static workspace corresponds to all the positions of the moving platform for which the tension of the cables are all positive. Moreover, the Jacobian matrix $\mathbf{M}$ can be simplified since the orientation of the platform is constant. First, the cross product $\boldsymbol{\rho}_i \times \mathbf{b}_i$ is computed as

$$\boldsymbol{\rho}_i \times \mathbf{b}_i = [\,-zb_{iy},\ zb_{ix},\ b_{iy}(x - a_{ix}) - b_{ix}(y - a_{iy})\,]^T. \tag{5}$$

Therefore, Eq. (2) can be rewritten as

$$\begin{bmatrix} x + b_{1x} - a_{1x} & \cdots \\ y + b_{1y} - a_{1y} & \cdots \\ z & \cdots \\ -zb_{1y} & \cdots \\ zb_{1x} & \cdots \\ b_{1y}(x - a_{1x}) - b_{1x}(y - a_{1y}) & \cdots \end{bmatrix} \begin{bmatrix} t_1 \\ t_2 \\ t_3 \\ t_4 \\ t_5 \\ t_6 \end{bmatrix} = \begin{bmatrix} 0 \\ 0 \\ g \\ 0 \\ 0 \\ 0 \end{bmatrix} \begin{matrix} \text{(I)} \\ \text{(II)} \\ \text{(III)} \\ \text{(IV)} \\ \text{(V)} \\ \text{(VI)} \end{matrix}. \tag{6}$$

Factoring out z in Eqs. (IV) and (V) yields the following expressions

$$b_{1y}t_1 + b_{2y}t_2 + b_{3y}t_3 + b_{4y}t_4 + b_{5y}t_5 + b_{6y}t_6 = 0 \tag{7}$$
$$b_{1x}t_1 + b_{2x}t_2 + b_{3x}t_3 + b_{4x}t_4 + b_{5x}t_5 + b_{6x}t_6 = 0 \tag{8}$$

which can then be used to simplify Eqs. (I), (II), (VI) since the sums introduced by Eqs. (7) and (8) appear in these equations. Thus, the system of Eq. (6) can be rewritten as

$$\begin{bmatrix} x - a_{1x} & x - a_{2x} & x - a_{3x} & x - a_{4x} & x - a_{5x} & x - a_{6x} \\ y - a_{1y} & y - a_{2y} & y - a_{3y} & y - a_{4y} & y - a_{5y} & y - a_{6y} \\ z & z & z & z & z & z \\ b_{1y} & b_{2y} & b_{3y} & b_{4y} & b_{5y} & b_{6y} \\ b_{1x} & b_{2x} & b_{3x} & b_{4x} & b_{5x} & b_{6x} \\ c_1 & c_2 & c_3 & c_4 & c_5 & c_6 \end{bmatrix} \begin{bmatrix} t_1 \\ t_2 \\ t_3 \\ t_4 \\ t_5 \\ t_6 \end{bmatrix} = \begin{bmatrix} 0 \\ 0 \\ g \\ 0 \\ 0 \\ 0 \end{bmatrix} \tag{9}$$

with

$$c_i = b_{ix}a_{iy} - a_{ix}b_{iy}, \quad i = 1, \ldots, 6 \tag{10}$$

which corresponds to the last component of the cross product $\mathbf{b}_i \times \mathbf{a}_i$. It can be observed that, as expected, the Jacobian matrix $\mathbf{M}$ is singular for $z = 0$, which occurs when the position of the reference point of the platform is in the plane defined by the fixed base. A potential boundary of the static workspace is found when one of the cable tensions is equal to zero. Setting one tension t_i to zero in Eq. (9) deletes the i-th column of the Jacobian matrix $\mathbf{M}$ and the i-th component of vector $\mathbf{t}$, which are denoted, respectively, matrix $\mathbf{M}_i$ with six rows and five columns and vector $\mathbf{t}_i$, which contains the remaining five tensions. The resulting system of equations is

$$\mathbf{M}_{i\ [6\times 5]}\mathbf{t}_{i\ [5\times 1]} = \mathbf{g}. \tag{11}$$

Then, one can define the i-th augmented matrix $\mathbf{M}_{a-i}$ based on matrix $\mathbf{M}_i$ and the wrench load $\mathbf{g}$, namely

$$\mathbf{M}_{a-i} = \begin{bmatrix} \mathbf{M}_i \mid \mathbf{g} \end{bmatrix} \tag{12}$$

which is a six-by-six matrix. For Eq. (11) to have a solution, vector $\mathbf{g}$ must be in the range of matrix $\mathbf{M}_i$, i.e., vector $\mathbf{g}$ must not be independent from the columns of $\mathbf{M}_i$. Therefore, the over-determined system in Eq. (11) yields a solution if the augmented matrix $\mathbf{M}_{a-i}$ has linearly dependent columns, which occurs when its determinant is zero, namely

$$\det(\mathbf{M}_{a-i}) = 0. \tag{13}$$

Equation (13) provides the positions $[x, y, z]$ where the tension of the i-th cable is zero, which corresponds to a potential limit of the static workspace. By inspection of Eq. (9), one can notice that expanding the determinant of Eq. (13) using the pivot placed on the third row and the last column of the augmented matrix $\mathbf{M}_{a-i}$ factors out variables g and z, which can then be eliminated because the determinant is set to 0. Thus, the static workspace of the effector for constant orientation is independent from the z coordinate of the platform and from the magnitude of the gravitational load. Moreover, subtracting the first column from columns 2 to 5 of the determinant of matrix $\mathbf{M}_{a-i}$ and expanding the determinant leads to the scalar equation

$$G_i x + H_i y + K_i = 0 \tag{14}$$

where coefficients G_i, H_i and K_i are functions of the architectural parameters of the mechanism only, defined by the components of vectors $\mathbf{a}_i$ and $\mathbf{b}_i$. Setting each cable tension to zero yields equations for six lines in the horizontal plane that define six half-planes when projected perpendicularly to the horizontal plane. The intersection of these six half-planes constitutes the static equilibrium workspace.

3.3 Application to a Three-DOF Cable-Suspended Mechanism with Crossing Cables

The architecture with crossing cables has particular properties that further simplify the system of equations presented in Eq. (9). First, one can define one pair of cables with vectors $\mathbf{a}_1, \mathbf{a}_2, \mathbf{b}_1, \mathbf{b}_2$ as shown in Fig. 4 and assume that the other two pairs are obtained with rotations of $\pm 2\pi/3$ around the centre of the base, referred to as the rotation matrices $\mathbf{Q}_1$ and $\mathbf{Q}_2$. For the particular architecture

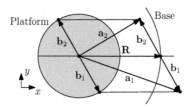

Fig. 4. Details of the architecture with crossing cables. The base lies in the plane $z = 0$ and the platform lies in a parallel plane at the z-coordinate of position vector $\mathbf{p}$, as defined in Fig. 3.

with crossing cables, the expressions for vectors $\mathbf{a}_1$ and $\mathbf{a}_2$ can be simplified as follows

$$\mathbf{a}_1 = \mathbf{R} + \mathbf{b}_1, \quad \mathbf{a}_2 = \mathbf{R} + \mathbf{b}_2 \tag{15}$$

where $\mathbf{R} = [R, 0, 0]^T$ is the position vector from the origin of the base frame to the midpoint of the line segment joining the attachment points on the base for the first pair of cables. Thus, the expressions for $\rho_1, \ldots, \rho_6$ from Eq. 1 become

$$\rho_1 = \rho_2 = \mathbf{p} - \mathbf{R}, \quad \rho_3 = \rho_4 = \mathbf{p} - \mathbf{Q}_1 \mathbf{R}, \quad \rho_5 = \rho_6 = \mathbf{p} - \mathbf{Q}_2 \mathbf{R}. \tag{16}$$

Moreover, the expression of the coefficients c_i in Eq. (9) is obtained from the last component of the cross product $\mathbf{b}_i \times \mathbf{a}_i$, whose expression can be written as

$$[\mathbf{b}_1 \times \mathbf{a}_1]_z = \underbrace{[\mathbf{Q}_1 \mathbf{b}_1 \times \mathbf{Q}_1 \mathbf{a}_1]_z}_{\mathbf{b}_3 \times \mathbf{a}_3} = \underbrace{[\mathbf{Q}_2 \mathbf{b}_1 \times \mathbf{Q}_2 \mathbf{a}_1]_z}_{\mathbf{b}_5 \times \mathbf{a}_5} = [\mathbf{b}_1 \times \mathbf{R}]_z = -R b_{1y}, \tag{17}$$

$$[\mathbf{b}_2 \times \mathbf{a}_2]_z = \underbrace{[\mathbf{Q}_1 \mathbf{b}_2 \times \mathbf{Q}_1 \mathbf{a}_2]_z}_{\mathbf{b}_4 \times \mathbf{a}_4} = \underbrace{[\mathbf{Q}_2 \mathbf{b}_2 \times \mathbf{Q}_2 \mathbf{a}_2]_z}_{\mathbf{b}_6 \times \mathbf{a}_6} = [\mathbf{b}_2 \times \mathbf{R}]_z = -R b_{2y}. \tag{18}$$

The results given in Eqs. (17) and (18) are obtained based on the fact that rotations $\mathbf{Q}_1$ and $\mathbf{Q}_2$ are performed in the same plane as vectors $\mathbf{b}_i$ and $\mathbf{R}$ and that the cross product is invariant to rotations performed in the plane defined by the two vectors to be multiplied. Using parameters $\mathbf{b}_i$ defined in Table 1 for the architecture with crossing cables, parameter r can be factored out from Eq. (9), which can then be further simplified after linear combinations among the last three rows, to become

$$\begin{bmatrix} x-R & x-R & x+\tfrac{1}{2}R & x+\tfrac{1}{2}R & x+\tfrac{1}{2}R & x+\tfrac{1}{2}R \\ y & y & y-\tfrac{\sqrt{3}}{2}R & y-\tfrac{\sqrt{3}}{2}R & y+\tfrac{\sqrt{3}}{2}R & y+\tfrac{\sqrt{3}}{2}R \\ z & z & z & z & z & z \\ 1 & -1 & 0 & 0 & 0 & 0 \\ 0 & 0 & 1 & -1 & 0 & 0 \\ 0 & 0 & 0 & 0 & 1 & -1 \end{bmatrix} \begin{bmatrix} t_1 \\ t_2 \\ t_3 \\ t_4 \\ t_5 \\ t_6 \end{bmatrix} = \begin{bmatrix} 0 \\ 0 \\ g \\ 0 \\ 0 \\ 0 \end{bmatrix}. \tag{19}$$

Expanding the determinant of the six augmented matrices $\mathbf{M}_{a-i}$ yields the three equations

$$x - R - \sqrt{3}y = 0, \qquad x - R + \sqrt{3}y = 0, \qquad x + \frac{1}{2}R = 0. \tag{20}$$

The equations for each cable in a given pair in Eq. (19) describe the same half plane, with the three half-planes then enclosing a triangular-prism-shaped volume. Furthermore, the static workspace of the architecture with crossing cables depends only on the radius R of the base frame and is independent from the size of the moving platform r.

4 Workspace Comparison for Cable-Suspended Parallel Mechanisms with Three Translational DOFs

The determination of the line equations that limit the static workspace for constant orientation is based on the calculation of the determinant of each augmented matrix $\mathbf{M}_{a-i}$, as shown in Subsect. 3.2. This method is now applied to different architectures for cable-suspended robots.

4.1 Mechanisms with Three DOFs

The analytical technique presented above is first applied to the two three-DOF architectures of Fig. 2. The static workspaces for both architectures, which are depicted with shaded areas, have triangular shapes, which means that the line equations that define the boundaries of the workspace are identical by pairs. For the trivial architecture, these equations are

$$x + \frac{1}{2}R - \frac{\sqrt{3}}{4}r - \sqrt{3}y = 0, \quad x + \frac{1}{2}R - \frac{\sqrt{3}}{4}r + \sqrt{3}y = 0, \quad x - \frac{1}{4}R + \frac{\sqrt{3}}{8}r = 0. \tag{21}$$

Figure 5 shows the static workspace delimited by the line equations of Eqs. (20) and (21), with unitary base radius R and platform radius $r = \frac{1}{2}R$. The static workspace for the trivial architecture shown in Fig. 5a is much smaller than that of the crossing-cables architecture shown in Fig. 5b. Indeed, crossing the cables over the mobile platform *significantly* increases the static workspace of the robot. The vertices of the static workspace for the crossing-cables architecture are located on the base circle of radius R, which represents an area about 12.5 times bigger than the area of the workspace of the trivial architecture.

4.2 Comparison with Six-DOF Architectures

Figure 6 shows the architecture and the workspace of a six-DOF cable mechanism actuated with six motors used for large-scale 3D printing [10] and for appearance modelling of objects [8]. This architecture is used as a reference for

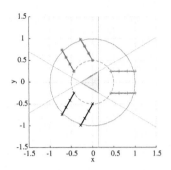
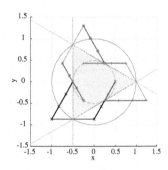

(a) Static workspace of the trivial architecture.

(b) Static workspace of the architecture with crossing cables.

Fig. 5. Static workspace of the centre point of the platform for two architectures of a three-DOF cable-suspended robot with parallelogram architecture. Distances in x and y are normalized by R, the radius of the base circle.

assessing the performance of the three-DOF cable-suspended architecture for translational motions proposed in this paper. The mobile platform for the six-DOF mechanism possesses three attachment points, with two cables connected to each, as shown in Fig. 6a. The line equations that define the static workspace for this mechanism are

$$x \pm R - \sqrt{3}y = 0, \quad x \pm R + \sqrt{3}y = 0, \quad x \pm \frac{1}{2}R = 0. \qquad (22)$$

The resulting hexagonal-prism-shaped workspace, depicted in Fig. 6b, is contained inside the static workspace shown in Fig. 5b for the proposed three-DOF architecture, which is about 1.5 bigger. Thus, the proposed architecture, shown in Fig. 2b, is more effective at generating a large static workspace. However, one can notice that upper bounds on the cable tensions were not taken

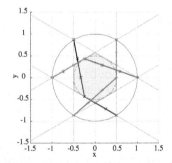

(a) Architecture of a six-DOF cable-suspended robot used for large-scale 3D printing [10].

(b) Corresponding static workspace of the six-DOF robot, for the centre point of the platform.

Fig. 6. Architecture and static workspace of a six-DOF cable-suspended robot.

into consideration during the analysis. Indeed, for the proposed architecture, only three motors support the weight of the effector whereas the wrench load is distributed among the six actuators for the configuration in Fig. 6a. Also, the wrench load is assumed to be applied at the centre of mass of the platform, which might not be the case in a real application.

5 Conclusion

This paper proposes the design of a three-DOF translational cable-suspended mechanism whose cables are arranged as parallelograms with cables crossing over the moving platform. The main motivation of the work is to reduce the number of actuators needed in a translational parallel cable-suspended robot while ensuring a large workspace. One of the applications of such a mechanism is large-scale 3D printing, which typically requires positioning an end-effector with a constant orientation. One particular arrangement of the proposed architecture was described and its static equilibrium workspace was determined analytically and compared to that of a six-DOF cable-suspended robot. It was shown that the proposed architecture produces a larger translational workspace, therefore justifying its potential to replace the system actuated with six motors. Future work involves the evaluation of the kinematic sensitivity of the three-DOF mechanism as well as the development of a prototype for experimentally assessing its static workspace and thus its capabilities for 3D printing and/or other applications.

Acknowledgements. Comments provided by Xiaoling Jiang and Pascal Dion-Gauvin were much appreciated during the research. This work was supported by The Natural Sciences and Engineering Research Council of Canada (NSERC), by the Fonds de la Recherche du Québec sur la Nature et les Technologies (FRQNT) and by the Canada Research Chair Program.

References

1. Riechel, A.T., Ebert-Uphoff, I.: Force-feasible workspace analysis for underconstrained, point-mass cable robots. In: IEEE International Conference on Robotics and Automation, pp. 4956–4962 (2004)
2. Albus, J., Bostelman, R., Dagalakis, N.: The NIST robocrane. J. Robot. Syst. **10**(5), 709–724 (1993)
3. Pusey, J., Fattah, A., Agrawal, S., Messina, E.: Design and workspace analysis of a 6–6 cable-suspended parallel robot. Mech. Mach. Theory **39**(7), 761–778 (2004)
4. Perreault, S., Cardou, P., Gosselin, C.M., Otis, M.J.-D.: Geometric determination of the interference-free constant-orientation workspace of parallel cable-driven mechanisms. J. Mech. Robot. **2** (2010)
5. Bosscher, P., Ebert-Uphoff, I.: Wrench-based analysis of cable-driven robots. In: IEEE International Conference on Robotics and Automation, pp. 4950–4955 (2004)
6. Stump, E., Kumar, V.: Workspaces of cable-actuated parallel manipulators. J. Mech. Des. **128**(1), 159–167 (2006)

7. Gouttefarde, M., Merlet, J.P., Daney, D.: Wrench-feasible workspace of parallel cable-driven mechanisms. In: IEEE International Conference on Robotics and Automation, pp. 1492–1497 (2007)
8. Gosselin, C., Bouchard, S.: A gravity-powered mechanism for extending the workspace of a cable-driven parallel mechanism: application to the appearance modelling of objects. Int. J. Automat. Technol. **4**(4), 372–379 (2010)
9. Pott, A., Bruckmann, T., Mikelsons, L.: Closed-form force distribution for parallel wire robots. In: Computational Kinematics, pp. 25–34 (2009)
10. Barnett, E., Gosselin, C.: Large-scale 3D printing with A cable-suspended robot. Addit. Manufact. **7**, 27–44 (2015)
11. Bosscher, P., Williams, R.L., Tummino, M.: A concept for rapidlydeployable cable robot search and rescue systems. In: International Design Engineering Technical Conferences and Computers and Information in Engineering Conference, pp. 1–10 (2005)
12. Saber, O.: A spatial translational cable robot. J. Mech. Robot. **7**(3) (2015)
13. Alikhani, A., Behzadipour, S., Vanini, S.A.S., Alasty, A.: Workspace analysis of a three DOF cable-driven mechanism. J. Mech. Robot. **1**(4), 041005-1–041005-7 (2009)
14. Lefrançois, S., Gosselin, C.: Point-to-point motion control of a pendulumlike 3-DOF underactuated cable-driven robot. In: IEEE International Conference on Robotics and Automation, pp. 5187–5193 (2010)
15. Gouttefarde, M., Collard, J.F., Riehl, N., Baradat, C.: Geometry selection of a redundantly actuated cable-suspended parallel robot. IEEE Trans. Robot. **31**(2), 501–510 (2015)
16. Otis, M.J.-D., Perreault, S., Nguyen Dang, T.-L., Lambert, P., Gouttefarde, M., Laurendeau, D., Gosselin, C.M.: Determination and management of cable interferences between two 6-DOF foot platforms in a cable-driven locomotion interface. IEEE Trans. Syst. Man Cybern. Part A Syst. Hum. **39**(3), 528–544 (2009)

On Improving Stiffness of Cable Robots

Carl A. Nelson(✉)

University of Nebraska-Lincoln, Lincoln, Nebraska, USA
cnelson5@unl.edu

Abstract. Stiffness of cable-driven parallel manipulators is dependent on the cable stiffnesses, cable tensions, and kinematic characteristics of the manipulator. In this paper, a general approach to stiffening this type of robot is put forward. The approach is based on two main principles: making constructive use of pulley-based force amplification, and adjusting the geometry of kinematic constraints. The approach is illustrated with an example.

1 Introduction

In cable-driven parallel manipulators, the cables can only carry tensile loads. These tensile force constraints inherent in cable-driven robots make their analysis and synthesis somewhat more complicated than for rigid-link robots. For example, this has bearing on the determination and optimization of feasible workspace [1] and the ability of the robot to interact meaningfully with its surroundings (apply loads) [2, 3]. It also requires special treatment of controller design [4]. Furthermore, cable interference with obstacles in the environment can be an issue, which has led to designs integrating movable cable anchor points [5].

In this paper the main focus is on robot stiffness. The stiffness behavior in these robots is known to be anisotropic and pose-dependent, and in addition to designing for a minimum desired stiffness map, it can also be desirable to optimize against failure of one or more cables/actuators to maintain a minimal stiffness threshold [6]. Cable mass is generally not negligible, and dynamic response is another aspect of stiffness behavior which is worthy of attention [7]. Increasing stiffness through increasing pretension in cables, though a popular solution, can also be problematic, as in some cases this can lead to robot instability [8]. Achievement of stiffness behavior which can be adjusted "on the fly" has also been a topic of interest for certain applications [9].

The remainder of the paper is organized as follows. In Sect. 2, a pulley-based method is presented for reducing tensile forces in the cables. Then in Sect. 3, this is enhanced by separating the spool/anchor locations, and the resulting pose dependency of stiffness behavior based on kinematic constraints is examined. An example is presented in Sect. 4, and conclusions are presented in Sect. 5.

2 Increasing Stiffness with Static Forces

One principle that can be brought to bear on the problem of robot stiffness is that higher cable tension produces higher stiffness of the end effector. Practically speaking, pretension in the cables is limited by the cables' strength properties. Therefore, one approach is to increase the number of cables (assuming the cable material and diameter to be fixed). However, in a general sense, adding cables adds constraints and therefore changes the mobility of the system; this can also exacerbate issues with cable interference [5].

In order to increase the effective number of cables and thereby alter the stiffness, without altering the kinematic characteristics of the system, the layout in Fig. 1 is proposed. By adding a single pulley, for a given cable tension the effective force on the end effector is doubled, whereas the kinematic constraint of a circular arc remains unchanged. Cable stiffness (considered as a lumped parameter in units of force per distance) can therefore be reduced by a factor of four(a factor of two due to each cable segment carrying half the load, and another factor of two due to the effective lengthening of cable being doubled) while retaining the same overall stiffness effect; another way of thinking of this is that, with lumped stiffness k equal to EA/L, and L effectively being doubled, either the cable cross-sectional area or the Young's modulus of the cable material can be reduced by half. Since motors tend to operate more efficiently at higher speed and lower torque, this allows overall improvements in robot performance (increased effective stiffness along with more frequent operation in the most efficient operating range of the actuator). One potential drawback is the need for larger spools, since the cable length doubles, but this is also offset by using smaller-diameter cables.

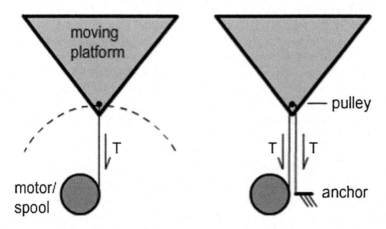

Fig. 1. Typical cable constraint (left); pulley-based cable constraint (right).

3 Increasing Stiffness by Modifying Geometry of Kinematic Constraints

Aside from cable tension and stiffness, the other main influence on manipulator stiffness is the geometry of the kinematic constraints. The Jacobian can be thought of as a way of expressing these constraints as a function of the robot's pose:

$$\delta \mathbf{P} = \mathbf{J}\delta \mathbf{q} \tag{1}$$

and

$$\tau = \mathbf{J}^T \mathbf{F} \tag{2}$$

where $\mathbf{q}$ is the vector of joint variables, $\mathbf{P}$ is the pose of the robot end effector, and $\mathbf{J}$ is the Jacobian matrix that couples the two; the Jacobian also couples the generalized joint efforts τ and the generalized forces applied at the end effector $\mathbf{F}$. Introducing an elastic actuator and/or cable,

$$\tau = \mathbf{K}\delta \mathbf{q} \tag{3}$$

results in a stiffness at the end effector involving both the Jacobian $\mathbf{J}$ and the matrix of stiffnesses $\mathbf{K}$:

$$\delta \mathbf{P} = \mathbf{J}\mathbf{K}^{-1}\mathbf{J}^T \mathbf{F} \tag{4}$$

Therefore the system stiffness depends on the component stiffnesses ($\mathbf{K}$) and the kinematic constraints (described in $\mathbf{J}$).

Without delving into the problem of singularities and directionality of stiffness properties, we can generalize for a single cable constraint as follows. For a given infinitesimal perturbation displacement of the end effector, a change in cable length (stretch) will occur, seen as a change in tensile force. The change in force divided by the change in length can be thought of as the cable's contribution to the manipulator stiffness in the direction of the perturbation displacement. It is apparent that the effective radius of curvature of the kinematic constraint imposed by the cable has an influence on the stiffness, particularly when the perturbation displacement is tangent to the curve of the constraint (see Fig. 2). Smaller radius of curvature leads to stiffer system behavior. Specifically, with reference to the nomenclature of Fig. 2, the dimensionless stretch in the cable is

$$dL/L = \left[(L + dy)^2 + dx^2\right]^{1/2}/L - 1 \tag{5}$$

and the larger the displacements relative to the unstretched cable length, the larger the resulting cable tension generated.

Based on this principle, the layout in Fig. 3 is proposed. In this configuration, the pulley principle of Fig. 1 is enhanced by allowing the radius of curvature to vary based

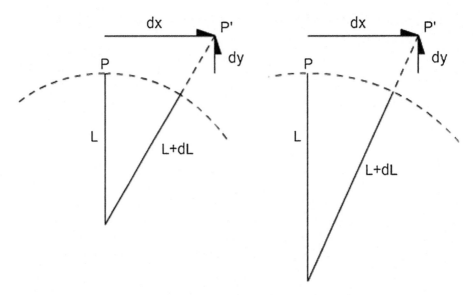

Fig. 2. Stiffness effect based on constraint path curvature – larger cable strain for smaller nominal constraint radius of curvature, for a given displacement.

on the pose of the end effector. This is done by locating the fixed cable anchor separately from the actuated cable spool, resulting in an elliptical constraint path as opposed to a circular one [10]. This can be considered a superior configuration in areas of the workspace for which the radius of curvature of the elliptical constraint path is smaller than that of the single-cable constraint (i.e., similar to Fig. 1) that would otherwise be used to achieve the given pose. For an ellipse with major and minor axes a and b described by

$$x^2/a^2 + y^2/b^2 = 1 \qquad (6)$$

or

$$x = a * \cos\theta, \quad y = b * \sin\theta \qquad (7)$$

and the location of its focus along the x-axis given by

$$c = (a^2 - b^2)^{1/2} \qquad (8)$$

The inverse kinematics are given by converting a position (x, y) or (r, θ) in the local coordinate frame of the ellipse to a cable length L (connecting the two foci to the point (x, y)) using the above equations given a fixed focal distance c, which results in

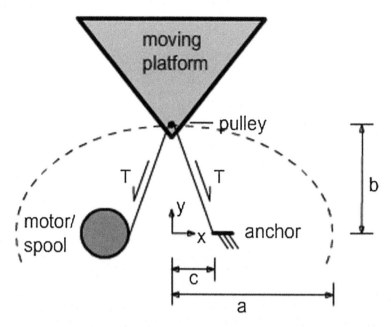

Fig. 3. Pulley-based cable constraint with separate anchor and spool locations, producing an elliptical constraint path.

$$r^2 = x^2 + y^2 \qquad (9)$$

$$a = (r^2 + c^2\sin^2\theta)^{1/2} \qquad (10)$$

and

$$L = L_1 + L_2 = 2a \qquad (11)$$

For an ellipse, it is known that the evolute (locus of centers of curvature) is an astroid. The local radius of curvature(length of the normal from a point on the ellipse to the corresponding point on the astroid) can be expressed as

$$r_c = (a^2\sin^2\theta + b^2\cos^2\theta)^{3/2}/ab \qquad (12)$$

and the most useful range of θ is the solution of these equations such that r_c is less than a characteristic radius of the system. Taking this characteristic radius to be the average of a and b, we have

$$(a^2\sin^2\theta + b^2\cos^2\theta)^{3/2}/ab < (a+b)/2 \qquad (13)$$

which gives the limiting value of θ in the vicinity of $\pi/4$ for a range of ellipse parameters, as shown in Fig. 4.

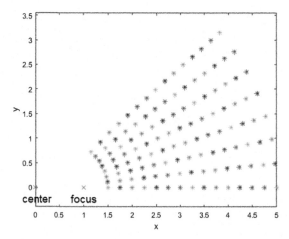

Fig. 4. High-stiffness portion of workspace for a single cable (fixed anchor and spool locations, but variable ellipse size, i.e., variable cable payout).

Because it may be difficult in practice to allow cables to swing past the pivot locations (anchor and spool), it is preferable to keep the usable workspace on one side of the ellipse's major axis. Therefore, the high-performance portion of the workspace (from the standpoint of a single cable) is limited to a segment of the ellipse, or range of angles, as shown in Fig. 4. A cable-actuated robot that leverages these principles on a system level by overlapping the cable-wise stiff areas of the workspace is shown qualitatively in Fig. 5.

4 Example

Consider the cable robot shown in Fig. 6. It is obvious from inspection that in this symmetric position the stiffness in the horizontal direction is inferior to that in the vertical direction. The stiffness value can be approximated numerically by assuming a small displacement from the nominal position and calculating the resultant restoring force. With $L = 1$ m, $dx = 0.01$ m, and cable stiffness of $K = 100000$ N/m, the cable stretches in the left and right cables are approximately 0.005 m and -0.005 m, inducing a net horizontal restoring force of approximately 1000 N. Substituting the cable-pulley arrangement of Fig. 1 (right) results in identical stiffness behavior using a cable stiffness of only 25000 N/m. The stiffness behavior can be further improved by adding the modification of Fig. 3 and optimizing the relative orientations α of the anchor and spool points (the ellipse foci). This was simulated in MATLAB with anchor/spool configurations symmetric about the line of displacement dx. For anchor/spool half-spacing $c = 0.3$ m, stiffness at this pose is optimized for left and right ellipse orientations of 1.48 and -1.57 rad, respectively, producing a 4% increase in stiffness compared to the baseline. Increasing the ellipse focal spacing to $c = 0.5$ m, the optimal left and right ellipse orientations are 1.40 and -1.57 rad, respectively, producing a stiffness improvement of more than 10% relative to the baseline. The stiffness

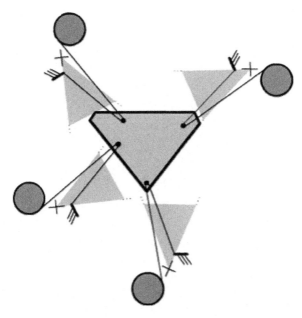

Fig. 5. Cable-based parallel manipulator configured for high stiffness in a portion of the workspace (zones of high stiffness shown for the respective pulley locations on the end effector).

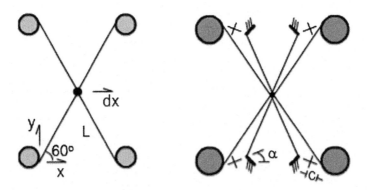

Fig. 6. Example of increasing robot stiffness using elliptical cable constraint paths.

improvement map as a function of the left and right ellipse orientations for $c = 0.5$ m is presented in Fig. 7. The general trend is that stiffness improvements increase with c (since increasing c decreases the effective radius of curvature of the constraint path), at the expense of a bulkier robot and possibly smaller usable workspace. It should be noted that, consistent with the principle illustrated in Fig. 4, there is a distinct "region" of anchor/spool orientations within which the stiffness behavior is improved, and outside of which the stiffness decreases with respect to the baseline. Therefore these orientations should be chosen carefully to correspond to the task requirements.

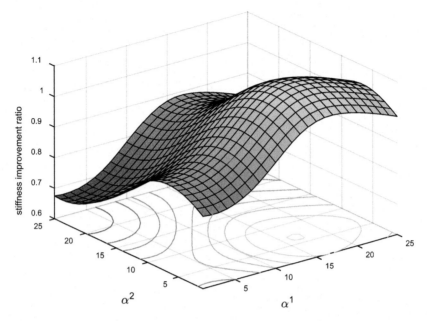

Fig. 7. Stiffness improvement map as a function of anchor/spool orientations for spacing of 0.5 m.

5 Discussion and Conclusions

The example just presented illustrates the optimization of stiffness for a specific pose. Just as manipulability is a local index of kinematic performance, and its associated principles have been adapted to develop global performance indices, this approach to stiffness analysis requires further extension in order to be applied as a global index. In particular, Eq. (4) can be explored in more detail to elucidate the combined effects of multiple elliptical kinematic constraints on overall manipulator stiffness. Although the principles outlined in this paper are illustrated in a planar sense, they are applicable to spatial cable-driven parallel robots as well. The principle of pulley amplification of tension is easily extended out of the plane, and the elliptical constraint presented here in the planar case becomes an ellipsoidal surface in the spatial case. Effectiveness of using pulleys to generate elliptical constraint paths for cable robots has been demonstrated through principle and example. Future work includes further analysis of appropriate stiffness optimization methods to accompany this general approach, development of more formal analytical approaches for dealing with these systems, and practical implementation in prototypes. This may also lead to new ways of adjusting stiffness and feasible workspace "on the fly" in combination with other design approaches previously mentioned [5, 9].

References

1. Fattah, A., Agrawal, S.K.: On the design of cable-suspended planar parallel robots. J. Mech. Des. **127**, 1021–1028 (2005)
2. Bosscher, P., Riechel, A.T., Ebert-Uphoff, I.: Wrench-feasible workspace generation for cable-driven robots. IEEE Trans. Robot. **22**(5), 890–902 (2006)
3. Bouchard, S., Gosselin, C.: On the ability of a cable-driven robot to generate a prescribed set of wrenches. J. Mech. Robot. **2**, 011010-1–011010-10 (2010)
4. Oh, S.-R., Agrawal, S.K.: Cable suspended planar robots with redundant cables: controllers with positive tensions. IEEE Trans. Robot. **21**(3), 457–465 (2005)
5. Bosscher, P., Williams II, R.L., Bryson, L.S., Castro-Lacouture, D.: Cable-suspended robotic contour crafting system. Autom. Constr. **17**, 45–55 (2007)
6. Moradi, A.: Stiffness analysis of cable-driven parallel robots, Ph.D. thesis, Queen's University, Kingston, Ontario, Canada (2013)
7. Yuan, H., Courteille, E., Deblaise, D.: Static and dynamic stiffness analyses of cable-driven parallel robots with non-negligible cable mass and elasticity. Mech. Mach. Theor. **85**, 64–81 (2015)
8. Behzadipour, S., Khajepour, A.: Stiffness of cable-based parallel manipulators with application to stability analysis. J. Mech. Des. **128**, 303–310 (2006)
9. Zhou, X., Jun, S.-K., Krovi, V.: Planar cable robot with variable stiffness. In: Hsieh, M.A., et al. (eds.) Experimental Robotics. Springer Tracts in Advanced Robotics, vol. 109 (2016). doi:10.1007/978-3-319-23778-7_26
10. Zhou, X.: Towards cooperative manipulation using cable robots, Ph.D. thesis, University at Buffalo, State University of New York (2014)

Optimal Design of a High-Speed Pick-and-Place Cable-Driven Parallel Robot

Zhaokun Zhang[1], Zhufeng Shao[1(✉)], Liping Wang[1], and Albert J. Shih[2]

[1] Beijing Key Lab of Precision/Ultra-Precision Manufacturing Equipments and Control, State Key Laboratory of Tribology, Tsinghua University, Beijing 100084, China
zzk15@mails.tsinghua.edu.cn,
{shaozf,lpwang}@mail.tsinghua.edu.cn
[2] College of Engineering, University of Michigan, Ann Arbor, MI 48109, USA
shiha@umich.edu

Abstract. The booming industrial demand has resulted in high-speed pick-and-place robots receiving increased attention from industry and academia. High-speed rigid parallel robots that comprise active pendulums and passive parallelograms limit high efficiency because of their complex structure and high cost. The cable-driven parallel robot (CDPR), which comprises parallel cables and tension branch, offers a promising new method. This study develops an optimal design of a CDPR with proposed novel transmission indices, which is normalized finite, dimensionally homogeneous, frame-free, and intuitive. The optimized result is verified with a numerical simulation. Given the optimized parameters, the vertical and physical prototypes of the CDPR are provided by considering the industrial application. The performance indices and optimal design metrology of this study can be further adopted in the optimal design of other CDPRs.

Keywords: Cable-driven parallel robot · Kinematic optimization · Mechanical design

1 Introduction

Robots are extensively used in industry, particularly for sorting and packaging operations, because of eternal pursuit of speed and efficiency. The serial mechanism with SCARA (selective compliance assembly robot arm) motion [1] is used as the pick-and-place manipulator firstly. Serial manipulators possess large workspace, but bulky serial open-chain configuration and large moving inertia limit high-speed performance. As the complementary to the serial mechanism, parallel manipulators possess inherent advantages in terms of low inertia of moving components, high speed, and good dynamic performance with closed kinematic chains. Parallel manipulators have been successfully adopted in various fields [2, 3], particularly as high-speed pick-and-place robots. For pick-and-place operations, 3 degrees of freedom (3-DoFs) translational motion is basically required. The Delta robot [4] is the most popular and extensively used one. In addition, high-speed parallel manipulators with SCARA motion are proposed, such as I4 [5], H4 [6], Par4 [7], and X4 [8], with one additional rotational DoF about the vertical direction. Above high-speed parallel manipulators have a common structural feature, that is, each limb comprises an active pendulum and a passive parallelogram, thereby limiting unnecessary terminal rotations. Lightweight materials, such as carbon fiber and aluminum alloy, are utilized to reduce the moving inertia, thereby leading to the complexity of structure and high cost.

When replacing the heavy rigid links of parallel robots with cables, cable-driven parallel robots (CDPRs) are proposed. Since the cable mass is negligible, the moving inertia of CDPRs can be reduced to the minimum, and extremely high speed and acceleration could be achieved with a simple structure and low cost [9, 10]. This condition makes CDPR a promising high-speed manipulator. Additional forces should be adopted in CDPRs to maintain all cables in tension because of the unilateral actuation property of cables. For high-speed CDPRs, the additional force is usually provided in two ways. Firstly, an additional driven cable is used. The FALCON-7 robot [11] is a typical example of the first class, which adopts seven cables to implement a 6-DoFs movement, and the maximum acceleration of 43 g can be achieved. The other approach is proposed by Landsberger [12]. A central extensible branch is employed, thereby creating a considerably simple structure and improved controllability. Furthermore, parallel cables are utilized to limit the end-effector's rotation, and a 3-DoFs translational CDPR was introduced [13] with the concept design. The current study executes the kinematic optimal design of the high-speed pick-and-place CDPR. Moreover, virtual and physical prototypes are deduced with detailed mechanical design, thereby promoting the industrial application of the CDPR.

General concerns toward CDPRs in terms of optimal design lie in "tension ability" [14] and stiffness [15]; these issues could be solved by the central extensible branch for our study object. Since transmitting force and motion is the nature of mechanisms, the kinematic optimal design of the object CDPR is carried out considering transmissibility. Although extensive studies in this field have been conducted, only a few existing indices can comply with all the requirements, such as simple, intuitive, and universal. Conditioning indices based on the Jacobian matrix are frame-related [16] and could cause resolution failure when applied to the translational parallel manipulator [17]. Angle-based transmission indices that are mainly used for planar manipulators, such as pressure and transmission angles [18], become difficult and complex in spatial analysis [19]. Another class of indices is deduced using screw theory, which is powerful but complicated and non-intuitive [20]. Thus, transmission index, which is frame-free, intuitive, and easy to calculate, is proposed and adopted in the current study for the CDPR based on matrix orthogonal degree.

The main methodologies for the optimal design can be divided into two categories. The first one is referred to as the algorithm-based optimum design, which converts the optimum design issue into a multi-objective optimization problem, and solved using complex nonlinear algorithms [21]. The second category is the atlas-based optimum design [22]. Optimum parameters can be determined accurately and intuitively based on the drawn performance atlases. Both methods are equivalent in deducing the optimal result. However, the atlas-based optimum design can illustrate the relationship between design parameters and manipulator performances, as well as immediately and conveniently adapt to the target changes.

This study employs atlas-based optimum design metrology to implement the kinematic optimal design of the object CDPR with proposed transmission index. The rest of this paper is organized as follows. Section 2 presents the virtual prototype, including the mechanical design details, of the CDPR and deduces the kinematic model. Section 3 focuses on the new transmission index and optimization of the CDPR. Section 4 analyzes and simulates the kinematic performance of the CDPR. Lastly, Sect. 5 presents the conclusions and prospects for this study.

2 Virtual Prototype and Kinematics

2.1 Virtual Prototype

The virtual prototype of the high-speed pick-and-place CDPR (see Fig. 1) comprises the base (1), actuation units (2), guiding pulleys (3), three groups of parallel cables (4), rigid extensible limb (5), and end effector (6). Partial design details of the virtual prototype are illustrated in Fig. 2. The actuation unit comprises the servomotor, reducer, coupling, and winch in series. Six cables are assigned into three groups, and two cables in each group are adjacent and parallel. Initially, the parallel cables are uniformly distributed circumferentially. Two cables of each group share the same actuation unit. Two identical spiral grooves are machined out on the surface of the winch to ensure the synchronized motion of the parallel cables.

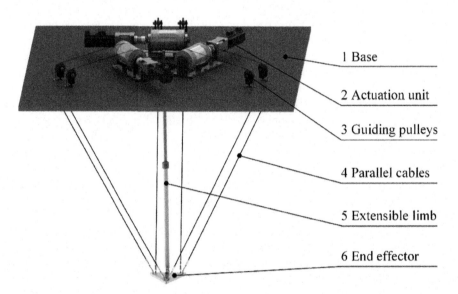

Fig. 1. Virtual prototype of the CDPR

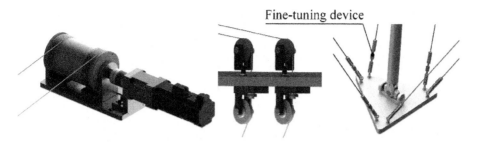

Fig. 2. Details of the actuation unit, guiding pulleys and end effector

Several measures are considered to ensure the parallel constraints of each pair of cables. Guiding pulleys are used to avoid cable wear caused by intense friction under high-speed motion. The lower pulleys are able to rotate vertically, and the rotation axis is coaxial with the corresponding cable. In practice, the lengths of the parallel cables may be dissimilar due to manufacturing and assembly errors, which may introduce unbalance loads of the cables. A fine-tuning device is attached to each cable in series near the end effector and utilizes the left-hand and right-hand threads at the ends. The cable length can be adjusted through the fine-tuning device, making the tension of parallel cables substantially the same to eliminate unbalanced loads. Given the parallelogram constraints, the unnecessary rotational DoFs are constrained and the purely translational motions of the end effector are guaranteed.

The rigid extensible limb is another important component of the CDPR. The extensible limb connects the base and end effector at the geometric centers through universal joints. The pitch and yaw angles of the limb can reach $\pm 45°$. Considering cost and energy consumption, spring instead of cylinder is employed in the prototype to apply appropriate pressure to the end effector, thereby maintaining all the cables in tension. Of course, the use of spring will inevitably bring a series of dynamic problems, such as effects on load capacity, acceleration capacity as well as vibration. These problems are left to be discussed later in the dynamic studies, while this article only focuses on the kinematics problems. Moreover, the limb could be utilized as the medium to implement terminal measurement and vibration suppression, thereby improving terminal accuracy and dynamics.

2.2 Inverse Kinematics

Although six cables connect the end effector and the base, the parallel cables of a group move synchronously and counteract the torque. In the inverse kinematic analysis, a group of parallel cables can be simplified into a single middle cable. The simplified kinematic model is shown in Fig. 3.

The upper triangle is the base and labeled as A_1, A_2 and A_3; and the feature size is

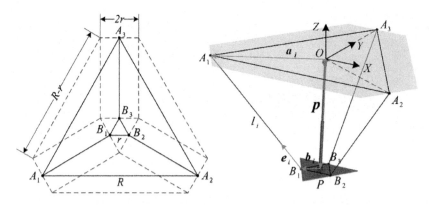

Fig. 3. Simplified kinematic model of the object CDPR

$|A_1A_2| = |A_2A_3| = |A_3A_1| = R$. The lower triangle is the end effector and labeled as B_1, B_2 and B_3. The size of the end effector is $|B_1B_2| = |B_2B_3| = |B_3B_1| = r$. P represents the center of the end effector. The global frame $\{O\text{-}XYZ\}$ is established at the geometric center of the base. The X-axis is parallel to A_1A_2, Y-axis points from O to A_3, and Z-axis is perpendicular to the base upward. Vectors $\boldsymbol{a}_i$ (vector connecting point O to point A_i) and $\boldsymbol{b}_i$ (vector connecting point P to point B_i) are defined as shown in the figure in the global frame. The end effector translation vector $\boldsymbol{p}$ connects point O to point P. $\boldsymbol{e}_i$ is the unit vector along the ith cable and the length of the ith cable is l_i. At the initial moment, branch OP is vertically downward with the maximum length h. Inverse kinematics, which is the basis of the optimal design and control, is the process to deduce the input variables (i.e., length, velocities, and accelerations of the three cables) with the given terminal trajectory.

The position vector of point P can be deduced with the vector chain as follows (see Fig. 3):

$$\boldsymbol{p} = \boldsymbol{a}_i - l_i\boldsymbol{e}_i - \boldsymbol{b}_i, \ (i = 1, 2, 3). \tag{1}$$

In addition, the length of each cable can be written as:

$$l_i = \|l_i\boldsymbol{e}_i\| = \|\boldsymbol{a}_i - \boldsymbol{b}_i - \boldsymbol{p}\|. \tag{2}$$

The unit vector along ith cable can be expressed as:

$$\boldsymbol{e}_i = (\boldsymbol{a}_i - \boldsymbol{b}_i - \boldsymbol{p})/l_i. \tag{3}$$

Taking the derivative of Eq. (1) with respect to time yields the velocity mapping function as follows:

$$\dot{\boldsymbol{p}} = -\dot{l}_i\boldsymbol{e}_i - l_i(\boldsymbol{\omega}_i \times \boldsymbol{e}_i), \tag{4}$$

where $\boldsymbol{\omega}_i$ is the angular velocity of the ith cable. By taking the dot product of Eq. (4) with unit vector $\boldsymbol{e}_i$ at both sides, the cable velocity can be obtained as:

$$\dot{l}_i = -\boldsymbol{e}_i \cdot \dot{\boldsymbol{p}}, \tag{5}$$

and

$$\dot{\boldsymbol{L}} = \begin{bmatrix} \dot{l}_1 & \dot{l}_2 & \dot{l}_3 \end{bmatrix}^{\mathrm{T}} = -\begin{bmatrix} \boldsymbol{e}_1 & \boldsymbol{e}_2 & \boldsymbol{e}_3 \end{bmatrix}^{\mathrm{T}} \dot{\boldsymbol{p}} = \boldsymbol{J}\dot{\boldsymbol{p}}, \tag{6}$$

where $\boldsymbol{J} = -\begin{bmatrix} \boldsymbol{e}_1 & \boldsymbol{e}_2 & \boldsymbol{e}_3 \end{bmatrix}^{\mathrm{T}}$ is the inverse Jacobian matrix of the robot.

Taking the derivative of Eq. (5) with respect to time yields the cable acceleration expression as follows:

$$\ddot{l}_i = d(\boldsymbol{e}_i \cdot \dot{\boldsymbol{p}}) = \dot{\boldsymbol{p}}(\boldsymbol{\omega}_i \times \boldsymbol{e}_i) - \boldsymbol{e}_i \cdot \ddot{\boldsymbol{p}} = -\dot{\boldsymbol{p}}(\dot{\boldsymbol{p}} + \dot{l}_i \boldsymbol{e}_i)/l_i - \boldsymbol{e}_i \cdot \ddot{\boldsymbol{p}}. \tag{7}$$

3 Optimal Design

3.1 Performance Index

Assuming that $X_{m \times n} = \begin{bmatrix} x_1 & x_2 & \cdots & x_n \end{bmatrix}$ is a real matrix composed of n real column vectors. The matrix orthogonal degree of $X_{m \times n}$ is defined as follows [23]: If $\min(\|x_i\|) = 0 \ (i = 1, \cdots, n)$, the matrix orthogonal degree $\mathrm{ort}(X)$ is zero. Else:

$$\mathrm{ort}(X) = \mathrm{vol}_n(X) \Big/ \prod_{i=1}^{n} \|x_i\| = \sqrt{\det(X^{\mathrm{T}} X)} \Big/ \prod_{i=1}^{n} \|x_i\|, \tag{8}$$

where $\mathrm{vol}_n(X) = \sqrt{\det(X^{\mathrm{T}} X)}$ is the volume of matrix $X_{m \times n}$ in the Euclidean space. When all the vectors in the matrix are unit-column vectors, the matrix orthogonal degree represents the volume of the n-dimensional parallel polyhedron with the column vector as the edge. Evidently, the range of the matrix orthogonal degree is [0, 1]. Only when column vectors are orthogonal to each other can the maximum value of 1 obtained. On the contrary, when the multi-collinearity of column vectors appears, the minimum value of 0 can be deduced. Extreme values of matrix orthogonal degree indicate two important geometric properties, i.e., multicollinearity and orthogonality of multiple vectors. Matrix orthogonal degree is a promising mathematical tool to study spatial force/motion vectors of mechanisms.

For the object CDPR, the middle branch ensures that all cables are in tension, and that cables jointly supply the force output of the end effector. When forces exerted by driven cables are completely orthogonal, the output performance of the arbitrary force is the same for the end effector. In addition, the end-effector transmissibility is best with isotropy. On the contrary, if the driving force vector collinear happens, then the dimension of the output force space is reduced. The end effector is unable to support the force in some particular direction; hence, a kinematic singularity occurs.

The preceding analysis indicates that the unit driving force vector of each cable is along the cable and equals the unit cable vector $\boldsymbol{e}_i$. Thereafter, the Orthogonal degree-based Local Transmission Index (OLTI) of the CDPR can be defined as:

$$\Gamma_{\text{OLTI}} = \text{ort}(U) = \sqrt{|U^{\text{T}}U|}, \tag{9}$$

where $U = [\,e_1\ e_2\ e_3\,]$. According to preceding definitions, the proposed transmission index is based on vector computation, and is frame-free with evident significance. The value range is finite and normalized as [0, 1]. Besides, the calculation is simple and concise.

3.2 Design Space

The theoretical workspace of the object CDPR is a hemispherical space (disregarding the limb length and swing angle) [11]. However, the swing limit of the extensible limb is $\pm 45°$ by considering universal joints, and the scalable range of the extensible limb is from $0.6\,h$ to h. The reachable workspace of the CDPR is the lower area of a cone (see Fig. 4). The required regular workspace of the pick-and-place robot is usually a flat cylinder. The diameter is D and the height is H. Considering the practical application, the diameter of the cylinder is set to four times of its height, i.e. $D = 4H$. The equation of geometric parameters is described as follows:

$$(0.6h + H)^2 + 4H^2 = h^2, \tag{10}$$

and

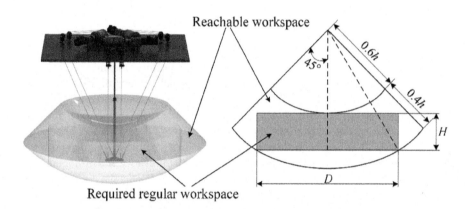

Fig. 4. Reachable workspace and required regular workspace

$$H = \frac{\sqrt{89} - 3}{25} h = 0.25736 h. \tag{11}$$

Thus, the regular workspace is determined by the middle extensible branch.

The configuration of the object CDPR can be expressed by three parameters: size of the end effector r, size of the base R, and maximum length of the extensible limb h. The impacts of these parameters on the performance of the CDPR need to be considered together during the optimization. The atlas-based kinematic optimum design method [24] is adopted because of its accuracy and intuitiveness. To embody all the possible combinations of these three parameters in a finite area, parameters should be normalized and physical dimensions must be eliminated. A dimension factor η is defined as:

$$\eta = (r + R + h)/3. \tag{12}$$

Then, the three dimensionless parameters $\lambda_i (i = 1, 2, 3)$ can be obtained as follows:

$$\lambda_1 = r/\eta, \; \lambda_2 = R/\eta, \; \lambda_3 = h/\eta. \tag{13}$$

The proportional relationship between the parameters can be expressed through λ_i. To facilitate further analysis, the parameters are illustrated in a two-dimensional plane (see Fig. 5). The λ_1-axis is vertical, λ_2-axis and λ_3-axis are perpendicular to each other, and the angle between λ_1-axis and λ_2-axis is $3\pi/4$. Theoretically, these three dimensionless parameters can assume any value from 0 to 3. However, the end effector for the object CDPR should be no larger than the base. Thus, the available design space is under the line of $R = r$ (marked with red lines in Fig. 5).

3.3 Optimization Process and Results

The Orthogonal degree-based Global Transmission Index (OGTI) can be defined to

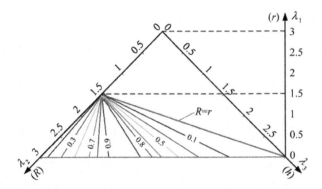

Fig. 5. Design space and atlas of OGTI within it

evaluate the global behavior in the entire regular workspace with a given configuration. OGTI can be expressed as follows:

$$\Gamma_{\text{OGTI}} = \int_W \Gamma_{\text{OLTI}} dW \bigg/ \int_W dW, \tag{14}$$

where W is the regular workspace of the CDPR. In general, a considerably large OGTI value presents an improved transmissibility and kinematic performance. Thereafter, the OGTI atlas for our object robot is obtained (see Fig. 5).

Given the performance atlas, we can clearly see the performance trends and realize the performance of the CDPR with any combination of r, R, and h. In the design space, the value of OGTI is substantially large in the middle area and decreases toward two sides. The candidate region for the optimal design can be deduced with the desired performance. For the optimization instance of this article, we assume that OGTI value of the desired CDPR is greater than 0.9. Taking into account the size of the regular workspace, $h(\lambda_3)$ should be as large as possible, and λ_3 is set not less than 1. In addition, the minimum value of $r(\lambda_1)$ also needs to be limited. If $r(\lambda_1)$ is too small, then the distance between the parallel cables is too small, and it cannot exert an effective limit to rotations and counteract the torque. In this example, the minimum value of λ_1 is set to 0.12. Based on the above conditions, the candidate area meeting our requirements can be determined, as shown in Fig. 6.

A set of non-dimensional optimum parameters can be determined by the most desired performance within the obtained candidate region. In this case, we want to get a manipulator with maximum regular workspace, and the maximum value of

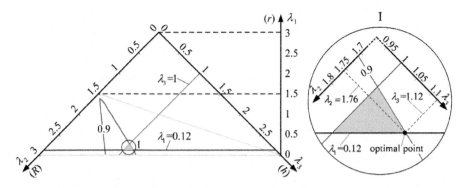

Fig. 6. Candidate region and the desired optimal point of the optimal design

$\lambda_3(h)$ should be adopted. Thus, $\lambda_1 = 0.12$, $\lambda_2 = 1.76$ and $\lambda_3 = 1.12$.

Lastly, the dimension factor η is determined, thereby converting the non-dimensional parameters into practical dimensional ones. If the maximum length of the extensive branch is $h = 1.12$ m, then the other two optimal parameters can be yielded as $R = 1.76$ m and $r = 0.12$ m. The physical prototype with the optimal parameters is manufactured and assembled, as shown in Fig. 7.

Fig. 7. Physical prototype of the high-speed pick-and-place CDPR

4 Simulation

In this section, the performance of the high-speed pick-and-place CDPR is analyzed by numerical simulation with deduced optimal parameters. The value distribution of OLTI within the regular workspace is shown in Fig. 8 on the left. In general, the OLTI value is relatively large in the regular workspace, which indicates that the object CDPR possesses good transmission performance. Moreover, OLTI value distributions on the *OXZ* and *OYZ* planes are drawn to explore the performance change (see Fig. 8 on the right). The region where $\Gamma_{OLTI} \geq 0.9$ occupies most areas of the regular workspace. The OLTI value at the top center of the regular workspace is largest and nearly 1 with the best transmission performance. The distribution of OLTI values could be fully utilized to the trajectory planning and control, to give full play to the potential performance.

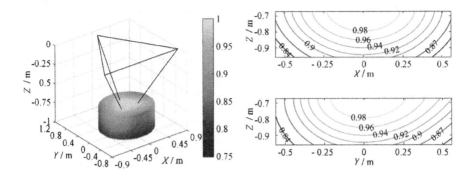

Fig. 8. OLTI distribution in the regular workspace of the optimized CDPR

With the established kinematic model and optimal parameters, the input motion of each cable could be deduced based on the given terminal trajectory. The center of the end effector is assumed to move along the X-direction on the horizontal plane of $z = -0.8$ m. In addition, the trajectory function of $x_P(t)$ is described as follows:

$$x_P(t) = 0.5 \times \sin(2\pi t - \frac{\pi}{2}), 0 \leq t \leq 1s. \tag{15}$$

The time-history graphs of the cable velocities and accelerations of the CDPR are deduced and illustrated in Fig. 9.

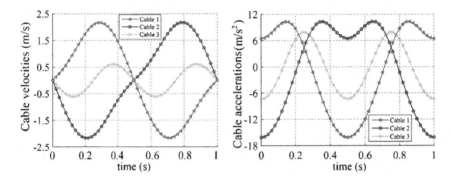

Fig. 9. Cable velocities and accelerations with the given terminal trajectory

5 Conclusions

This study uses the matrix orthogonal degree as basis to propose new transmissibility indices of OLTI and OGTI for the high-speed pick-and-place CDPR, which possess such advantages as being frame-free, clear significance, and concise calculation. The atlas-based optimum design method is adopted and the optimal dimension parameters of CDPR are deduced by considering the regular workspace and transmission performance in the defined design space. Numerical simulations are developed to verify the transmission performance of the optimized CDPR and its input variables with the given trajectory. A practical prototype of the high-speed pick-and-place CDPR is completed and the detailed mechanical designs of the implement of the prototype are illustrated. The deduced performance atlas substantially illustrates the effect of the parameters on performance. Moreover, the proposed indices and optimization metrology can be further adapted to other CDPRs. The rigid-flexible coupling structure of the CDPR robot could introduce vibration problem, which will be further studied with the dynamics.

Acknowledgments. This work is supported by the National Natural Science Foundation of China (No. 51575292), National Key Technology Research and Development Program of the Ministry of Science and Technology of China (No. 2015BAF19B00), and National Scholarship Fund of the Ministry of Education of China (No. 201606215004).

References

1. Xie, F.G., Liu, X.J., Zhou, Y.H.: A parallel robot with SCARA motions and its kinematic issues. In: Proceedings of the 3th IFToMM International Symposium on Robotics and Mechatronics, Singapore, 2–4 October 2013, pp. 53–62 (2013)
2. Shao, Z.F., Tang, X., Wang, L., et al.: A fuzzy PID approach for the vibration control of the FSPM. Int. J. Adv. Robo. Syst. **10**(1), 59 (2013)
3. Shao, Z.F., Tang, X., Wang, L.: Dynamic verification experiment of the Stewart parallel manipulator. Int. J. Adv. Robot. Syst. **12**(10), 144 (2015)
4. Clavel, R.: Delta: a fast robot with parallel geometry. In: Proceedings of 18th International Symposium on Industrial Robots, pp. 91–100 (1988)
5. Krut, S., Company, O., Benoit, M., et al.: I4: a new parallel mechanism for SCARA motions. In: Proceedings of the 2003 IEEE International Conference on Robotics and Automation, vol. 2, pp. 1875–1880 (2003)
6. Company, O., Marquet, F., Pierrot, F.: A new high-speed 4-DOF parallel robot synthesis and modeling issues. IEEE Trans. Robot. Autom. **19**(3), 411–420 (2003)
7. Nabat, V., De La Rodriguez, M.O., Company, O., et al.: Par4: very high speed parallel robot for pick-and-place. In: IEEE/RSJ International Conference on Intelligent Robots and Systems, pp. 553–558 (2005)
8. Mo, J., Shao, Z.F., Guan, L., et al.: Dynamic performance analysis of the X4 high-speed pick-and-place parallel robot. Robot. Comput.-Integr. Manuf. **46**, 48–57 (2017)
9. Gosselin, C.: Cable-driven parallel mechanisms: state of the art and perspectives. Mech. Eng. Rev. **1**(1), 1–17 (2014)
10. Shao, Z.F., Tang, X., Yi, W.: Optimal design of 3-DOF cable-driven upper arm exoskeleton. Adv. Mech. Eng. **6**, 1–8 (2014)
11. Kawamura, S., Choe, W., Tanaka, S., et al.: Development of an ultrahigh speed robot FALCON using wire drive system. In: IEEE International Conference on Robotics and Automation, vol. 1, pp. 215–220 (1995)
12. Landsberger, S.E.: Design and construction of a cable-controlled parallel link manipulator. Dissertation, Massachusetts Institute of TechnolTIy (1984)
13. Behzadipour, S., Khajepour, A.: A new cable-based parallel robot with three degrees of freedom. Multibody Syst. Dyn. **13**(4), 371–383 (2005)
14. Pham, C.B., Yeo, S.H., Yang, G., et al.: Workspace analysis of fully restrained cable-driven manipulators. Robot. Auton. Syst. **57**(9), 901–912 (2009)
15. Behzadipour, S., Khajepour, A.: Stiffness of cable-based parallel manipulators with application to stability analysis. J. Mech. Des. **128**(6), 303–310 (2006)
16. Liu, X.J., Wu, C., Wang, J.: A new approach for singularity analysis and close-ness measurement to singularities of parallel manipulators. J. Mech. Robot. **4**(4), 1–10 (2012)
17. Wang, J., Wu, C., Liu, X.J.: Performance evaluation of parallel manipulators: motion/force transmissibility and its index. Mech. Mach. Theory **45**(10), 1462–1476 (2010)
18. Balli, S.S., Chand, S.: Transmission angle in mechanism (triangle in mesh). Mech. Mach. Theory **37**(2), 175–195 (2002)
19. Zhang, L., Mei, J., Zhao, X., et al.: Dimensional synthesis of the delta robot using transmission angle constrains. Robotica **30**, 343–349 (2012)
20. Chen, X., Liu, X.J., Xie, F.G., et al.: A comparison study on motion/force transmissibility of two typical 3-DOF parallel manipulators: Sprint Z3 and A3 tool heads. Int. J. Adv. Robot. Syst. **11**(5), 1–10 (2014)

21. Fahham, H.R., Farid, M.: Optimum design of planar redundant cable-suspended robots for minimum time trajectory tracking. In: International Conference on Control Automation and Systems, pp. 2156–2163 (2010)
22. Shao, Z.F., Tang, X., Wang, L.P.: Optimum design of 3-3 Stewart platform considering inertia property. Adv. Mech. Eng. **5**, 1–10 (2013)
23. Wang, X., Dang, Y., Xue, S.: A numerical value index applied to diagnosis of ill-conditioning problems: orthogonal degree of matrices. Geomatics Inf. Sci. Wuhan University **34**(2), 244–247 (2009)
24. Shao, Z., Tang, X., Wang, L., et al.: Atlas based kinematic optimum design of the stewart parallel manipulator. Chinese J. Mech. Eng. **28**(1), 20–28 (2015)

On the Improvements of a Cable-Driven Parallel Robot for Achieving Additive Manufacturing for Construction

Jean-Baptiste Izard[1(✉)], Alexandre Dubor[2], Pierre-Elie Hervé[1], Edouard Cabay[2], David Culla[3], Mariola Rodriguez[3], and Mikel Barrado[3]

[1] Tecnalia France, 950 Rue Saint-Priest, 34070 Montpellier, France
jeanbaptiste.izard@tecnalia.com
[2] Institute of Advanced Architecture of Catalonia,
Pujades 102, 08005 Barcelona, Spain
[3] Tecnalia, Paseo Mikeletegi 2/7 - Parque Tecnologico,
20009 San Sebastian, Spain

Abstract. Generalization of additive manufacturing has led to consider this technological solution for more and more challenging use cases. Porting this technology to construction industry is a major step to overcome. Most of the recent research deal with materials, construction and extrusion techniques. Positioning of the extruder or material handler is mostly carried out by standard anthropomorphic robots or large-scale gantries. Cable-driven parallel robots (CDPR) can be an efficient alternative to these positioning solutions, being capable of automated motions in six degrees of freedom and easily relocated.

The combination of the Cogiro CDPR (Tecnalia, LIRMM-CNRS, 2010) with the extruder and material of the Pylos project (IAAC, 2013), open the opportunity to a 3D printing machine with a workspace of $13.6 \times 9.4 \times 3.3$ m. Two prints, with different patterns, have been achieved with the Pylos extruder mounted on Cogiro, drawing a wire of material of 11 m in width and 3 mm in height: the first spanning 3.5 m in length, the second, reaching a height of 0.86 m.

The motivation of this paper is to give an insight to the necessary technical implementations on a CDPR for dealing with additive manufacturing process relevant for construction, in particular acute modelling of the cable and its extension under load, and to showcase the experimental prints carried out by the authors.

1 Introduction

The recent developments in Additive Manufacturing (AM) have led to consider ever more various materials for printing parts. Cement and construction materials such as clay are some of them. With AM technology, based on CAD design, architects get the possibility of manufacturing quickly and with precision optimized complex geometries for parts, offering free shape design for serving specific purposes and refinement in material distribution. A clear potential have been identified by industries, contractors and architects to reduce the cost of customized fabrication, and therefore create a change of paradigm from the twentieth century standardized architecture of mass production toward the contemporary digital architecture of mass customization and site specific adaptation [1].

The optimal material for printing strong parts with total freedom of design and with a proven printing process is yet to be found, and this is a subject of intensive research. Typically, printing is carried out through brick laying and joining with cement [2–4], fused deposit modelling [5–7] or inkjet-like process [8], with in some cases shaping tools [7]. Positioning of the printing process head is however a much less investigated area. Some of the processes focus on the printing of individual parts of smaller scale for architecture in a workshop, to be brought later on at a construction site, and therefore use anthropomorphic arms for moving the printing head [2–5, 9]. Alternatively, using gantry robots for carrying the head is fairly common practice [7, 8]. Another disclosed solution [6] has been to extend the dimensions of a linear delta robot design to reach a cylindrical workspace with a 6 m diameter and height.

These solutions are typically unfit for printing directly on-site, a mandatory option as printed parts get larger, up to printing a whole building in AM. Anthropomorphic robots require a strong base for operation, and have limited workspace of a couple of meters at maximum, which can however be extended using mobile robots [3, 4]; anthropomorphic architectures with longer reach could be used and can be automatized, as they already have been for brick laying applications for example, but they require expensive sensors for compensating for their flexibility [2]. Gantry robots [6, 7] can be installed on site with however important impacts in terms of logistics and price for large-scale structures.

Use of more novel solutions such as cable-driven parallel robots (CDPR) for AM in construction have already been investigated [10, 11] but are yet to be proven in this specific field, which is the purpose of the experiment disclosed in this paper. CDPR offer the advantage of having its longest elements being cables, which can be reeled in for minimum storage space; its frame can be limited to a series of posts for the drawing points, more easily displaceable than the supporting beams and rails of a gantry robot. Such capability has been shown in foreseen applications for emergency deployable robots [12]; outdoor operation following autonomously detailed trajectories is a common operation for aerial cameras [13] and radio telescope receivers [14]. An example of 3D printing over a workspace spanning almost a meter has also been achieved [15].

AM for construction is a process that requires the printing machine to be able to follow trajectories with a tolerance of half the width of the material wire on horizontal directions, and one third of the thickness of the wire, for the print to be considered satisfactory for the current experiment, using clay as the manufacturing material. The objective is to print a part that is sufficiently close to its CAD design. It can be noted that imprecision within these limits can provide a texture to the material which can be appreciated in specific situations.

This paper will first focus on the enhancements on the Cogiro CDPR [16] that have been required to achieve properly AM. The integration of the Pylos extruder [17] using clay on Cogiro is then discussed. A sample of the test prints is finally disclosed, with a discussion on the facts witnessed during this printing experiment.

2 Enhancement of Vertical Precision of Cogiro

Cogiro is a CDPR owned by Tecnalia and CNRS-LIRMM [18]. Its original point of design resides in the way the cables are connected to the frame, called the configuration of the CDPR, which makes it a very stable design [16]. With a footprint of 15 × 11 m, 6 m high, it is capable of holding a load up to 500 kg over more than 80% of the footprint. Several demonstrations of industrial scenarios have already been run using Cogiro [18, 19] picturing the versatility of a CDPR.

Cogiro winches are capable of pulling the 4 mm diameter cables at up to 4150 N, or 1/3 of their calculated break strength. These cables are non-rotating cables manufactured in steel. They show a weight per length of 68.5 g/m and an EA factor (calculated as per ISO 12076) of 860 kN.

Advances in the control of the robot have allowed to reach repeatability in the millimeter range and precision in the low centimeter range [20], depending on the position. Most of the precision loss is today to be accounted on mostly the vertical direction at the center of the workspace, and progressively horizontal directions as the platform is brought closer to the sides. Yet higher positions will however show higher error in precision because of the geometry and the forces in the cables. This imprecision will therefore mostly bring an error on the height of the part to be printed. The position of the platform being reconstructed in the control scheme exclusively through the use of the angular position of the winches. It therefore appears that this imprecision would be due to error in the models, being geometric or due to the stiffness of the structural elements as the cables or the structure. Similar approaches in other CDPR [21] suggest that the modelling of the drawing points and the yield of the cables under stress would be the main factors of this imprecision.

Calibration of the robot is carried out using a laser tracker capable of very high resolution (10 µm) over a large span (50 m). A global frame is reconstructed each time the laser tracker is powered on, using elements fixed to the floor. The positions of the fixing points at the platform are measured at the same time as a series of fixed elements on the platform to reconstruct a platform frame. The points measured for the drawing points, coined as reference points, are the intersection of the axis of rotation of the pulley with the plane normal to this axis passing through the center of the pulley, measured using features of the parts holding the pulleys. Finally, the position of the platform in the homing position of the CDPR is measured using the laser tracker as well, allowing for linking drums positions with cables lengths.

2.1 Basic Implementation

In this first implementation, taken as the basis for this study, cables are considered inextensible and massless. The length of the cables are computed being the straight distance between the fixing points with the desired position and the reference points for the pulleys. The desired length of each cable over the wished trajectory is then fed to the control scheme that computes the input for each winch.

2.2 Stretching Cables

In order to qualify the extension of the cables under various tensile forces, a stress test in accordance to norm ISO12076 has been carried out using a stretch test bench over a cable length of 600 mm for a force spanning from 0 to 5000 N (Fig. 1).

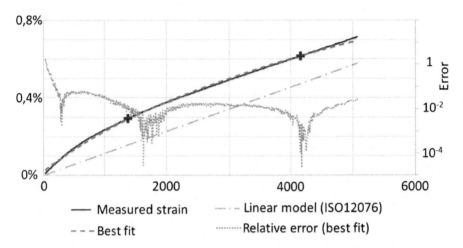

Fig. 1. Strain ϵ (%) vs Tensile force (N) F measured in the cable used in Cogiro. Crosses show the measure points for determination of the Young's modulus as per ISO12076.

It appears the number of 860 kN for EA factor is misguiding, as this is in fact a measurement of the slope between two single points, one at 10% of the break strength, and another at 30% of the break strength as per ISO12076. In reality the cable shows a faster extension at lower forces (EA factor 339 kN) followed by a stiffer behavior at higher forces. Best fit between strain ε and tensile force F has been found to be an offset squared logarithmic curve with the shape of $\varepsilon = a.\left(\ln^2(F - F_0)\right) + b$.

In addition to the basic implementation, at each point of the trajectory, a set of forces in the cables are computed using the inverse of the Jacobian of the system for balancing the expected weight on the platform. This requires the user to enter by hand the weight in the control interface. Each force is converted into a stretch, which is added to the target length for the control at each position.

2.3 Swiveling Pulleys

Another aspect taken into account is the model of the pulleys. Details of the model implementation has been developed in [20]. With this model, the length taken into account is the length over the circle of the pulley added to the length of the straight massless inextensible cable tangent to the pulley to the fixing point.

2.4 Test and Results

The test carried out for qualifying the improvement in accuracy is the measurement of the vertical position of the platform for various targets in position. These measures have been made at the center of the workspace, which is also the homing position of the CDPR. Platform actual position is measured every 0.1 m from $Z = 1.27$ m to $Z = 3.57$ m. Platform payload has been measured at 142 kg. The center of gravity of the platform is expected to be at the center of the cube.

Figure 2 shows the impact on the precision that the extension of the cable on one hand and the approximations in the cable model on the other had on the precision of the machine. From an original error of 0.25 m (at a very high position, where the cables are almost horizontal), the consideration of the real path of the cables and the extension in the cable, calculating the force in the cables, reduces the error to a few millimeters (max 6 mm, mean error 2.5 mm) in the measured conditions.

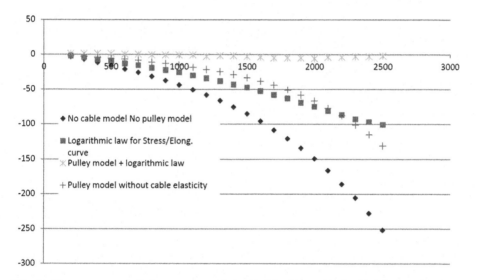

Fig. 2. Error measures in Z, in mm, versus desired height, in mm. Desired height is expressed in relative to the homing position, situated at $Z = 1.07$ m.

3 Combining Cogiro and Pylos

Pylos [17] is the result of a research action at IAAC on AM processes of large scale, using material with low ecological impact, 100% natural and biodegradable, for architecture. The material is a soil based mixture with natural additive specially tailored for AM process with an improved tensile strength and viscosity compared to the traditional soil used in construction. The extruder developed in order to test this material is composed of a canister with 15L of capacity for the material, compressed by a pneumatic cylinder. The extruder measures $0.3 \times 0.3 \times 2$ m, weighing between 65 kg empty and 85 kg full. It prints with a wire between 1 and 7 mm in thickness and 6 to 30 mm in width, at a speed between 0.05 and 1 m/s.

3.1 Mounting Pylos on Cogiro

The whole platform together with the extruder filled with clay weighs between 157 kg empty and 169 kg filled up. It features the standard cubic platform of Cogiro with the Pylos extruder mounted on it by the means of a connecting plate (Fig. 3). The extruder is controlled by an output of the controller of the CDPR through the switching on and off of a pneumatic valve.

Fig. 3. Assembly of Pylos extruder on Cogiro.

The original design of the Pylos extruder has been modified for its use on the CDPR. Originally, the material chamber was filled directly into the extruder. For this operation the extruder is placed upside down and the nozzle removed. This operation is not possible with a CDPR because of cable collisions. It is also unpractical for large volume prints such as envisioned for such a large machine, with down time between two prints being too important for an efficient printing.

For these reasons a system of canisters has been developed, through which the piston pushes the material. It offers the advantage for the team to be able to prepare the material in masked time, when the printing is being processed. This reduces the amount of clay that can be printed in a go from 20 kg in the original design to 12 kg.

With this payload, Cogiro is capable of moving the tip of the extruder with the platform in upright position inside a rectangle measuring 13.5×9.4 m, and up to 3.36 m above the floor. Figure 4 shows the volume within which the wrench due to the weight of the platform and extruder at the center of the cube, added to half of the weight of deployed cable at each fixing point, accordingly to the simplified cable model proposed by Irvine, can be balanced by the cables by taking into account their limitations. Coordinates show the position of the tip of the extruder. For balancing this wrench, when the tip is inside the pictured volume, there exist a set of forces in the cables, expressed along force lines set by fixing points at the platform and drawing points at the frame, which are all lower than 4150 N (1/3 of break strength of the cable) and higher than $20 \times$ the weight of the deployed cable.

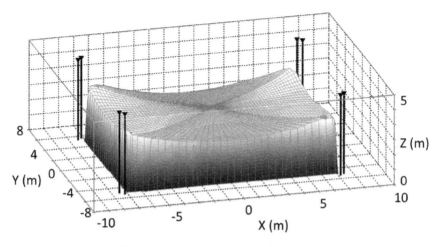

Fig. 4. Wrench-closure workspace of Cogiro with Pylos extruder mounted. Black triangles indicate the drawing points.

3.2 Trajectory Control of the CDPR

In order for Cogiro to follow precisely the trajectories, a G-Code postprocessor module has been integrated, based on CNC modules on the B&R software and hardware of the CDPR, running on Automation Studio 4.

Trajectories are designed using Rhinoceros 3D and Grasshopper 3D, using a custom script crafted by IAAC. This script computes the optimal path of the CDPR much like a CAM software. The output of the script is a G-Code file with the position and feed instructions for following the desired trajectory with a deviation of less than 2 mm. Each canister is able to print 100 m of wire; one file is calculated to correspond to one canister, and featured with a homing movement at the beginning and the end of the file, for approaching the printing position at first, and for going back to home position at last.

The G-Code postprocessor is a specific module integrated into the B&R Automation Studio suite enabling up to 5 axis of synchronous movement composed of lines, which are connected by circular portions to manage a continuous trajectory. Once a trajectory file is processed, it provides a timed output providing the position target that the CDPR should follow in order to follow the trajectory. This output is linked to the position target of the control of Cogiro until the end of the trajectory is reached. The G-Code also controls the output for controlling the starting and stopping of the extruder, so that the printing process starts and stops adequately.

Curvature has also proven to be a very important feature of the trajectory for printing with a CDPR. Preliminary tests on Cogiro with a print wire of 11×3 mm have proven that at the desired printing speed (0.15 m/s) the local radius of curvature should not be lower than 25 mm. Undesired vibrations exceeding the tolerances (5 mm horizontal, 1 mm vertical) show up when this limit is not respected. When a smaller radius is required, the printing speed is brought lower in order to limit the acceleration

from the trajectory. In addition, at the end of each layer, the robot has been programmed to make a start and stop at both the finishing point of the current layer and the starting point of the coming layer.

4 Test Prints

Several test prints have been achieved with Pylos and Cogiro. First tests involved the qualification of the best printing speed for the process, and the achievement of a 0.25 m high and 1 m long print with a sinusoidal shape (Fig. 5).

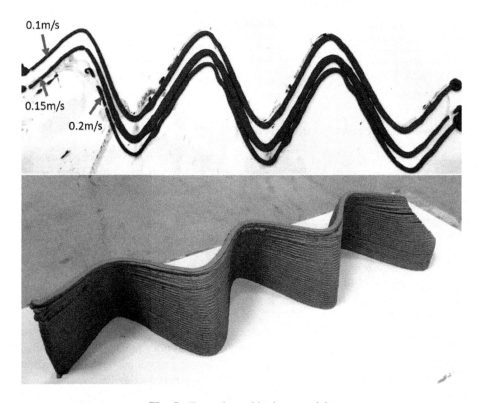

Fig. 5. Test prints with clay material.

The first demonstrating print for showing the CDPR capabilities for 3D printing is 0.2 m wide and 3.5 m long, in an overall straight line (Fig. 6). It features a special pattern that allows it to deal with the material shrinking under the process of drying, and improvements in the material composition for improving material strength. The pattern was repetitive which makes it more practical for the designer to design long structures. The coordinates for the center of the print were (X = 0, Y = 2.2 m, Z = 0). The print showed a steady height over its full length.

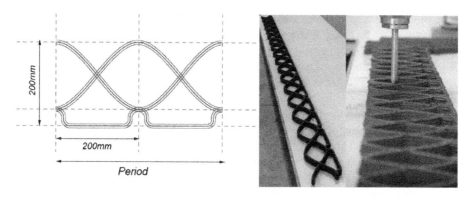

Fig. 6. Long wall print.

The pattern used for this first print led to further improvements in the pattern design, optimizing the crossing of the lines and the dimensions of free lengths of wall. Parametric design has also been implemented for a shape of the wall changing with the height, showing the full potential of the 3D printing process. The second printed piece targeted a 1.5 m height and consisted of two periods of 0.338 m long by 0.415 m wide

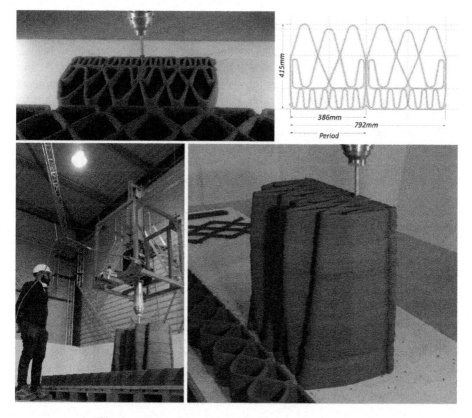

Fig. 7. Cogiro, Pylos and the high print being processed.

each, which could be extended endlessly (Fig. 7). All prints have been carried out at coordinates (X = 0, Y = 2.2, Z = 0).

As the part was getting printed, it appeared that the unloading of the clay from the CDPR was generating a steady vertical shift of about +5 mm. This made every stack of layers printed by a single canister 5 mm thicker than it should be. In order to compensate for that effect, after each canister the actual height of the print was measured and the Z positions in the G-Code file were offset such that the first layer of the new canister prints right on top of the last printed layer. Alternatively, other solutions were to modify the weight input of the control of the CDPR in function of time for the control to compensate for the weight loss; or use adaptive control [22] in order to compensate automatically such changes in the model.

At higher levels of print, the quality of the level of the layers had also degraded. The layers were showing oscillations on Z of ± 1 mm; they did not show any dilatation on the width however, because the oscillations were very repeatable from one layer to the next. This shows that the control of the CDPR needed finer tuning to achieve the wanted precision of ± 1 mm.

5 Discussion on the Results

Both prints are considered good results, and the test campaign discussed proved fruitful in experience. One of the main outcomes of the demonstration is to show that the Cogiro CDPR is capable of trajectories with millimeter precision, despite its workspace spanning meters long. It has been capable of drawing a material wire measuring 11 mm in width by 3 mm in height over long trajectories without substantial modifications on the height of the wire. This has been achieved only using the encoders of the motors for reconstructing the position of the extruder.

The need of acute control design for achieving such a precision has also been pointed out. Natural stiffness of the CDPRs played a role too: it clearly appears as the first source of errors for the estimation of the pose of the platform, and had an incidence on the print itself because of the unloading of the printing material. At this level of precision, a lot of detail is required on the models and parameters.

Future steps foresee the development of a continuous flow extruder, allowing to suppress down time from extruder loading, and print large scale elements of the size of a small building. The use of concrete as building material, requiring pumping of the material to the platform, will be investigated as well. Further research is also required on how to achieve high levels of precision with a CDPR in outdoor conditions, towards the application of on-site robotics for the construction industry.

Acknowledgements. Presented work is the fruit of a collaboration between IAAC and Tecnalia. It is part of the Open Thesis Fabrication program of IAAC, with the support and expertise of A. Markopoulou, S. Gianakopoulos, D. Stanojevic, K. Chadha and M. Marengo. The help of the students of the program, S. Chukkapalli, I. Giraud, A. Ibrahim, C. Raaghav, L. Ratoi, L. Tayefi, N. Abadi and T. Thomas, in both physical and intellectual details of the experiments, is greatly acknowledged.

References

1. Carpo, M., Kholer, M.: Challenging the thresholds. In: Fabricate: Negotiating design and making, pp. 12–21 (2014)
2. Fastbrick Robotics: June 2015. http://bit.ly/2lC850w
3. Peters, S., Podkaminer, N., Coller, T.: Brick Laying System. US Patent US8965571 B2 (2011)
4. Helm, V.: In-situ fabrication: Mobile robotic units on construction sites. Architectural Des. **3**(84), 100–107 (2014)
5. Laarman, J., Jokic, S., Novikov, P., Fraguada, L., Markopoulou, A.: Antigravity additive manufacturing. In: Fabricate: Negotiating Design and Making, pp. 192–197 (2014)
6. Carlucci, V.: World Advanced Saving Project. Archeomatica, vol. 6(4) (2016)
7. Khoshnevis, B.: Automated construction by contour crafting—related robotics and information technologies. Autom. Constr. **13**(1), 5–19 (2004)
8. Cesaretti, G., Dini, E., De Kestelier, X., Colla, V., Pambaguian, L.: Building components for an outpost on the Lunar soil by means of a novel 3D printing technology. Acta Astronaut. **93**, 430–450 (2014)
9. Gosselin, C., Duballet, R., Roux, P., Gaudillière, N., Dirrenberger, J., Morel, P.: Large-scale 3D printing of ultra-high performance concrete – a new processing route for architects and builders. Mater. Des. **100**, 102–109 (2016)
10. Izard, J.-B., Gouttefarde, M., Baradat, C., Culla, D., Sallé, D.: Integration of a parallel cable-driven robot on an existing building façade. In: Cable-Driven Parallel Robots, pp. 149–164 (2013)
11. Bosscher, P., Williams, R., Bryson, L., Castro-Lacouture, D.: Cable-suspended robotic contour crafting system. Autom. Constr. **17**(1), 45–55 (2007)
12. Merlet, J.-P., Daney, D., Winch, A., Needs, A.: A portable, modular parallel wire crane for rescue operations. In: IEEE International Conference on Robotics and Automation, Anchorage, AK, USA (2010)
13. SkyCam. http://skycam.tv/
14. Nan, R., Li, D., Jin, C., Wang, Q., Zhu, L., Zhu, W., Zhang, H., Yue, Y., Qian, L.: The five-hundred-meter aperture spherical radio telescope (FAST) project. Int. J. Modern Phys. D **20**(6), 989–1024 (2011)
15. Barnett, E., Gosselin, C.: Large-scale 3D printing with a cable-suspended robot. Addit. Manuf. **7**, 27–44 (2015)
16. Gouttefarde, M., Collard, J., Riehl, N., Baradat, C.: Geometry selection of a redundantly actuated cable-suspended parallel robot. IEEE Trans. Robot. **31**(2), 501–510 (2015)
17. Giannakopoulos, S.: Pylos. IAAC (2015). https://iaac.net/research-projects/large-scale-3d-printing/pylos/
18. Tecnalia: Cable Driven Parallel Robotics for industrial applications (2015). https://youtu.be/px8vwNerkuo
19. Tecnalia: COGIRO; Cable Robot for agile operation in large workspaces, February 2016. https://youtu.be/6RjBKvoc6N4
20. Nguyen, D.: On the Study of Large-Dimension Reconfigurable Cable-Driven Parallel Robots, Montpellier: Université Montpellier II - Sciences et Techniques du Languedoc (2014)
21. Schmidt, V., Pott, A.: Increase of position accuracy for cable-driven parallel robots using a model for elongation of plastic fiber ropes. Mech. Mach. Sci. **43**, 335–343 (2017)
22. Lamaury, J., Gouttefarde, M., Chemori, A., Hervé, P.: Dual-space adaptive control of redundantly actuated cable-driven parallel robots. In: 2013 IEEE/RSJ International Conference on Intelligent Robots and Systems, Tokyo (2013)

Concept Studies of Automated Construction Using Cable-Driven Parallel Robots

Tobias Bruckmann[1](✉), Christopher Reichert[1], Michael Meik[1],
Patrik Lemmen[1], Arnim Spengler[2], Hannah Mattern[3], and Markus König[3]

[1] Chair of Mechatronics, University of Duisburg-Essen, Duisburg, Germany
{bruckmann,reichert,meik,lemmen}@imech.de
[2] Institute of Construction Management,
University of Duisburg-Essen, Duisburg, Germany
arnim.spengler@uni-due.de
[3] Chair of Computing in Engineering, Ruhr University Bochum, Bochum, Germany
hannah.mattern@rub.de, koenig@inf.bi.rub.de

Abstract. The construction industry is still dominated by manual processes. This is both due to conventional planning tools that do not allow for a complete digital representation of a building and the limitation of the workspace of conventional robots that is far below the dimensions of most buildings. With the introduction of Building Information Modeling (BIM), the first is subject to change. BIM will allow for a holistic representation of any building and thus allow for digital workflows that enable efficient automation. Second, the development of cable-driven parallel robots meanwhile allows to create large manipulators that even cover the volume of a construction site. This allows for automated construction processes. The paper addresses BIM as a base for motion planning, investigates workspace aspects and site layout and introduces an experimental setup for feasibility studies. Initial experimental results are presented and discussed.

1 Introduction

Since the beginning of 1980s, automated robotic systems have been a great success in most fields of mass production. To both reduce costs and increase precision in repeated processes, robots have replaced human workers in the production of goods in nearly all industrial branches.

While this process is highly established in mass production, the production of single or individual items is still challenging in most cases. In the field of mechanical engineering, the introduction of Industry 4.0 is intended to enable robot systems for made-to-order production.

Contrarily, in the field of civil engineering and construction of buildings, manual processes are still dominating and the success of robots is limited to rare special cases. Amongst others, two factors can be identified:

- First, the programming of a robot – at least as long as manual teaching is not employed – is based on digital data. Usually, in mechanical engineering this

data is created upon CAD models. As a CAD model is a digital model of the goods to be produced, modern software tools provide a complete tool chain to derive the robot programming from most popular CAD environments. In contrast, in construction conventional 2D planning is still popular for the detailed planning of most buildings and divisions.
- Second, conventional robots are limited in their range and therefore have a limited workspace. For sure, the workspace even of large serial robots does not allow to cover the size of most buildings. Generally, this workspace is extendable by additional motion capabilities. Technically, this might be realized e.g. by a linear rail (which is expensive) or a wheeled mobile platform (which is challenging regarding accuracy and payload). Therefore, an extension of the workspace is not trivial.

These two factors are currently dramatically changing.

- First, the construction industry is facing a digital revolution called Building Information Modeling (BIM). As mentioned, the planning of buildings conventionally is conventionally realized per discipline (e.g. masonry, sanitary, electrics,...). This might be done in specialized CAD models per subsection, but as there was no common and parametric description language before the introduction of BIM, all models were isolated per domain and planned mainly individually based on 2D drawings. In consequence, there was no complete model of the building that contained each and every part in a consistent way. BIM is about to introduce complete and detailed digital models of a building and thus opens the door for efficient automation. All data required to feed automated and especially robotic systems can be derived. Section 2 contains a detailed introduction to BIM.
- Second, the research activities in the field of cable-driven parallel robots (or simply cable robots) meanwhile has reached a mature status. Especially in the field of control, in the last years both the implemented methods as well as the required hardware allow to realize large cable robots that are reliable and cost-effective. Thus, now for the first time in the history of robots, the realization of huge manipulators with a workspace covering the volume of a building is reasonable both under economic and technological aspects.

This contribution focuses on describing the concept of cable robots for construction, based on construction data derived from BIM models. As most steps in construction currently are manually done, production paradigms as well as site layout need to be changed. This is detailed in Sect. 2. The mechatronic design as well as initial experimental results from a feasibility study are presented in Sect. 4.

2 BIM, Robots and Path Planning

In mechanical production, automated systems like robots have increased productivity over the last decades. This is a process driven also by digitalization and digital planning tools like CAD, as already introduced. In contrast, according to [11], data from the U.S. Department of Commerce and the Bureau of Labor Statistics indicate that the construction industry nowadays is even less productive than it was in the 1960s. This might appear unexpected. But construction projects became more complex and subject to extended documentation regulations. On the other hand, they are still governed by manual steps, printed construction plans and lack of communication platforms between the designers and the contractors who realize the buildings [1].

Here, BIM is expected to be a key for future economic and technical improvements. In the early 2000s, the concept of BIM was created, derived from the concept of CAD. But while CAD basically represents geometric information, the BIM methodology extends this by parametric properties and attributes for the planning and realization, but also the operation and maintenance of a building. BIM standards like Industry Foundation Classes (IFC) provide standards beyond proprietary formats. Even specific information like e.g. the heating power of a radiator, the set values for valves, or even maintenance manuals can be integrated. Using open data formats allows to create a single model of a building that all disciplines can use, edit and harmonize. In countries like the U.S. or The Netherlands, BIM-based planning is mandatory for all new public buildings. It becomes even best practice for private buildings. In Germany, the government has set a roadmap to introduce BIM as a standard [4].

Additionally, BIM is *the* key for holistic digital site planning. A BIM model contains the data required to plan all process steps to realize a building. As this data is digitally available, it can be used to derive a trajectory planning for automated systems like robots. For many process steps, the usage of robots may be very efficient to decrease costs. In addition, robots open a totally new dimension of available precision on site. Even extremely complicated shapes and structures can be realized, using motion data provided by the BIM model. Still, the BIM model needs to be processed to derive the trajectories which is subject to ongoing research [7]. In the context of additive technologies like 3D concrete printing, this post processing requires the slicing of all printed elements like walls [3]. As 3D concrete printing for building structures is currently subject to extensive R&D activities, but difficult in terms of legal situation, norms and customer acceptance in most western countries, conventional materials are focused within the described concept. In case of traditional processes like bricking, composed elements like walls need to be disassembled into bricks that can be picked and placed [7]. Other parts like precast elements need to be integrated. This requires the following properties to be included by the path planning. Most of them are subject to future research [2]:

- Completeness: For each single part to be moved, a trajectory must be generated that provides transportation from its storage or delivery location to its assembly pose
- Efficiency: To reasonably automate the construction phase, it must become cheaper than the equivalent conventional process. Up to a certain degree, this correlates with the reduction of the construction time, i.e. the robot must be

fast and should move large material elements and parts. Accordingly, the sum of the generated trajectory times must be subject to minimization, e.g. by numerical optimization.
- Sequences: The path planning must include the pose of all material elements and parts at all times of the construction phase. This is required due to two facts:
  – Most material elements are delivered in a stacked manner, e.g. palletized. Thus, only the top layer of the material is accessible and can be picked by the robot.
  – All stacked structures can only be created bottom-up in layers. Accordingly, a production sequence needs to be created and considered. This step is a key for efficient building production in the future and may lead to paradigm changes.
- Collisions: During the construction phase, the building grows and thus intrinsically changes its shape and volume occupation. The wire robot and its payload may not collide with storage, delivered parts, the building, other machinery or itself (self-collisions). As most of these obstacles may also change their size and/or pose during the construction phase, both offline trajectory planning and online collision prevention are demanding. The latter is also essential to any human-machine interaction and safety considerations. Still, it is subject to ongoing investigations whether human workers are involved in specific processes (semi-automated) or not (fully automated). Clearly, this challenge must also be addressed by the employed control concepts.

For initial feasibility studies presented here, most of these requirements can be relaxed and define future activities. However, and as already mentioned, both simulation and experimental steps require the decomposing of larger structures into smaller units that can be picked and placed. Obviously, this holds especially for walls made from bricks. Unfortunately, the IFC standard does not provide rules

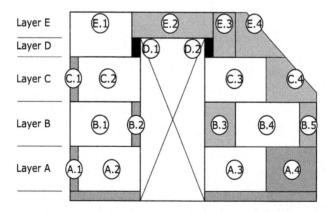

Fig. 1. Decomposed wall. Bricks are entitled as "Layer.Sequence". Image based on [5, 7]

for the wall decomposing procedure. Currently, proprietary solutions of the building materials suppliers close this gap. The resulting part list can be used as a base to create wall layers and identify the parts delivery and installation sequence (see Fig. 1). For architects and designers, the automated derivation of single brick pose information opens a new horizon for facade and wall design. As both the pose offset and the vertical brick alignment can be precisely set by the wire robot, sophisticated structures that are extremely challenging regarding optical quality are not longer limited by manual measurements which are prone to error.

In the future, close collaboration between construction engineers and robot researchers is required to develop optimized work processes, including (semi-)automated approaches and manual steps. This will identify the required dynamic capabilities of the cable robot, the needed workspace and the associated optimal site layout. First ideas for the latter are introduced in the next section.

3 Robot Geometry

For initial theoretical investigations of the concept, a certain geometry of a platform and a huge cuboid frame was set, intended to cover the volume of a small house. The frame needs to be mobile and is erected on site. For high buildings, it might be advantageous if the frame can climb as the building grows. To avoid collisions between the robot and objects or workers on the site, a suspended design according to the French CoGiRo layout was chosen [9], see Table 1. To increase the stiffness of the cable robot, it may be very beneficial to apply movable basepoints for the pulleyes that can slide along the vertical columns of the frame [10].

Table 1. Geometrical parameters

Winch №	Platform connection point coordinates (w.r.t. platform center) [m]	Pulleye coordinates (w.r.t. frame bottom center) [m]
1	$[0.125, 0.125, 0.125]^T$	$[15, -15, 20]^T$
2	$[0.125, -0.125, 0.125]^T$	$[-15, -15, 20]^T$
3	$[-0.125, -0.125, -0.125]^T$	$[15, -15, 20]^T$
4	$[-0.125, 0.125, -0.125]^T$	$[-15, -15, 20]^T$
5	$[-0.125, 0.125, 0.125]^T$	$[15, 15, 20]^T$
6	$[-0.125, -0.125, 0.125]^T$	$[-15, 15, 20]^T$
7	$[0.125, -0.125, -0.125]^T$	$[15, 15, 20]^T$
8	$[0.125, 0.125, -0.125]^T$	$[-15, 15, 20]^T$

Based upon that, a workspace computation was performed (see Fig. 2). The total mass of the end effector including payload is 250 kg. The lower wire force limit was 100 N for the lower force limit to avoid slackness where the maximum force provided by the winches is 10.000 N.

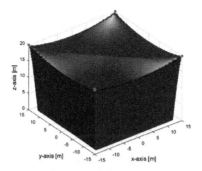

Fig. 2. Workspace of cable robot using the parameters in Table 1

Note, that in future investigations also additional parts like precast elements or ceilings might be installed by a cable robot. Accordingly, the payload needed would be higher and therefore require additional workspace syntheses.

After defining the robot geometry, the resulting footprint can be used to realize concepts for site layout and possible material logistics and delivery (see Fig. 3).

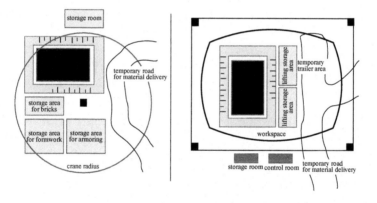

Fig. 3. Site layout conventionally (left) and automated using a cable robot (right)

The site layout is of special interest for any future research in automated construction. Initial investigations have shown that paradigm changes are likely to occur, starting with the prefabrication of tailored bricks that are cut to exactly fitting dimensions. These bricks have to be delivered in the right sequence. Accordingly, buffers and storage space both accessible for delivery trucks and the cable robot must be available. Here, lifting storage devices will be needed to allow for both truck delivery on road level and picking of bricks within the collision-free part of the cable robot workspace.

4 Experimental Setup

Based upon an existing cable robot prototype of 12 m length, 6 m height and 2 m depth, the experimental bricking of a wall was planned to confirm the technical feasibility of the concept (see Fig. 4). The goal of the study is to investigate the practical implementation of the brick transportation process, including applicability of distance sensors, influence of oscillations and the reliability of a gripping device (which was still to be developed).

Fig. 4. Experimental feasibility study

As for any robot, the end effector of the cable robot needs to be adapted for the described task. The following process steps need to be realized by the end effector:

1. The bricks are delivered on site, e.g. stacked. The next brick to be moved must be detected and its pose must be measured with respect to the end effector.
2. The brick needs to be gripped in a safe and reliable manner.
3. Transportation of the bricks from the place of delivery to their destination needs to be performed with a certain speed. Collisions with any object must be avoided during the motion.
4. When approaching the destination, the wall needs to be detected. A pose measurement relative to the intended dropping location must be performed.
5. The bricks need to be dropped with a certain accuracy.

Under these requirements, an end effector was realized that includes several mechatronic components (see Fig. 5). It carries a gripper that can pick a brick by pressing steel plates with spikes against the flanks of a brick. The steel plates are in parallel with the wall surface and actuated by motors and springs. This means that two opposite sides of the brick need to be accessible which has to be considered especially for the delivery of bricks. Rotation of the brick is not considered for the feasibility study, but subject to extended experiments.

In the horizontal direction and aligned with the wall, a linear rail with a spring allows for small movements. This enables for a smooth brick alignment.

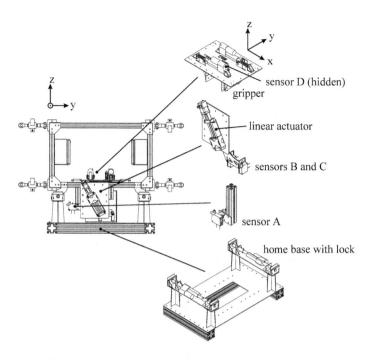

Fig. 5. End effector for bricking. Image based on: [8]

Furthermore, four laser distance sensors have been installed. While sensor A is pivot-mounted to be able to both measure downwards and horizontally towards the gripped brick, sensors B and C are mounted on a linear actuator to measure the alignment between end effector and wall plus the air gap between the contact surfaces of brick and wall. Sensor D measures the distance to the brick's top face relative to the end effector.

A programming scheme (see Fig. 6) which includes a state machine has been implemented (see [6]) to enable the following automated bricking process (states with dashed lines are not yet implemented). Encircled numbers in the following description illustrate the state of the state machine (see Fig. 6) and the recorded measurements (see Figs. 7 and 8) that are fed to the machine. Note, that an Augmented PD controller is employed. For all positioning tasks, a nominal-actual value comparison is performed in an iterative scheme. If the calculated error exceeds a defined tolerance range, an adjustment is executed. As it can be observed from the measurements (see Fig. 8), the maximum error observed after the first iteration is typically below 1 mm and therefore acceptable.

- The process starts with the robot at the home base. After the initializing routine the robot moves to the gripping pose with safety offsets in the horizontal and vertical position to avoid collision in case of inaccuracy. The linear actuator carrying sensors B and C is retracted. Both sensors B and C measure horizontal distances (x direction) to the brick side surface to ensure the

alignment of gripper and brick ②. Sensor D measures the distance to the brick's top face (z direction) relative to the end effector ③. Sensor A is aligned to measure horizontally (y direction) towards the front surface of the brick ④.
- When the intended gripping pose is reached, the gripper is closed ⑤.
- The robot moves to the calculated destination position ⑥, along with the wall and with a small height and horizontal offset with respect to the last brick layer and the last placed brick, respectively. As the sensors are covered by the gripped brick, sensors A, B and C are realigned ⑦. Sensor A is rotated downwards to measure the vertical distance to the last brick layer (z direction). The linear actuator for sensors B and C is extended to monitor the alignment of wall and end effector (x direction). Sensor D is static and therefore does not allow repositioning.
- The end effector moves downwards to the last brick layer ⑧. When the intended position is reached, sensors B and C measure the distance (x direction) to the last brick layer to align the gripped brick and the wall ⑨. As it can be seen from (Fig. 7), the robot must adjust twice in the given run to fulfill the tolerance criteria. At the current state, the rotation around the z axis is not yet implemented.
- After reaching the final position in x direction, the robot makes use of the measurement of sensor A to move downwards ⑩. At the final position the gripper opens ⑪ to place the brick. In case of the given run from (Fig. 7), the process has finished. While moving to the home position the linear actuator retracts and the rotational actuator rotates to get the sensors to a safety configuration.

Table 2 summarizes a set of experimental results on picking and placing six bricks. Note, the applied tolerance values were set according to discussion results with practitioners:

In the first row of Table 2, the remaining positioning errors in x direction are measured with sensor B during positioning of the bricks on the wall. For this direction, the requirements derived from the application allow for a tolerance of 2 mm.

The values of the second row are measured with sensor A during the gripping process and show the accuracy in y direction. Again, a tolerance of 2 mm is allowed. Note, that a direct measurement during the placing process in this direction is not possible due to the employed sensor arrangement.

The third row shows the values measured with sensor D in z direction during the gripping process. Here, required accuracy for the gripping process is set to 1 mm.

Accordingly, the average accuracy in y and z directions is good and absolutely satisfactory for the intended application. However, the flat design of the employed prototype led to oscillations in the x direction that led to increased average errors. Still, the damping of the system, the iterative positioning

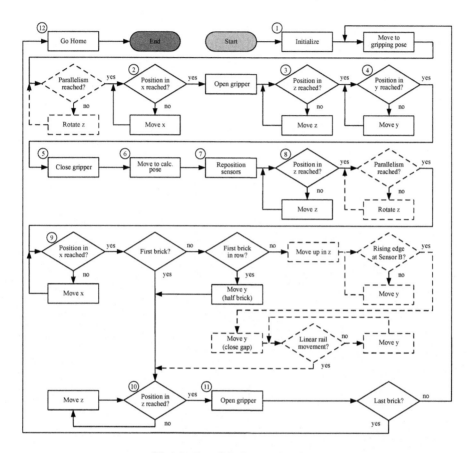

Fig. 6. Simplified state machine

Table 2. Accuracy measurements

Mean error in positioning	Brick 1	Brick 2	Brick 3	Brick 4	Brick 5	Brick 6	Total average
Sensor B (x direction) [mm] 2 mm tolerance	1.746	0.279	1.475	0.362	1.323	0.537	0.954
Sensor A (y direction) [mm] 2 mm tolerance	0.266	0.277	0.011	0.146	1.011	1.058	0.462
Sensor D (z direction) [mm] 1 mm tolerance	0.101	0.121	0.102	0.184	0.088	0.070	0.111

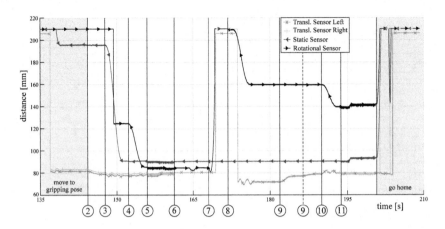

Fig. 7. Distance measurements

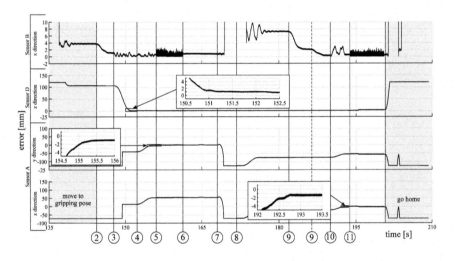

Fig. 8. Measurement errors

approach and its waiting periods led to satisfactory results. Concluding, the accuracy requirements for translations in all dimensions are met.

5 Summary and Outlook

The paper discussed multiple aspects on the automation of construction using cable robots. Based on Building Information Modeling, the digital construction plan of a building can serve as a base to feed automated processes using robots. Here, cable-driven robots can serve as universal handling devices that might enable the automation of many process steps. Currently, bricking has been subject to initial investigations on workspace, mechatronic design of the end effector and an experimental feasibility study. During the experiments it was found out that the distance measurements using laser devices is reliable and allows for an acceptable accuracy of the bricking process. Still, the development of a large spatial cable robot for more complex wall geometries, the rotation of bricks, the bricking of multiple walls and the application of mortar are open and subject to extended investigations.

References

1. Arch2O Internship Team: "Robotization" of BIM: How robots could improve BIM workflow. http://www.arch2o.com/robotization-bim-robots-improve-bim-workflow/
2. Bruckmann, T., Mattern, H., Spengler, A., Reichert, C., Malkwitz, A., König, M.: Automated construction of masonry buildings using cable-driven parallel robots. In: Proceedings of 33rd International Symposium on Automation and Robotics in Construction (ISARC 2016) (2016)
3. Ding, L., Wei, R., Che, H.: Development of a BIM-based automated construction system. Procedia Eng. **85**, 123–131 (2014)
4. Federal Ministry of Transport, Digital Infrastructure: Road map for digital design, construction: introduction of modern, IT-based processes and technologies for the design, construction and operation of assets in the built environment (2015). http://www.bmvi.de/SharedDocs/DE/Publikationen/DG/roadmap-stufenplan-in-engl-digitales-bauen.html?nn=214524
5. KS-PLUS Wandsystem GmbH: Maßgeschneiderte Lösungen aus Kalksandstein. Verlag Bau+Technik GmbH (2014)
6. Lemmen, P.: Entwurf einer Steuerungsstruktur für einen Seilroboter nach IEC 61131. Master's thesis, University of Duisburg-Essen (2016)
7. Mattern, H., Bruckmann, T., Spengler, A., König, M.: Simulation of automated construction using wire robots. In: Roeder, T.M.K., Frazier, P.I., Szechtman, R., Zhou, E., Huschka, T., Chick, S.E. (eds.) Proceedings of the 2016 Winter Simulation Conference, pp. 3302–3313 (2016)
8. Meik, M.: Entwurf eines Endeffektors und Implementierung einer Regelung für einen Seilroboter. Master's thesis, University of Duisburg-Essen (2016)
9. Michelin, M., Baradat, C., Nguyen, D.Q., Gouttefarde, M.: Simulation and control with XDE and Matlab/Simulink of a cable-driven parallel robot (CoGiRo). In: Pott, A., Bruckmann, T. (eds.) Cable-Driven Parallel Robots: Proceedings of the Second International Conference on Cable-Driven Parallel Robots, pp. 71–83. Springer International Publishing, Cham (2015)
10. Reichert, C., Glogowski, P., Bruckmann, T.: Dynamische Rekonfiguration eines seilbasierten Manipulators zur Verbesserung der mechanischen Steifigkeit. In: Bertram, T., Corves, B., Janschek, K. (eds.) Fachtagung Mechatronik (2015)
11. Whaley, M.: How BIM, robotics and 3-D printing will improve efficiencies on the jobsite. Construction Business Owner Magazine, Issue, August 2014

Inverse Kinematics for a Novel Rehabilitation Robot for Lower Limbs

Abdelhak Badi[1(✉)], Maarouf Saad[1], Guy Gauthier[1], and Philippe Archambault[2]

[1] ETS, University of Quebec, 1100, Notre-Dame West, Montreal, QC H3C 1K3, Canada
abdelhak.badi.1@ens.etsmtl.ca, {maarouf.saad,guy.gauthier}@etsmtl.ca
[2] School of Physical and Occupational Therapy,
McGill University, 3654 Promenade Sir William Osler, Montreal, Canada
philippe.archambault@mcgill.ca

Abstract. We present in this paper a new structure of cable robot for rehabilitation of lower limbs. The proposed concept is distinguished by the ability to synchronize and coordinate the joints of the hip, knee and foot. It explains a modeling approach to find explicit relations expressing the relationship between the desired trajectories in the physiological limb and the articular variables to be applied to the robot motors. The proposed robot (KINECAB) can be deployed in two configurations. In this work, we study a configuration that allows the two lower limbs to be manipulated to reproduce planar movements helping human walk or similar exercises. The inverse kinematics developed will be analyzed on the basis of a kinematic reference model of the physiological members. Validation tests by simulation and experimentation are also proposed. A patent application is deposited by ETS, University of Quebec (Application Number PCT/CA2016/051376).

1 Introduction

According to the Canadian Institute for Health Information, the client groups admitted to inpatient rehabilitation are ever on the rise. For cerebrovascular accidents alone, close to 45,741 cases were recorded between 2008 and 2015. These deficiencies present in patients locomotor and/or major neurological dysfunctions in patients [1]. Rehabilitation aims to partially or sometimes completely recover, the motor abilities by coming up with appropriate and adapted exercises.

In this work, we present a new robotic structure to reproduce the movements useful for the functional rehabilitation of the lower limb. First, we focus on a first configuration for walking and in another article, we discuss a second configuration capable of affecting all the joint movements of the legs.

Studies are currently ongoing aimed at optimizing physiotherapy techniques using a treadmill [2,3], through research aimed at ensuring adequate training of the joints [4] or by combining several techniques [5]. These techniques aim primarily to: (1) reproduce the correct/healthy joint movement pattern; (2) avoid inhibition of mobility caused by the residual use of prostheseis; (3) help synchronize the two legs and coordinate the phases of the walk cycle, and finally, (4) ensure a large number of repetitions of walk cycles. In this study, we present a new cable-driven robotic system, which provides interesting advantages when compared to other currently available robotic devices.

The cable-driven robot presented in this paper is able to perform the physical exercises necessary for the execution of rehabilitation protocols requiring the hip, knee and ankle. In order to perform a wide range of exercises, we plan to have the robot adopt two possible configurations. A configuration in which the individual is in a standing position targets movements assisting human walking. A second configuration, in which the person is in a lying position, and likely to reproduce all movements of which the physiological joints are capable. It is used on patients, with more severe neurological conditions, who are unable to actively move their own legs. In this paper, we cover only the standing posture configuration. Therefore, the target movements are limited to the flexion/extensions of the hip, knee and ankle in both legs. The robotic system is composed of a fixed frame serving as a base for actuators, together with two mobile orthoses (or platforms for parallel robots) to support the lower limbs. Each orthosis has a passive rotary joint to accommodate movements of the ankle joints, independently of the other articulations. Each platform (or orthosis) is manipulated by four cables in order to perform movements in the X-Y plane and in a Z-axis rotation. The total system therefore includes eight cables, arranged such as to ensure a coordinated and synchronized movement of the two leg orthoses, both having three degrees of freedom.

Our work focusses on the study of a particular problem. Since each orthosis is made of two segments interconnected by a passive rotary joint, we need to control a trajectory composed of two independent motions. We also need to develop explicit models to facilitate future work: the elaboration of a trajectory generator and the design of an efficient human-machine interface with the clinician.

The proposed approach aims to explicitly formulate the kinematic models of the system in order to facilitate the movements to be reproduced based on the clinicians' needs. At this stage of the project, the physical system is realized. Experimental and simulation validations of the model have already been completed.

The rest of the paper is organized as follows: In Sect. 2, a description of the system is given. This system is composed of the leg segments and the cable robot. Section 3 covers the kinematic model of both the leg and the robot. Simulation and experimental results are presented in Sect. 4. Section 5 contains a conclusion and the interpretation of the results obtained.

2 Description of the System

2.1 Physiological Members

The lower-limb part of the system is modeled as two parallel kinematic chains linked to a rigid frame. Each of these kinematics chain is composed of three links and three joints as shown in Fig. 1A. In this paper, we only consider the flexion/extension movements of these articulations. The reference frame is placed at the midpoint between the two joints of the hip. The Z_0-axis is oriented in the same direction as Z_{3r}, as shown in Fig. 1B. The kinematic model uses the Denavit-Hartenberg convention [7] (see Table 1 for the link parameters).

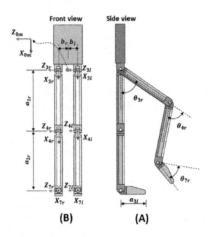

Fig. 1. A: The referentiel and geometric parameters B: Lower-limb articulation rotations

Since there is a symmetry between the two kinematic chains (Fig. 1A), we use subscript k in the equations. This subscript must be replaced by r in the equations with respect to the right kinematic chain. Similarly, the subscript l refers to the left kinematic chain. There is only one exception to this notation: the distance b_k is replaced by b_r for the right side and $-b_l$ for the left side.

Figure 1B shows that these three angles move the leg segments only in the X-Y plane[1], while the other physiological angles create movements outside this plane.

Table 1. Link parameters of the k-th kinematic chain

Joint	α_k	a_k	d_k	θ_k
3	0	0	b_k	θ_{3k}
4	0	a_{1k}	0	θ_{4k}
7	0	a_{2k}	0	$\theta_{7k} + \pi/2$

---

[1] The plane can extend to three dimensions in the second configuration of this proposed robot. This configuration, where the patient is in a supine position, is treated in a second paper which will shortly be submitted.

2.2 Cable Robot

KINECAB is a parallel cable-robot to be used for lower-limb rehabilitation. Figure 2A shows the robot in the configuration with the subject standing. There are two mobile orthoses (one per leg) composed of two parts connected by a passive joint, one to hold the leg and the other to hold the foot. One end of the cables is attached to the member support and the other end side is wound around motor-driven winders. The trajectories executed by the engines will manipulate the leg supports which will manipulate the members. The patient is held standing by a harness, the first part of support will hold a leg and a second part will hold a foot. The cables will be attached after patient positioning. The support can be adjusted to fit for different morphologies of the subjects. Four cables are needed to control the trajectory of an orthosis and the corresponding leg. Thus, the robotic system contains a total of eight cables, since we control the movements of both legs. The cables are attached to the supports after positioning the patient. The position and the orientation of the orthosis segments are obtained by adjusting the cable length designated by variables ρ_{1k}, ρ_{2k}, ρ_{3k} and ρ_{4k}. The length ρ_{jk} of cable j of the side k is the distance between its attachment

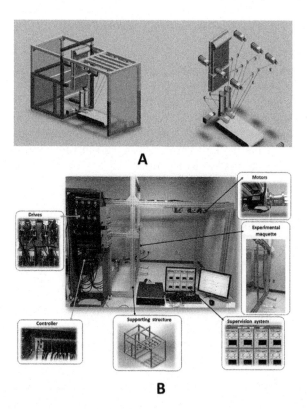

Fig. 2. A: Preliminary design of KINECAB robot — standing configuration B: Built academic prototype KINECAB

point P_{jk} at the fixed structure and the attachment point V_{jk} on the orthosis. The attachment points P_{jk} are defined relative to the reference frame $\{X_{0m}, Y_{0m}, Z_{0m}\}$:

$$P_{1k} = \begin{bmatrix} -a_{4k} & e_{4k} & b_k \end{bmatrix}^T$$

$$P_{2k} = \begin{bmatrix} -a_{4k} & e_{4k} + e_{5k} & b_k \end{bmatrix}^T$$

$$P_{3k} = \begin{bmatrix} -a_{4k} & e_{4k} + e_{5k} + e_{6k} & b_k \end{bmatrix}^T$$

$$P_{4k} = \begin{bmatrix} e_{8k} - a_{4k} & -e_{7k} & b_k \end{bmatrix}^T$$

Furthermore, the attachment points V_{jk} are defined relative to the orthosis frame $\{X_{c1}, Y_{c1}, Z_{c1}\}$:

$$V_{1k} = \begin{bmatrix} -e_{2k} & e_{1k}/2 & 0 \end{bmatrix}^T$$

$$V_{2k} = \begin{bmatrix} e_{3k} - e_{2k} & e_{1k}/2 & 0 \end{bmatrix}^T$$

$$V_{4k} = \begin{bmatrix} e_{3k}/2 - e_{2k} & -e_{1k}/2 & 0 \end{bmatrix}^T$$

except for the attachment point V_{3k} where the frame used is $\{X_{c2}, Y_{c2}, Z_{c2}\}$:

$$V_{3k} = \begin{bmatrix} -e_{9k}\sin(\theta_{7k}) & e_{9k}\cos(\theta_{7k}) & 0 \end{bmatrix}^T$$

$e_{1k}, e_{2k}...e_{9k}$ represent the geometric parameters of the robot shown in Fig. 3.

Each cable is passed through a pulley of radius r_{jk}, where $j \in \{1,2,3,4\}$ and k indicate the right and left side, respectively.

Each orthosis is divided into two parts, articulated around a passive rotary joint. In cartesian space, the robot has to realize a movement consisting in the translation and rotation of the entire orthosis and in the rotation of the second segment of the orthosis relative to the first.

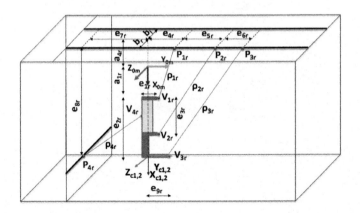

Fig. 3. Parametrization and placement of frames on the cables mechanism (right hand side)

Both parts of the orthosis translate in their X-Y plane and rotate around their Z-axis. The reference frames $\{X_{c1}, Y_{c1}, Z_{c1}\}$ and $\{X_{c2}, Y_{c2}, Z_{c2}\}$ of both parts of the orthosis are located at the rotary joint with their origin at the same location and the Z axis in the same direction.

Note that Fig. 3 shows only the right-side orthosis, in order to allow a clearer picture of the system. The left-side orthosis is placed symmetrically at distance b_l from the system reference frame.

The origins of the leg segment frames (one for the left leg and, one for the right leg) are located exactly at the reference frame $\{X_{0m}, Y_{0m}, Z_{0m}\}$.

3 Kinematic Model

In this section, we begin by defining the reference model of the robot. At this level, the direct and inverse kinematics of the leg segments are calculated. Their velocities and accelerations are also obtained. Then, we evaluate the inverse and differential kinematics of the cable mechanism, by using the relationship between the desired physiological joints angle ($\theta_{3k}, \theta_{4k}, \theta_{7k}$) and the cable real drum angle ($q_{1k}, q_{2k}, q_{3k}, q_{4k}$).

3.1 Physiological Members Kinematic [7]

The direct kinematic of both chains are defined by the following homogeneous transformation matrix:

$$_{7k}^{0}\mathbf{H} = {}_{3k}^{0}\mathbf{H}\, {}_{4k}^{3k}\mathbf{H}\, {}_{7k}^{4k}\mathbf{H} = \begin{bmatrix} -s_{3k+4k+7k} & -c_{3k+4k+7k} & 0 & \beta_1 \\ c_{3k+4k+7k} & s_{3k+4k+7k} & 0 & \beta_2 \\ 0 & 0 & 1 & b_k \\ 0 & 0 & 0 & 1 \end{bmatrix} \quad (1)$$

where $s_\bullet$ and $c_\bullet$ correspond respectively to $\sin(\bullet)$ and $\cos(\bullet)$. Recall that k is r or l — right or left.

Also, $\beta_1 = a_{2k} c_{3k+4k} + a_{1k} c_{3k}$ and $\beta_2 = a_{2k} s_{3k+4k} + a_{1k} s_{3k}$.

The coordinates of the point $\mathbf{C}_k$ (at the ankle) relative to the reference frame is:

$$\mathbf{C}_k = \begin{bmatrix} \beta_1 \\ \beta_2 \\ b_k \end{bmatrix} = \begin{bmatrix} a_{1k}\cos(\theta_{3k}) + a_{2k}\cos(\theta_{3k}+\theta_{4k}) \\ a_{1k}\sin(\theta_{3k}) + a_{2k}\sin(\theta_{3k}+\theta_{4k}) \\ b_k \end{bmatrix}$$

At the end of the chain, the direct kinematics is given by:

$$_{Fk}^{0}\mathbf{H} = {}_{7k}^{0}\mathbf{H}\, {}_{Fk}^{7k}\mathbf{H} = \begin{bmatrix} -s_{3k+4k+7k} & -c_{3k+4k+7k} & 0 & \beta_3 \\ c_{3k+4k+7k} & s_{3k+4k+7k} & 0 & \beta_4 \\ 0 & 0 & 1 & b_k \\ 0 & 0 & 0 & 1 \end{bmatrix} \quad (2)$$

where $\beta_3 = -a_{3k} s_{3k+4k+7k} + a_{2k} c_{3k+4k} + a_{1k} c_{3k}$ and $\beta_4 = a_{3k} c_{3k+4k+7k} + a_{2k} s_{3k+4k} + a_{1k} s_{3k}$.

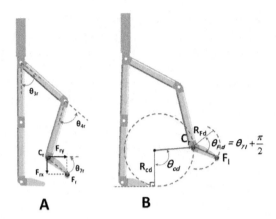

Fig. 4. A: Representation of points $\mathbf{C}_r$ and $\mathbf{F}_r$ ($k = r$ for right side) B: Desired trajectory used in the simulation

We now analyze the inverse kinematics of both chains.

Since we are interested in obtaining the independent relationships between the movements of the foot and the movements of the leg, we need to find the inverse relationship of the points $\mathbf{C}_k$ expressed in the reference frame from the angles $(\theta_{3k}, \theta_{4k})$ — see Fig. 4A. Then we determine the inverse relationship of point $\mathbf{F}_k$, at the end of the foot, relative to the point $\mathbf{C}_k$. The X and Y coordinates of point $\mathbf{C}_k$ are given respectively by the following equations:

$$C_{xk} = a_{2k}\cos(\theta_{3k} + \theta_{4k}) + a_{1k}\cos(\theta_{3k}) \tag{3}$$

$$C_{yk} = a_{2k}\sin(\theta_{3k} + \theta_{4k}) + a_{1k}\sin(\theta_{3k}) \tag{4}$$

where:

$$\theta_{4k} = \operatorname{atan2}(\alpha_1, \alpha_2) \tag{5}$$

In the three previous equation, we need to evaluate:

$$\alpha_2 = \frac{C_{xk}^2 + C_{yk}^2 - a_{2k}^2 - a_{1k}^2}{2a_{2k}a_{1k}} \text{ and } \alpha_1 = \pm\sqrt{1 - \alpha_2^2}$$

From (3) and knowing that:

$$\cos(\theta_{3k} + \theta_{4k}) = \cos(\theta_{3k})\cos(\theta_{4k}) - \sin(\theta_{3k})\sin(\theta_{4k})$$

And we find:

$$\theta_{3k} = \operatorname{atan2}(C_{yk}, C_{xk}) - \operatorname{atan2}(a_{2k}\alpha_1, a_{2k}\alpha_2 + a_{1k}) \tag{6}$$

and:

$$\theta_{7k} = \operatorname{atan2}(F_{yk}, F_{xk}) - \frac{\pi}{2} \tag{7}$$

Finally we conclude this subsection, by analyzing the differential kinematics. The vector of linear and angular velocities at point C_k (respectively $\boldsymbol{\nu}_k \in \mathbb{R}^3$ and $\boldsymbol{\omega}_k \in \mathbb{R}^3$) with respect to the framework referential is given by:

$$\mathbf{v}_k = \begin{bmatrix} \boldsymbol{\nu}_k \\ \boldsymbol{\omega}_k \end{bmatrix} = \mathbf{J}_k \dot{\boldsymbol{\theta}}_k \tag{8}$$

where $\mathbf{J}_k \in \mathbb{R}^{3\times 3}$ is:

$$\mathbf{J}_k = \begin{bmatrix} -a_{2k}s_{3k+4k} - a_{1k}s_{3k} & -a_{2k}s_{3k+4k} & 0 \\ a_{2k}c_{3k+4k} + a_{1k}c_{3k} & a_{2k}s_{3k+4k} & 0 \\ 0 & 0 & 0 \\ 0 & 0 & 0 \\ 0 & 0 & 0 \\ 1 & 1 & 1 \end{bmatrix} \tag{9}$$

The combined acceleration vector is given by:

$$\dot{\mathbf{v}}_k = \begin{bmatrix} \dot{\boldsymbol{\nu}}_k \\ \dot{\boldsymbol{\omega}}_k \end{bmatrix} = \dot{\mathbf{J}}_k \dot{\boldsymbol{\theta}}_k + \mathbf{J}_k \ddot{\boldsymbol{\theta}}_k \tag{10}$$

From (8) and (10) we obtain the inverse relations as:

$$\dot{\boldsymbol{\theta}}_k = \mathbf{J}_k^{-1} \mathbf{v}_k \tag{11}$$

and:

$$\ddot{\boldsymbol{\theta}}_k = \mathbf{J}_k^{-1} \left(\dot{\mathbf{v}}_k - \dot{\mathbf{J}}_k \mathbf{J}_k^{-1} \mathbf{v}_k \right) \tag{12}$$

3.2 Cable Robot Kinematics [12–14, 17]

The direct kinematics of the robot is expressed (for each side) by:

$$\mathbf{C}_{ck} = \boldsymbol{\rho}_{ikv} + \mathbf{P}_{ik} - \mathbf{Q}_k \mathbf{V}_{ik} \tag{13}$$

where $i \in \{1, 2, 3, 4\}$ is the cable number and $\mathbf{Q}_k \in \mathbb{R}^{3\times 3}$ is the rotation matrix for the $\{X_{Ck}, Y_{Ck}, Z_{Ck}\}$ orthosis frame of side k:

$$\mathbf{Q}_k = \begin{bmatrix} \cos(\theta_{3k} + \theta_{4k}) & -\sin(\theta_{3k} + \theta_{4k}) & 0 \\ \sin(\theta_{3k} + \theta_{4k}) & \cos(\theta_{3k} + \theta_{4k}) & 0 \\ 0 & 0 & 1 \end{bmatrix} \tag{14}$$

From (13) we obtain:

$$\boldsymbol{\rho}_{ikv} = \mathbf{C}_{ck} + \mathbf{Q}_k \mathbf{V}_{ik} - \mathbf{P}_{ik} = \begin{bmatrix} \rho_{ikx} \\ \rho_{iky} \\ \rho_{ikz} \end{bmatrix}$$

The inverse kinematics of both sides gives the cables length and is defined by:

$$\rho_{ik} = ||\mathbf{C}_{ck} + \mathbf{Q}_k \mathbf{V}_{ik} - \mathbf{P}_{ik}|| \tag{15}$$

At the initial condition shown in Fig. 3, the initial lengths of cables are defined by those vectors:

$$\begin{aligned}
\rho_{10k} &= (a_{1k} + a_{4k})^2 + (e_{4k} - e_{1k}/2)^2 \\
\rho_{20k} &= (a_{1k} + a_{4k} + e_{3k})^2 + (e_{4k} + e_{5k} - e_{1k}/2)^2 \\
\rho_{30k} &= (a_{1k} + a_{4k} + e_{2k})^2 + (e_{4k} + e_{5k} + e_{6k} - e_{9k})^2 \\
\rho_{40k} &= (e_{8k} - a_{1k} - a_{4k} - e_{3k}/2)^2 + (e_{7k} - e_{1k}/2)^2
\end{aligned}$$

After a displacement of the cables, their lengths become:

$$\boldsymbol{\rho}_k = \boldsymbol{\rho}_{0k} - \mathbf{r}_k \mathbf{q}_k$$

then we can write:

$$\mathbf{q}_k = \mathbf{r}_k^{-1} (\boldsymbol{\rho}_{0k} - \boldsymbol{\rho}_k) \tag{16}$$

where:

$$\mathbf{q}_k = \begin{bmatrix} q_{1k} & q_{2k} & q_{3k} & q_{4k} \end{bmatrix}^T$$

is the vector of the cable reel drum angle and:

$$\mathbf{r}_k = \text{diag}(r_{1k}, r_{2k}, r_{3k}, r_{4k})$$

is the diagonal matrix of the reel radius and r_{ik} is the radius of the i-th reel of side k.

The cable speeds are calculated from the derivative of the cable length with time. Then, from (15) we find:

$$\rho_{ik}\dot{\rho}_{ik} = (\mathbf{C}_{ck} + \mathbf{Q}_k \mathbf{V}_{ik} - \mathbf{P}_{ik})^T \dot{\mathbf{C}}_{ck} + + (\mathbf{Q}_k \mathbf{V}_{ik} \times (\mathbf{C}_{ck} - \mathbf{P}_{ik}))^T \boldsymbol{\omega}_k$$

which can be rewritten as:

$$\mathbf{A}_k \dot{\boldsymbol{\rho}}_k = \mathbf{B}_k \mathbf{t}_k \tag{17}$$

where $\mathbf{A}_k \in \mathbb{R}^{4 \times 4}$ is the following diagonal matrix of cable length:

$$\mathbf{A}_k = \text{diag}(\rho_{1k}, \rho_{2k}, \rho_{3k}, \rho_{4k})$$

$\mathbf{B}_k \in \mathbb{R}^{4 \times 6}$ is:

$$\mathbf{B}_k = \begin{bmatrix} \mathbf{b}_{1k}^T & \mathbf{b}_{2k}^T & \mathbf{b}_{3k}^T & \mathbf{b}_{4k}^T \end{bmatrix}^T$$

and each entry $\mathbf{b}_{ik} \in \mathbb{R}^6$ in $\mathbf{B}_k$ is:

$$\mathbf{b}_{ik}^T = \begin{bmatrix} (\mathbf{C}_{ck} + \mathbf{Q}_k \mathbf{V}_{ik} - \mathbf{P}_{ik})^T & (\mathbf{Q}_k \mathbf{V}_{ik} \times (\mathbf{C}_{ck} - \mathbf{P}_{ik}))^T \end{bmatrix}$$

Furthermore, the vector $\mathbf{t}_k \in \mathbb{R}^6$ is:

$$\mathbf{t}_k = \begin{bmatrix} \dot{\mathbf{C}}_{ck}^T & \boldsymbol{\omega}_k^T \end{bmatrix}^T$$

Then, we can obtain the derivative of (16):

$$\dot{\mathbf{q}}_k = -\mathbf{r}_k^{-1} \dot{\boldsymbol{\rho}}_k = -\mathbf{r}_k^{-1} \mathbf{A}_k^{-1} \mathbf{B}_k \mathbf{t}_k \tag{18}$$

We can also obtain the cable accelerations by taking the time derivative of (17):

$$\dot{\mathbf{A}}_k \dot{\boldsymbol{\rho}} + \mathbf{A}_k \ddot{\boldsymbol{\rho}}_k = \dot{\mathbf{B}}_k \mathbf{t}_k + \mathbf{B}_k \dot{\mathbf{t}}_k$$

and rewriting this equation as:

$$\ddot{\boldsymbol{\rho}}_k = \mathbf{A}_k^{-1} \left(\dot{\mathbf{B}}_k \mathbf{t}_k + \mathbf{B}_k \dot{\mathbf{t}}_k - \dot{\mathbf{A}}_k \mathbf{A}_k^{-1} \mathbf{B}_k \mathbf{t}_k \right)$$

where:

$$\dot{\mathbf{A}}_k = \mathrm{diag}(\dot{\rho}_{1k}, \dot{\rho}_{2k}, \dot{\rho}_{3k}, \dot{\rho}_{4k})$$

$\dot{\mathbf{B}}_k \in \mathbb{R}^{4 \times 6}$ is:

$$\dot{\mathbf{B}}_k = \begin{bmatrix} \dot{\mathbf{b}}_{1k}^T & \dot{\mathbf{b}}_{2k}^T & \dot{\mathbf{b}}_{3k}^T & \dot{\mathbf{b}}_{4k}^T \end{bmatrix}^T$$

where each entry $\dot{\mathbf{b}}_{ik} \in \mathbb{R}^6$ in $\dot{\mathbf{B}}_k$ is:

$$\dot{\mathbf{b}}_{ik}^T = \begin{bmatrix} \left(\dot{\mathbf{C}}_{ck} + (\boldsymbol{\omega}_k \times \mathbf{Q}_k \mathbf{V}_{ik}) \right) \\ \left(\boldsymbol{\omega}_k \times \mathbf{Q}_k \mathbf{V}_{ik} \times (\mathbf{C}_{ck} - \mathbf{P}_{ik}) + \mathbf{Q}_k \mathbf{V}_{ik} \times \dot{\mathbf{C}}_{ck} \right) \end{bmatrix}^T$$

Also, we have $\dot{\mathbf{t}}_k \in \mathbb{R}^6$ defined as:

$$\dot{\mathbf{t}} = \begin{bmatrix} \ddot{\mathbf{C}}_{ck}^T & \dot{\boldsymbol{\omega}}_k^T \end{bmatrix}^T$$

Then, the time derivative of (18) is:

$$\ddot{\mathbf{q}}_k = -\mathbf{r}_k^{-1} \ddot{\boldsymbol{\rho}}_k \tag{19}$$

Since the cable lengths are not large and their inertia is low, the cable flexion problem will not be addressed in this context.

4 Simulations and Experimental Results

The KINECAB robot prototype is presented in Fig. 2B. This preliminary academic version will help us to validate the kinematic models developed, as well as the feasibility of the idea. It will allow us to test the different works we will undertake in future studies.

To experimentally validate the explanatory model of the robot, we will perform two checks: The first consists in applying cable lengths, which are calculated by the model according to certain angular amplitudes selected by the robot, and then verifying the angular amplitudes produced by the robot. (Table 2) and (Fig. 6). Table 2 shows the cable lengths calculated based on the desired angular configuration (hip, knee, ankle) and angle values obtained by applying the calculated lengths. Figure 5 shows the desired configurations obtained by simulation and those obtained experimentally on the robot by applying the calculated lengths. The observed differences are due mainly to the measurement uncertainties. The second check is based on a PID applied to each drive system.

The validity of the kinematic model is evaluated by analyzing the position and orientation of a reference point of the leg (ankle joint), and a second frame on the cable point of the orthosis. The kinematic model is considered valid if the position and orientation of both reference points are identical.

To verify the models developed, we define first a circular trajectory with a radius R_{cd} at point **C**, where θ_{cd} goes from 0 to 2π (Fig. 4B). The foot is oriented at the angle $\theta_{Fd} = \theta_{7r} + \pi/2$. For the simulation, this movement is applied simultaneously to both legs.

Table 2. Verification of some angular amplitudes

Desired amplitudes (.^{0})			Cable lengths on robot (m)				Measured amplitudes (.^{0})		
θ_3	θ_4	θ_7	ρ_1	ρ_2	ρ_3	ρ_4	θ_3	θ_4	θ_7
0	0	0	1.45	1.98	2.05	0.52	0	0	1
45	−45	0	1.16	1.69	1.75	0.86	42	−41	1
90	−90	0	0.82	1.36	1.42	1.1	88	−93	48
45	0	30	1.11	1.39	1.37	1.05	44	2	26

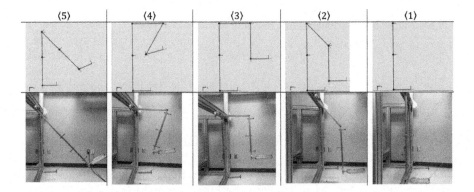

Fig. 5. Verification of some angular amplitudes

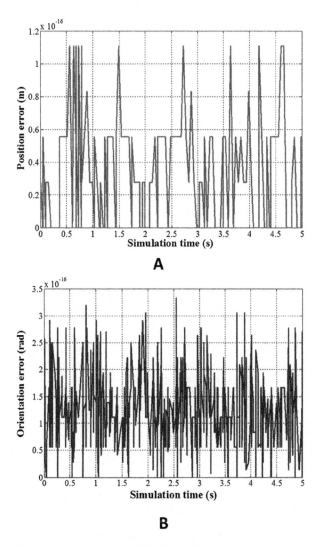

Fig. 6. A: Maximum position errors of the left side B: Maximum error of $\mathbf{QV}_1$ noted on the left side

5 Conclusion and Interpretation

This paper presented a new cable mechanism used for rehabilitation exercises for the lower limbs. The objective was to conduct a process to express the kinematic the cable mechanism based on reference models of leg segments. Its novelty lies in the movements the system introduces with respect to kinesitherapy by manipulating two parts of a leg-support connected by a rotating joint. This permits the three independent movements of the physiological joints. It then allows the manipulation of two effectors that can operate in synchronization.

The simulation results show the similarities between the trajectories resulting from the kinematic model of reference and those observed in the kinematic model of the cable robot. The experiments showed the validity of the models allowing the performance of rehabilitation exercises based on desired movements. Nevertheless, for the sake of accuracy, purely mechanical improvements are required in order to optimize the cable spooling mechanism and reduce the minimum gaps observed as a result of geometric parameter uncertainties. The trajectory tracking will be investigated in future studies addressing the question of appropriate control strategies for this type of structure. Moreover, other concepts associated with realizing control systems will be studied; particularly, the concept of continually ensure the positivity of tension cables and the evaluation of the biomechanical approach to the modeling of physiological member.

References

1. Schmidt, H., Werner, C., Bernardt, R., Hesse, S., Kruger, J.: Gait rehabilitation machines based on programmable footplates. J. Neuroeng. Rehabil. (2007)
2. Duncan, P.W., Sullivan, K.J., Behrman, A.L., Azen, S.P., Wu, S.S., Nadeau, S.E., Dobkin, B.H., Rose, D.K., Tilson, J.K., Cen, S., Hayden, S.K.: Body-weight-supported treadmill. Engl. J. Med. (2011)
3. Westlake, K.P., Patten, C.: Pilot study of Lokomat versus manual-assisted treadmill training for locomotor recovery post-stroke. J. Neuroeng. Rehabil. (2009)
4. Belda-Lois, J.-M., Mena-del Horno, S., Bermejo-Bosch, I., Moreno, J.C., Pons, J.L., Farina, D., Iosa, M., Molinari, M., Tamburella, F., Ramos, A., Caria, A., Solis-Escalante, T., Brunner, C., Rea, M.: Rehabilitation of gait after stroke: a review towards a top-down approach. J. Neuroeng. Rehabil. (2011)
5. Maple, F.W., Tong, R.K.Y., Li, L.S.W.: A pilot study of randomized clinical controlled trial of gait training in subacute stroke patients with partial body-weight support electromechanical gait trainer and functional electrical stimulation. J. Neuroeng. Rehabil. (2007)
6. Banala, S.K., Kim, S.H., Agrawal, S.K., Scholz, J.P.: Robot assisted gait training with active leg exoskeleton (ALEX). IEEE Trans. Neural Syst. Rehabil. Eng. **17**(1), 2–8 (2009)
7. Craig, J.J.: Introduction to Robotics, Mechanincs and Control. Wesley, Reading (2005)
8. Spong, M.W., Vidyasagar, M.: Robot Dynamics and Control. Wiley, New York (1989)
9. Merlet, J.-P.: Wire-driven parallel robot: open issues. Romansy 19-Robot Design, Dynamics and Control (2013)
10. Gosselin, C.: Global planning of dynamically feasible trajectories for three-DOF spatial cable-suspended parallel robots. In: Cable-Driven Parallel Robots Mechanisms and Machine Science, vol. 12, pp. 3–22 (2013)
11. Barrette, G., Gosselin, C.M.: Determination of the dynamic workspace of cable-driven planar parallel mechanisms. Trans. ASME J. Mech. Des. **127**(2), 242–248 (2005)
12. Trevisani, A., Gallina, P., William, R.L.: Cable-direct-driven robot (CDDR) with passive SCARA support: theory and simulation. J. Intell. Rob. Syst. **46**(1), 73–94 (2006)

13. Oftadeh, R., Aref, M.M., Taghirad, H.D.: Forward kinematic analysis of a planar cable driven redundant parallel manipulator using force sensors. In: 2010 IEEE/RSJ International Conference on Intelligent Robots and Systems (IROS), pp. 2295–2300, 18–22 October 2010
14. Tang, X., Tang, L., Wang, J., Sun, D.: Configuration synthesis for fully restrained 7-cable-driven manipulators. Int. J. Adv. Rob. Syst. **9**, 142 (2012). doi:10.5772/52147
15. Wernig, A., Muller, S., Nanassy, A., Cagol, E.: Laufband therapy based on 'rules of spinal locomotion' is effective in spinal cord injured persons. Eur. J. Neurosci. **7**(6), 823–829 (1995)
16. Lathrop, R.: Locomotor training: the effects of readmill speed and body weight support on gait kinematics. 23rd Hayes Graduate Research Forum at the Ohio State University, 25 April 2009. http://hdl.handle.net/1811/45253
17. Merlet, J.P.: On the inverse kinematics of cable-driven parallel robots with up to 6 sagging cables. In: 2015 IEEE/RSJ International Conference on Intelligent Robots and Systems (IROS), pp. 4356–4361, September 2015

On the Design of a Novel Cable-Driven Parallel Robot Capable of Large Rotation About One Axis

Alexis Fortin-Côté[1]([✉]), Céline Faure[3], Laurent Bouyer[3],
Bradford J. McFadyen[3], Catherine Mercier[3], Michaël Bonenfant[2],
Denis Laurendeau[2], Philippe Cardou[1], and Clément Gosselin[1]

[1] Department of Mechanical Engineering, Université Laval, Quebec City, Canada
alexis.fortin-cote.1@ulaval.ca
[2] Department of Electrical Engineering, Université Laval, Quebec City, Canada
[3] Department of Rehabilitation, CIRRIS, Université Laval, Quebec City, Canada

Abstract. This paper presents a novel architecture of cable-driven parallel mechanism that yields large orientation capability around one axis. Design decisions and justifications are presented with the aim of producing a human-scale haptic interface used for the rehabilitation of patients in a fully immersive virtual environment. The rehabilitation task to be reproduced by the haptic interface consists in carrying a crate from a shelf to a nearby table. This task requires a large rotation of the crate about the vertical axis, which cannot be achieved by conventional cable-driven parallel architectures. The novel architecture presented can generate Schönflies motion which should be useful in other tasks accomplished by cable-driven parallel robots.

1 Introduction

Cables-driven parallel robots (CDPRs) possess many proven qualities such as great stiffness, large payload capabilities for a given footprint and low inertia which results from the use of low mass cables instead of rigid links. These characteristics are highly desirable for a device used as a haptic interface. A common disadvantage of these parallel mechanisms is their limited range of motion. While CDPRs are often praised for their large translational workspace, their orientation workspace is always limited. This is a major obstacle to their use as a haptic interface, especially as six-degrees-of-freedom (6-DOFs) devices. For certain tasks, large orientation capabilities are required.

This drawback was of major concern in the realization the haptic interface used as a rehabilitation device presented in [3,4]. The task to be rendered requires large orientation capabilities about the vertical axis of rotation in addition to 6-DOF motion. Specifically, the task to be reproduced by the haptic interface consists in carrying crates between two surfaces, the crate impedance being reproduced by the interface. The end-effector in this case is a mockup crate with which the user interacts with the system. The goal of this task is to evaluate a patient on his capability of lifting a weight during a task involving a rotation of his trunk. The patient has to move the crate along a circular arc of at least 90° along the vertical axis, as shown in Fig. 1. Furthermore, a capability of at least 180° is desired so that the patient can move the crate to both his right and left.

A conventional cable-driven architecture such as those of the FALCON [8], the MARIONET family [13] or the IPAnema family [16] are not suitable. A fully constrained architecture, as opposed to a suspended mechanism, is required to maintain proper force capabilities in all directions. Of the existing architecture, the MACARM [12] comes close to what is desired with its gimbaled end-effector. The problem is that the gimbal is not actuated and so it does not render the rotational inertia or torque needed. A similar implementation exists for the mechanism used in [2], where the gravity powers the reorientation along the vertical axis. Few or none of the solutions proposed in the literature are then suitable for the intended haptic application.

A custom solution is thus needed. The solution presented here stems from ideas seen previously in the literature. The addition of a serial DOF is one of the most promising ideas [6]. Reconfigurable architectures [1] are an existing solution but require additional DOFs at the base, which increase complexity and cost. Adding kinematic redundancies [5,9] have proven able to increase orientation workspace notably but can also increase complexity. Since large reorientation capability along the vertical axis is desired, an architecture capable of Schönflies-like [14] movements would be adequate. To this end, a serial DOF could be added to the end-effector. Coupling a serial mechanism with a CDPR was already reported in the literature. One such example is the coupling of a passive planar serial arm to a CDPR [18]. Another recently proposed solution consists in doubling some of the cables allowing the addition and actuation of DOFs on the end-effector [10]. The *endless Z* architectures was presented in [15], which constitutes a promising architecture for large rotations about an axis, but this solution requires at least nine cables. The architecture presented in this paper has similar capabilities but uses only eight cables.

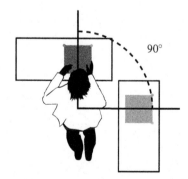

Fig. 1. Schematic of the crate moving task. The patient is asked to move the crate (blue) to the table at his right (pale blue). (Color figure online)

2 Design Criteria

From the previously presented task, the architecture should allow cables to produce forces and torques along the three main axes. It should have large translational capabilities and large rotational capability along the vertical axis of at least 180°. Since the mechanism is used as a haptic interface, its moving inertia should be kept to a minimum to maintain a high transparency. It should not impede user movements. The use of more than eight cables is discouraged because of implementation constraints.

3 Initial Concepts

Two concepts have initially been considered, both being based on the addition of a rotational DOF at the end-effector of a conventional CDPR. They differ by the way means of actuation of this joint. The first concept uses a rotational electric motor mounted on the mobile platform, as shown in Fig. 2a. The six DOFs of the mobile platform, which is of a conventional architecture, is then used to move and hold the motor, while the motor manages the last DOF which is the large rotation about the main axis. The mobile platform is then kept at a constant orientation around the main axis and only has to resist the torque transmitted by the motor. Since the interface is used to render the weight of a virtual crate, the added weight of the motor does not impact the usable workspace of the interface. On the other hand, a major drawback of this concept is the considerable increase of the mobile inertia due to the motor being on the mobile platform which would negatively affect haptic rendering by lowering the transparency. Another drawback is the need to route power to this motor. Electric wires are effectively needed to bring power to this motor and would add inertia and complexity since they should not imped motions of the end-effector or of the user. Due to these drawbacks, an alternative is considered.

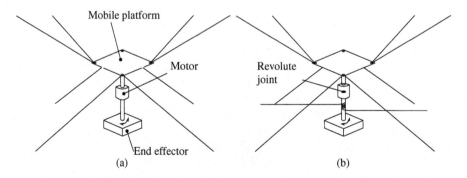

Fig. 2. Initial sketches of (a) a serial DOF actuated by a motor mounted on the end-effector and (b) a serial DOF actuated by cables.

While keeping the added serial DOF, it would be interesting to move the actuation to the base to alleviate the preceding drawbacks. By using two cables, it should be possible to actuate a pulley driving the last DOF, as shown in [10]. Letting the cable be wound on the end-effector would allow for a large reorientation, like presented in [11]. This would then greatly increase usable workspace. A sketch of the implementation is shown in Fig. 2b.

The implementation as sketched should be feasible, but requires ten cables, which, for practical purposes, is not desirable. With that in mind, it should still be feasible with eight cables since only six DOFs are desired, which for a fully constrained mechanism, requires a minimum of seven cables. Since a passive revolute joint between the end-effector and the mobile platform exists and the actuation of the serial DOF comes from the fixed base, no torque should be transmitted to the mobile platform in the direction of the main axis. A 5-DOFs arrangement of cables for the mobile platform is then sufficient.

An architecture inspired from the FALCON [8], but composed of six cables, was then developed for the mobile platform. The six cables are attached at two points on the mobile platform by groups of three. This mobile platform takes the form of a rigid rod. As such, two different configurations exist, one making the rod work under compression (a) and the other under tension (b) as shown in Fig. 3. Since the height clearance is limited, configuration (a) is preferred because it maximizes the workspace volume for a given height between anchoring points on the fixed base. Its drawback is an increased possibility of cable interferences, but since rotation capabilities around horizontal axes are limited and since there are only three cables per group instead of four, such interferences should be easily avoided.

Having made this design decision, the architecture of the anchoring points on the mobile platform is complete. The mobile platform has therefore five DOFs, three translations and two rotations (2R3T). The mobile platform can therefore balance external wrenches in all DOFs except around its main axis. The end-effector is attached to the mobile platform by a revolute joint so that forces and torques are transmitted, except for the torque about the main axis. The sixth DOF of the CDPR is actuated at the end-effector by a pair of cables wound on a drum so that they are able to balance an external moment around the main axis. Balancing an external moment around the main axis induces a parasitic resulting force at the end effector that will need to be balanced by the mobile platform. To reduce the impact of this resulting force the drum is placed as close as possible to the mobile platform. This architecture is presented in Fig. 4.

In a real implementation, a small parasitic moment might be transmitted through the revolute joint. This small moment is easily balanced by the mobile platform, which can withstand small moments about its main axis due to the cables not being exactly anchored on the main axis and therefore can keep the mobile platform at a relatively constant angle.

The drum can be chosen to allow as long a cable as needed to be wound on the end-effector. There is therefore a potential for many turns about the main axis. Since there are two actuators on a single cable, their action is coupled. A winding

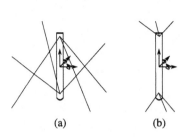
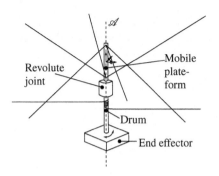

Fig. 3. Two possible configurations of the mobile platform.

Fig. 4. Schematic of the architecture of the mobile platform and the end-effector.

or unwinding of both winches allows for a translation of the end-effector. If the winding is opposite, the end-effector will rotate about its main axis. In theory, this rotation could be unlimited, but in practice it will be limited by the design of the drum assembly and by the sensor measuring the rotation angle. Indeed, the winding of the cable must be guided to avoid jamming and uneven winding. Furthermore, without guides, the effective anchoring points on the end-effector would be dependent on its pose, on the cable tensions and pre-existing winding. A guide is then necessary to ensure a proper winding and unwinding and a precise effective anchoring point. By using an eyelet, as used on most winches, it becomes simple to determine the pose of the effective anchoring point, and thence, compute the wrench applied by the cable on the end-effector or solve the forward kinematics. Figure 5 presents various concepts for the guides. The first one (a) affixes guides which do at the end-effector, which is a simple solution, but does not allow complete rotations and is thus discarded. Concept (b), which affixes the guides on the mobile platform, allows rotations but also transmits torque along the main axis to the mobile platform, which it can't resist. Final solution is then concept (c), where the guides are allowed to rotate with respect to the end-effector, using revolute joints, which allows rotations and prevent torque transmission along the main axis.

The effective position of the anchoring points is then only dependent on the mobile platform pose and is constrained on a plane normal to the main axis $\mathscr{A}$ at a constant distance from the centre of the rod. To be able to compute the wrench matrix, the position of each anchor point on the end-effector must be known. The six anchors to the mobile platform are constant in the mobile reference frame and are obtained through the architecture of the mobile platform. However, the other two depend on the pose, shown in Fig. 6. By projecting all points of interest on a plane normal to the axis $\mathscr{A}$, i.e., to the axis of the drum, it is possible to find the position of those anchoring points referred to as R_1 and R_2 geometrically. One way to obtain point R_1 is by first finding point S_1. This point is located

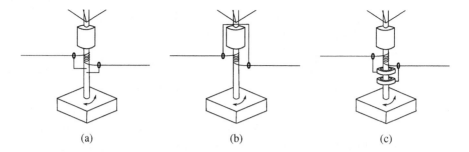

Fig. 5. Three possible choices of the placement of the cable guides.

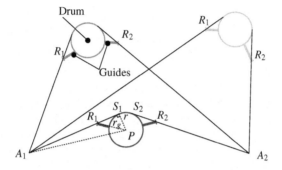

Fig. 6. Top view of different poses (one per colour) of the end-effector and its effect on the positions of the guides. (Color figure online)

at the intersection of the circle centred at point A_1 having a radius equal to the length of the segment $\overline{A_1P}$ and the circle centred at P having a radius of r, the radius of the drum. At most two intersections exist, but finding the one of interest should be simple since it is always on the same side of the segment $\overline{A_1P}$. Having found the position of point S_1, the right angled triangle PR_1S_1 is fully defined, the position of two points are known as well as the length of its hypotenuse, r_g which is the length of the guide arm, and one of its sides, r, which is the radius of the drum. The position of point R_1 can therefore be computed.

4 Drum and Guide Assembly Design

To ensure controlled and predictable winding of the cable on the end-effector, a proper drum design is crucial. One of the first design choices is the radius of the drum. This radius mainly affects the torque capability along the main axis. A rough estimate of the available torque in the most favourable condition is obtained by computing the moment about the main axis when one of the cables is under maximal tension and the other is at minimal tension. This is mathematically written as

$$\eta = r(\overline{\tau} - \underline{\tau}), \tag{1}$$

where η is the moment around the main axis, r is the radius of the drum and $\overline{\tau}$ and $\underline{\tau}$ are the maximum and minimum cable tensions respectively. To determine the actual torque capability, the wrench feasible workspace has to be traced [7]. There are two main downsides of having a large radius, one is that it reduces rotational speed around the axis, the other is that it results in a bulky end-effector.

Another important design decision concerns the measure of the angle of rotation between the mobile platform and the end-effector. This angle can be inferred using the measurement of the lengths of the cables, but the sensitivity of the angle estimate with respect to the cable-length measurements is high. This means that a small error on the length of a cable results in a large error in the angle measurement. This effect has been observed in the early implementations. In our case, a direct measurement of this angle is thus necessary, since this quantity is crucial to the haptic interface. It indeed has a great influence on the reading of the force/torque sensor installed on the end-effector and used in the control of the interface. A multi-turn potentiometer is used to directly measure this angle. It is coupled to a small and lightweight plastic gear assembly that increases its sensitivity by a factor of three. Since the potentiometer has a range of 10 turns, the range of rotation between the mobile platform and the end-effector is limited to 3.3 turns, which is sufficient, since the task at hand requires a minimum of half a turn. To avoid routing electric cables to the end-effector to measure the output of the potentiometer, a small battery-powered wireless transmitter is used.

With these considerations in mind, a prototype was assembled. For ease of fabrication and in order to reduce the weight, 3D printable plastic (ABS) is used for the drum and guide assembly. Aluminum is used as the material for the rod in the mobile platform and the end-effector. Two ball bearings are used to allow rotations along the main axis and to prevent movement in the other directions. Figure 7 shows a view of the CAD prototype with annotated parts. To mitigate possible incorrect winding of the cables, a double drum has been conceived, allowing each cable to wind on its own drum, which connects to the

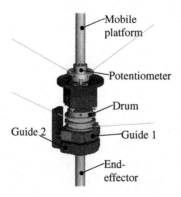

Fig. 7. Illustration of the revolute joint and drum assembly.

end-effector rod. Although winding a single cable on the end-effector would allow it to wind and unwind indefinitely, it is preferred to use two cables, which are wound a minimum of two times on their respective drums. This simplifies of the drum assembly, since it ensures that the cables are correctly wound without overlapping, as a single cable on a single drum. Winding each cable with a minimum of two turns on its drum ensures that the end-effector can rotate over the full range of 3.3 turns.

5 Implementation Considerations

With the general architecture in place, only the placement of the winches has yet to be determined. A smaller prototype of the mechanism was first built as a proof of concept. Because of its smaller size, interferences with the user were present, but the prototype worked as expected. For the full size mechanism, the geometric arrangement of the winches needs to be tailored to the task at hand and must account for environmental constraints, e.g., the size of the room, possible anchoring points and obstacles. It is also desired to maximize the workspace. For the targeted task, which is the carrying of a crate by a user, the user who is situated in the middle of the workspace is the greatest constraint because cables should not interfere user movements. To this end, the whole mechanism is moved above the user as shown in Fig. 8. The end-effector rod is then extended to reach the user. By elongating this rod, forces at the end-effector generate larger torques at the mobile platform. To counteract this effect, the length of the mobile platform rod can also be increased to provide a better lever effect. On the other hand, this increase has the effect of reducing the height of the usable workspace. Therefore, a compromise has to be made. The preferred approach is to find and use the minimum length of the end-effector rod needed to avoid interference with the user. A model of an average height person is then used to evaluate this geometrically. Figure 8 shows a representation of the CDPR along with the user in the reference position. The point in front of the user represents the bottom of the end-effector where the user interacts with the device. This visualization is then used to find the minimum length, where the user head is clear from interferences with the cables even when holding the box in a low position close to the ground.

As for the anchoring points on the fixed base, they are also placed to keep the cables away from the user. The three higher anchoring points are placed as high as the room ceiling allows. The three lower anchoring points are placed at the lowest point before interference with the user appears. With the use of ARACHNIS [17], software tool for evaluating the wrench feasible workspace, the constant orientation workspace is computed and used to assess the usable volume where the task can be accomplished. Figure 9 presents the constant orientation workspace of the mechanism for the set of external wrench needed by the task, 15 N along the Z axis, 5 N along the X and Y axes and 0.1 Nm around all three axes taking into consideration the force of all eight cables.

The final design decision to be made is the position of the anchoring points on the fixed base for the two winches driving the added serial DOF. Intuitively,

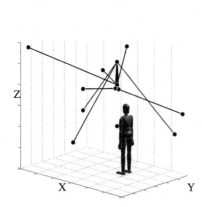 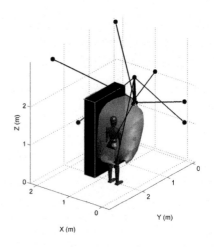

Fig. 8. Visualization of the user in the reference position along with the CDPR.

Fig. 9. Constant orientation wrench feasible workspace visualization (green volume) along with the user and a virtual shelf (black block) from which the crate is moved. (Color figure online)

these should be placed in a plane normal to the Z axis which corresponds to the main axis in the reference position. To reduce friction at the eyelet of the guides, the height of the winches should correspond to the height of the drum during typical use. This therefore leaves only one parameter to define, i.e., the angular position about the main axis at the reference position, as shown in Fig. 10. While three possible configurations are presented in this figure, the choice of the angle is effectively continuous in the range from 0° to 360°. The volume of the constant orientation workspace has been evaluated for the configuration with

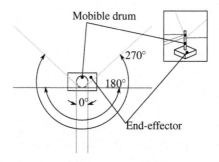

Fig. 10. Top view of the mechanism in its reference position, where the cable pair of interest is shown in three different configurations 0° in red, 180° in blue and 270° in green. (Color figure online)

Table 1. Geometric parameters of the architecture [m]. B_2 and B_6 are the attachment points of the two cables driving the added serial DOF so their position varies with the end effector pose.

(a) Fixed frame								
	A_1	A_2	A_3	A_4	A_5	A_6	A_7	A_8
X	2.09	0.02	0.02	2.08	−0.12	−0.14	−0.14	2.08
Y	0.07	−0.00	−0.00	1.58	1.35	3.22	2.95	2.58
Z	3.07	1.85	1.00	0.98	2.86	1.85	0.98	3.01
(b) End-effector								
	B_1	B_2	B_3	B_4	B_5	B_6	B_7	B_8
X	0.00	-	0.00	0.00	0.00	-	0.00	0.00
Y	0.00	-	0.00	0.00	0.00	-	0.00	0.00
Z	0.74	0.64	1.32	1.32	0.74	0.64	1.32	0.74

angles, 0°, 90°, 180° and 270°, which corresponds to possible anchoring points on the fixed frame. From the results obtained, the configuration with the angle equal to 270° was chosen, since it results in the largest workspace volume and facilitates anchoring. A more detailed investigation of this parameter should be considered in future work. Final geometric parameters are presented in Table 1.

6 Quantitative Results and Conclusion

With the prototype in place, as shown in Fig. 11, the desired capabilities are achieved. The end-effector can rotate about its main axis for 3.3 turns which is more than needed for the task that requires a range of ±90° from the reference position. The usable workspace of the robot covers the parts of a virtual shelf

Fig. 11. The haptic device prototype being used in a fully immersive environment.

and table, that are needed for the rehabilitation task. The user is therefore able to pick up a virtual crate from the shelf and place it on a table by his side, which involves a rotation of his trunk.

The novel architecture presented is a good candidate for tasks needing large orientation capabilities about one axis. Other possibilities exist for this kind of architecture of CDPR. With a type of movement related to Schönflies-like motions, screwing tasks as well as palletization tasks are potential candidates. Future work on this architecture could investigate the possibility of removing one cable while achieving a similar workspace. Indeed, only seven cables are needed to fully constrain a 6-DOF CDPR. Conversely, the addition of cables to the mobile platform cannot be ruled out as a way to increase the workspace or wrench capabilities.

Acknowledgements. The authors would like to thank Simon Foucault, Thierry Laliberté, Steve Forest and Nicolas Robitaille for their technical help in the making of the haptic interface. This work was supported by The Natural Sciences and Engineering Research Council of Canada (NSERC) as well as the Fonds de Recherche du Québec Nature et Technologie (FRQNT) 2015-PR-180481.

References

1. Bande, P., Seibt, M., Uhlmann, E., Saha, S.K., Rao, P.V.M.: Kinematics analyses of Dodekapod. Mech. Mach. Theory **40**(6), 740–756 (2005)
2. Deschênes, J.D., Lambert, P., Perreault, S., Martel-Brisson, N., Zoso, N., Zaccarin, A., Hébert, P., Bouchard, S., Gosselin, C.M.: A cable-driven parallel mechanism for capturing object appearance from multiple viewpoints. In: Sixth International Conference on 3-D Digital Imaging and Modeling (3DIM 2007), pp. 367–374. IEEE, Montreal (2007)
3. Faure, C., Bouyer, L., Mercier, C., Robitaille, N., J McFadyen, B., Fortin-Cote, A., Cardou, P., Gosselin, C., Bonenfant, M., Laurendeau, D.: Development of a virtual-reality system with largescale haptic interface and accessible motion capture for rehabilitation (under review). In: Proceedings of the International Conference on Virtual Rehabilitation, Montreal, QC, Canada (2017)
4. Fortin-Cote, A.: Développement d'un mécanisme parallèle entraîné par câbles utilisé comme interface à retour haptique visant la réadaptation physique en environnement immersif (to come). Ph.D. thesis, Université Laval (2017)
5. Gosselin, C., Schreiber, L.T.: Kinematically redundant spatial parallel mechanisms for singularity avoidance and large orientational workspace. IEEE Trans. Robot. **32**(2), 286–300 (2016)
6. Gosselin, F., Andriot, C., Bergez, F., Merlhiot, X.: Widening 6-DOF haptic devices workspace with an additional degree of freedom. In: Second Joint EuroHaptics Conference and Symposium on Haptic Interfaces for Virtual Environment and Teleoperator Systems (WHC 2007), pp. 452–457 (2007)
7. Gouttefarde, M., Merlet, J.P., Daney, D.: Wrench-feasible workspace of parallel cable-driven mechanisms. In: IEEE International Conference on Robotics and Automation, Roma, Italy, pp. 1492–1497 (2007)
8. Kawamura, S., Choe, W., Tanaka, S., Pandian, S.: Development of an ultrahigh speed robot FALCON using wire drive system. In: Proceedings of 1995 IEEE International Conference on Robotics and Automation, vol. 1, pp. 215–220. IEEE (1995)

9. Kotlarski, J., Abdellatif, H., Ortmaier, T., Heimann, B.: Enlarging the useable workspace of planar parallel robots using mechanisms of variable geometry. In: 2009 ASME/IFToMM International Conference on Reconfigurable Mechanisms and Robots, pp. 63–72 (2009)
10. Le, T.N., Dobashi, H., Nagai, K.: Configuration of redundant drive wire mechanism using double actuator modules. ROBOMECH J. **3**(1), 25 (2016)
11. Lei, M.C., Oetomo, D.: Cable wrapping phenomenon in cable-driven parallel manipulators. J. Robot. Mechatron. **28**(3), 386–396 (2016)
12. Mayhew, D., Bachrach, B., Rymer, W.: Beer R Development of the MACARM - a Novel cable robot for upper limb neurorehabilitation. In: 9th International Conference on Rehabilitation Robotics, ICORR 2005, pp. 299–302 (2005)
13. Merlet, J.P.: MARIONET, a family of modularwire-driven parallel. In: Lenarcic, J., Stanisic, M.M. (eds.) Advances in Robot Kinematics: Motion in Man and Machine, pp. 53–61. Springer, Dordrecht (2010)
14. Nabat, V., De La O Rodríguez, M.: Schoenflies motion generator: a new non redundant parallel manipulator with unlimited rotation capability. In: Proceedings - IEEE International Conference on Robotics and Automation, April 2005, pp. 3250–3255 (2005)
15. Pott, A., Miermeister, P.: Workspace and interference analysis of cable-driven parallel robots with an unlimited rotation axis. In: Proceedings of the 15th International Conference on Advances in Robot Kinematics, Grasse, France, pp. 345–353 (2016)
16. Pott, A., Mütherich, H., Kraus, W., Schmidt, V., Miermeister, P., Verl, A.: IPAnema: a family of cable-driven parallel robots for industrial applications. In: Bruckmann, T., Pott, A. (eds.) Cable-Driven Parallel Robots, Mechanisms and Machine Science. Springer, vol. 12, pp. 119–134. Springer, Berlin (2013)
17. Ruiz, A.L.C., Caro, S., Cardou, P., Guay, F.: ARACHNIS: analysis of robots actuated by cables with handy and neat interface software. Mech. Mach. Sci. **32**, 293–305 (2015)
18. Trevisani, A., Gallina, P., Williams, R.L.: Cable-direct-driven robot (CDDR) with passive SCARA support: theory and simulation. J. Intell. Robot. Syst. Theory Appl. **46**(1), 73–94 (2006)

Preliminary Running and Performance Test of the Huge Cable Robot of FAST Telescope

Hui Li[1,2(✉)], Jinghai Sun[1,2], Gaofeng Pan[1,2], and Qingge Yang[1,2]

[1] National Astronomical Observatories,
Chinese Academy of Sciences, Beijing 100012, China
lihui@nao.cas.cn
[2] Key Laboratory of Radio Astronomy,
Chinese Academy of Sciences, Beijing 100012, China

Abstract. The paper gives an overall introduction about the huge cable-driven parallel robot of FAST, including its mechanical design, control architecture, integrate system debugging and performance test. In Nov. of 2015, the cabin was successfully lifted airborne via robot operations, indicating that the cable robot is coming into being. As the concerning focus, three kinds of prototype tests are carried out to check cable tension, dynamics and control errors of the cable robot. Finally assessment and conclusion are made on the performance of the cable robot.

1 Introduction

The Five-hundred-meter Aperture Spherical radio Telescope, FAST, has been in construction since the spring of 2011 in a Karst depression of southwest China, which is expected to be complete in 2016. The huge radio telescope has a flexible and adjustable feed support system that distinguishes it obviously from many others [1, 2]. Because of its large size, it is very difficult and inconvenient to build a solid support structure between the feed and the reflector. A flexible mechanism, so called cable-driven parallel robot, is then designed to fulfill such functions as supporting and driving the feed airborne, and even preliminary pose controlling of the feed cabin. Another advantage is that the design greatly reduces both the size and weight of the cabin [1, 2].

FAST takes only a part of its active reflector as the illumination area that deforms from spherical dome to paraboloid to converge radio waves to its transient focus. Both the effective illumination aperture and the focal ratio are constants, but the position of illumination area can move continuously anywhere on the reflector dome. The fact indicates that in observation mode the movement of the feed as well as the cabin is limited only on a 206 m-aperture focal surface which contains all possible transient focuses. This may produce a set of requirements on the cable robot. First the great movement range requires that flexible cable be strong enough to support the 30-ton cabin anywhere. Second the cable robot should be capable of driving the feed cabin to track transient focus with enough accuracy. The maximal positioning error is less than 48 mm (or rms value less than 16 mm) and maximal pointing error less than 1°.

© Springer International Publishing AG 2018
C. Gosselin et al. (eds.), *Cable-Driven Parallel Robots*,
Mechanisms and Machine Science 53, DOI 10.1007/978-3-319-61431-1_34

Although feasibility of the conceptual design has been proved via a Sino-Germany end-to-end simulation in 2007 [3, 4] and demonstrations of a few downscale models [5], a lot of details are left to be testified on the prototype especially after the detailed design had been finished in 2013. In the middle of 2015 the construction of the prototype was nearly complete. In November a model cabin was successfully pulled airborne by six parallel steel cables. Both its size and weight as well as the mass distribution are quite similar to the real one. It drifted in the air along designated trajectory with small vibration due to some disturbances like wind. During the next two months, performance tests were tentatively done on the measurement of cable tensions, vibration experiments and analysis, and statistics and evaluation of positioning accuracy of the cable robot. Each of the tests aims to verify a set of key design parameters or to update them according to the real performance.

In the following pages, the author would like to introduce the huge cable robot. First attention is paid on the preliminary running of the cable robot, including its composition and functions, the preparatory integrate system debugging, the first lifting and tentative trajectory-tracking of the cabin. Then relevant instances of performance tests are given graphically with comparisons between actual and analytical data or between the prototype and down-scaled models. Finally assessment and conclusion are made on the performance of the cable robot.

2 Mechanical Design

The main structure of huge cable robot is made of 6 parallel flexible link drives, each of which includes a long steel cable and a capstan as the drive unit [6], as shown in Fig. 1. Six cables, each supported by a high-rise steel tower, are reeled coordinately by six capstans to drive the cabin to track with the desired speed and pose the transient focus. One end of steel cable is jointed to the cabin, while another end is connected to capstan drum. Six towers are equally spaced around a 600 m-diameter ring. Each steel cable carries quite a few trolleys under which electrical cable for power supply or optic fiber cable for signal transmission is hung up, as shown in Fig. 3. It can automatically adapt to the length change of steel cable due to movable trolleys. Furthermore each upper pulley on a tower is installed on an omnidirectional slewing gear to adapt to the change of cabin position.

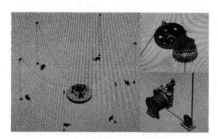

Fig. 1. Overall view of the cable robot of FAST

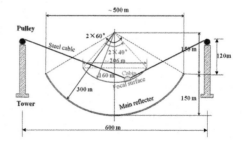

Fig. 2. Sectional dimension

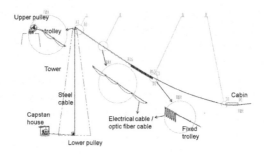

Fig. 3. Flexible link drive and its suspensions

Each of 6 capstans with additional electric equipments is installed in a steel house. The layout guides the cable vertical to the drum axis for reliable coiling and uncoiling of the cable. Each capstan is composed of servomotor, gearbox, clutch, drum and 2 working brakers and a safety braker. An asynchronous servomotor with integrated multi-turn incremental encoders is coupled to a multi-stage gearbox. Working brakers are installed between motor and gearbox for stopping high-speed shaft if necessary. Safety braker is installed on one end of drum which works to stop drum rotation only under emergence that high-speed shaft is unexpectedly broken.

Table 1 gives a brief description of the main mechanical parts and a set of technical requirements or specifications that have to be carefully considered in the detailed design and simulations.

Table 1. Description and requirement of mechanical parts of the cable robot

Mechanical part	Description	Technical requirement/Specification
Cabin (end-effector)	Carries the secondary tuning system, powering/wiring system and the receivers for astronomical observation	Weight in total: ≤ 30 tons; Movement range: A calotte, about 38 m high and 206 m in aperture in working status, which shares with the same center of the spherical reflector. It may also descend/ascend vertically to/from the bottom harbor for maintenance/normal running Maximal tilt: $15°$ Maximal speed: 400 mm/s; Normal speed: $0 \sim 24$ mm/s inn working status

(*continued*)

Table 1. (*continued*)

Mechanical part	Description	Technical requirement/Specification
6 Steel cables	6 cables support and drag the cabin airborne. They also provide support of the cable suspensions	Cable length: about 600 m for each; Normal tension range: 140~400 KN; Maximal change of length: ~200 m; Minimal safety factor in design of cable force: 3.5; Sectional diameter: 46 mm; Broken load: 1900 KN; Weight of wire suspension: about 2 tons
6 Capstans	As the main driving unit, they wind/unwind the steel cables coordinately so that the cabin can follow specially planned track and speed. Each is made up of motor, reducer, shaft and drum	Maximal power of motor: 257 KW; Winding/unwinding accuracy: ≤ 2 mm; Transmission ratio of gear box: up to 345.5; Dimension of drum: 2.3 m long and 2.4 m in diameter
Pulleys and omnidirectional slewing gears	Include a pair of upper pulley and lower pulley to guide the steel cable	Diameter: about 2 m; The ratio of pulley diameter to rope: 45 The upper pulley can turn left or right due to eccentric cable by as much as 20° in a round

3 Control Architecture

The system architecture is decomposed into 3 layered network structures, namely the topmost software control, intermediate programmable logical control (PLC) and 6 parallel bottommost electric drives. The software control system mainly devotes to the kernel algorithms, such as astronomical trajectory planning, computation of 6 coordinate drum positions for cable coiling or uncoiling, optimization of cabin pose and cable tensions, and dynamic stability. It is made up of a few modules including user interface, intercommunication and database management. The software is implemented into a PC-based real-time operating system. The intermediate PLC, composed of AC500-PM590-ETH controller and engineer workstation, works to gather process parameters, carry out operation commands and complete system maintenance. Each bottommost electric drive belongs to ACSM1-04AM-580A-4. The control system and motor drives can command either position, velocity or torque set values for the motors. The inner level torque and current control loops are embedded in standard motor drives (Fig. 4).

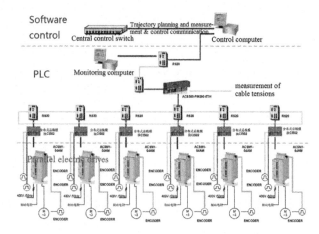

Fig. 4. Network of PLC and bottommost electric drives

4 System Debugging

System debugging of the cable robot took up the main working stage after equipment installment. The integrate system performance can be further confirmed via debugging. It includes unloaded test and loaded test. During the unloaded test steel cables are dismounted from capstan drums. Each single link drive is first tested, checking component status, circuit connection, PLC commands and logic relations. Then coordination performance is tested for the six parallel link drives. At the operating station, the robot controller is also under testing, including its human machine interface (HMI), control commands, logic function and data communication with both the main controller and the PLC. At this stage system safety and its effective self-protection are highly concerned under unexpected situations like communication interrupt, power off or equipment malfunction. Another key task is to verify the feasibility of control algorithm via a hardware-in-the-loop simulation of trajectory tracking.

The following loaded test of the robot involves lifting the cabin airborne. Before that test the model cabin is mounted on a ring-shaped harbor located on the bottom of the telescope depression, as shown in Fig. 5(1). Safety check is still the upmost focus, such as brakers, mechanical connection, operation procedure, emergent stop and interference with other equipments. Then the cabin continues to move upward until it stops at a position 20 m high where the cabin orientation is checked. After that the cabin returns to the harbor under inverse operations. Figure 5(2) shows the airborne cabin.

Fig. 5. View of the model cabin (in a left-right order): (1) Model cabin and its harbor; (2) Airborne cabin

5 Performance Test

At this stage the cable robot can lift the model cabin airborne 140 m high to reach a calotte-shaped zone that is enclosed by the focal surface and its upper aperture plane. The related position and dimension of the zone are shown in Fig. 2. A series of tests are done here to get or verify key robot performances, such as kinematics, dynamics, control parameters and system braking. Here three kinds of tests are introduced, namely the control test of astronomical tracking, force measurement and verification of steel cable, and vibration test and damping identification of the flexible cabin-cable suspension system.

Several kinds of sensors are installed on the model cabin to complete measurement of such tests, including force sensors, accelerators, targets of laser total station and anemoclinograph. Their installations are shown in Fig. 6(1). The most important is the laser total station whose pose feedback is the key information of control test. Three leica TS30 total stations are set up on measurement bases scattered on the telescope dish, as shown in the photo of Fig. 6(2). Six targets are installed on the bottom of the cabin. At least three targets are necessary in measurement during which spatial coordinates are transformed into the cabin pose. Its nominal accuracy under tracking mode is 1 mm + 1 ppm and 0.5 angular seconds in rms value. Tentative site tests show that the actual accuracy of spatial coordinates is below 6 mm rms.

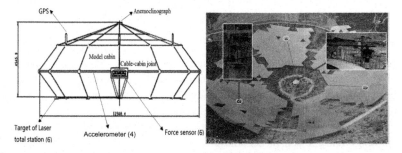

Fig. 6. Sensors in the performance test: (1) Sensors on the cabin; (2) Laser total station and its position (yellow rings). (Color figure online)

5.1 Tension Verification

Tensions of 6 steel cables are the key performance in the mechanical design of the cable robot. The end-to-end simulation based on Sino-Germany cooperation in 2007 first gave calculation results of cable tensions [3, 7]. The simulation furthers that one-to-one mapping exists between cabin pose and 6 cable tensions [8, 9], so the planning of cabin orientation can be determined via optimization of cable tension once the cabin path is given.

In the verification test 6 steady state cable tensions are measured near 6 cable-cabin joints. The maximal measuring range of the force sensor is 50t with the minimal resolution equal to 0.1t. Its nominal error is 1% of the measuring range but the actual value may reach about 3% via site test. Movement path of the cabin is predefined during the test. The cabin first follows a ring on the edge of the focal surface. Then it goes down along a straight line to the bottom of the surface. Along such a path cable tension goes through all possible values between the maximum and minimum.

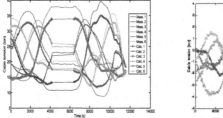
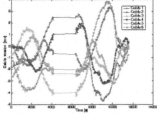

Fig. 7. Comparison of cable tension between calculation and measurement (in a left-right order): (1) Tension-time curves; (2) Tension error (measured-analytical)-time curves

Figure 7 (1) and (2) shows the comparison between theoretical calculation and measurement for each of 6 cable tensions. The theoretical calculation is based on a united static equilibrium analysis of both the cabin and 6 sagged cables, as demonstrated in the references [7, 10]. The 12 curves keep horizontal due to an operational pause during the period from about 4000 s to 6000 s. Obviously each pair of the theoretical and measured curves shows a similar shape. As the cabin follows the ring, they fluctuate periodically with nearly the same phase difference. When the cabin descends to the bottom, the curves go down and converge to nearly the same value. The error of each cable tension is relatively small as it varies in the range of about [15t, 30t]. However it goes up fast outside this range and reaches about 4t at the maximal tension. The authors estimate that the error may come from the modeling of sagged cable carrying wire suspensions of the cabin. The wire suspensions are not distributed smoothly and they may vary nonlinearly. Such modeling error is too difficult to include in the analytical model. Nevertheless the error is allowable compared with the measurement range and the sensor accuracy.

5.2 Vibration Test and Damping Identification

Cabin-cable suspension system of the robot has structurally weak-stiffness dynamics with low damping performance, which makes it quite sensitive to disturbance-induced vibrations. Certain damping can increase the additional energy dissipation of the system and therefore suppress such disturbance-induced vibrations. Based on the results of series of model tests, the end-to-end simulation assumes the damping ratio to be 0.22% [5]. Nevertheless the key parameter is hardly testified other than prototype test.

The dynamics of cabin-cable suspension system changes as the cabin moves. Therefore several typical points of the focal surface are selected as the working positions (WP) in the test. The selections include the bottom point of the surface (WP1), an edge point nearest to a tower (WP2) and an edge point between two neighbor towers (WP3). The test results on these points can hold enough information of the system dynamics.

Artificial impact is made to excite vibration of the flexible suspension system. Let the cabin run to the working position at the maximal speed (100 mm/s) in x/y/z direction respectively. Then the six capstan drums are synchronously braked to stop urgently at the position to induce free decaying vibration. Two highly sensitive SLJ-100 accelerators on the cabin can record the weak spatial acceleration response of the cabin with a sampling rate of 200 Hz. Each gives data in three directions. Figure 8 shows an example of vibration measurement at WP1. The vibration is induced when the cabin goes down vertically to WP1 and is then stopped.

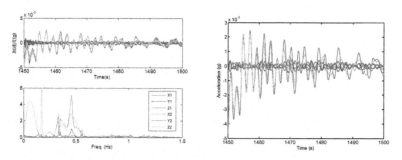

Fig. 8. Acceleration of the cabin at WP1: (1) Original data (2) After low-pass filtering

The vibration lasted for several minutes before it completely decayed. First, the data on the starting stage are discarded because they are by far higher than the normal vibration level. Similarly the data on the end are also neglected due to lower signal-noise ratio. Second, the measured data are processed via low-pass filtering to further delete high-frequency noise. The cut-off frequency is about 0.6 Hz. Finally, the method of Ibrahim Time Domain (ITD) [11, 12] is applied to identify natural frequency and relating damping ratio, as shown in Table 2. It is worth noting that damping ratio is much more sensitive to measurement error than natural frequency. In many occasions artificial impact may induce multiple frequencies of the system. The author advices that measured signals be preprocessed via band-pass filtering to separate several frequency bands for a better identification accuracy.

Table 2. Identified natural frequency and damping ratio of the suspension system

Working position	Coordinates			Natural freq. (Hz)	Damping ratio
	X(m)	Y(m)	Z(m)		
WP1	0	0	−161.70	0.1711	0.0035
				0.4662	0.0125
WP2	72.202	74.767	−123.17	0.1460	0.0058
				0.5407	0.0067
WP3	99.912	28.649	−123.17	0.1438	0.0063
				0.1803	0.0040
				0.5225	0.0035
WP4	38.418	39.783	−151.25	0.1437	0.0046
				0.1799	0.0039

Obviously the first-order natural frequency falls in an interval of [0.14 Hz, 0.18 Hz], in good accordance with the results of the end-to-end simulation [3, 4]. However the identified damping ratio scatters in a broad range. Generally it is low according to the first-order frequency. The lowest value is 0.35%, quite satisfying result compared with the previous assumption in the end-to-end simulation. Higher damping ratio means more energy dissipation and better control stability of the flexible suspension system.

5.3 Control Test

In the control test the robot drives its cabin to track astronomical trajectories. Tracking error is always the concerning focus. Taking the reference origin on the rotational center of the cabin, the technical specification requires that translational error is less than 48 mm and the orientation error less than 1 angular degree. Meanwhile system braking and stable control are also tested in regard of safety. So the optimization of control algorithm and parameters goes all over the test. Figure 9 shows the basic logic diagram of the control system of cable robot. Here the robot controller has a feed forward algorithm other than feed backward, so that a smooth control of the flexible system can be anticipated. The feedback controller receives two kinds of feedbacks:- cabin pose and cable tensions. So it is necessary to set pose feedback as the predominant control considering accuracy and compulsory requirement of large movement of the cabin. It also avoids a potential conflict due to different feedbacks. Actually for the sake of simplicity we set that 6 tensions are usually under monitoring and it works only if any of 6 cable tensions deviates significantly from the predetermined tension range between 140 KN and 400 KN.

Six working modes are tested, namely slewing, tracking of single source, basket-weaving, drift-scan, on-the-fly (OTF) mapping and user-defined mapping. Accurate positioning is required for most of the working modes except for slewing, also called observation modes. In the slewing mode the cabin is required to move fast to the designated starting point of observation path within limit time, normally 10 min. The path may not be restricted on the focal surface, but a straight line instead. On the

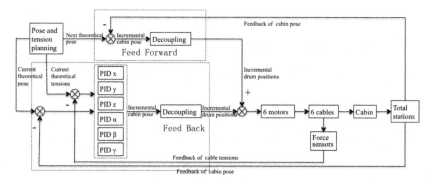

Fig. 9. Control loop of cable robot

contrary, in observation modes most of trajectories are sectionally continuous arcs on the focal surface.

An operational scenario of cabin position with constant acceleration and deceleration can be planned for most of the modes, as shown in Fig. 10. The only exception is the drift-scan mode when the cabin is fixed to a point. During a working mode the scenario is made of two stages of constant acceleration/deceleration m, t_1, and a long period of constant speed, t_2. The maximal speed, V_0, may reach 400 mm/s with the acceleration/deceleration time equal to 20 s in the slewing mode. But in observation modes the respective maximal values are 24 mm/s, doubling the self-rotation speed of the earth, and 2 s. When the planned path is quite short, the stage of constant speed may disappear. Similarly an operational scenario of cabin orientation can also be planned. The only difference is that orientation planning is not directly determined by working mode and its parameters, but based on the analysis of optimal tension distribution of the cable robot [7, 10] along a given cabin path.

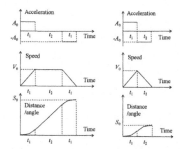

Fig. 10. Operational scenario of cabin pose

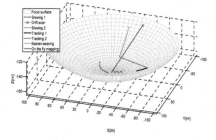

Fig. 11. Some trajectories in control test

Figure 11 shows some testing trajectories. Control errors of some trajectories are plotted as the time-error curves, as shown in Fig. 12. The blue and red horizontal lines in each plot are the required spatial and angular limits respectively. Several significant phenomena can be observed in the figure. First, for all observation modes the control errors are below the required limit in the most part of period. The errors may further decrease as the time goes long. At the starting point, however, large initial errors may exist. The author advices that positioning control should be turned on earlier before the starting of observation mode so that initial error may be reduced greatly in advance. Second, spatial error deceases fast than angular error which fluctuates obviously with very low frequency. It sometimes approaches to or even surpasses the limit. The possible reason may be relatively larger rotational inertia of the cabin and smaller stiffness. It may be improved via optimizing the control parameters in the next tests. Finally, periodical jumping of spatial error can be observed in the mode of on-the-fly mapping. The jumps appear synchronously with the sharp turns of the trajectory. This is the problem on how to keep good positioning control at such unsmooth transitions between two continuous paths. Based on dynamics of the flexible suspension system, optimization of the control parameters or kinematic planning may be a solution.

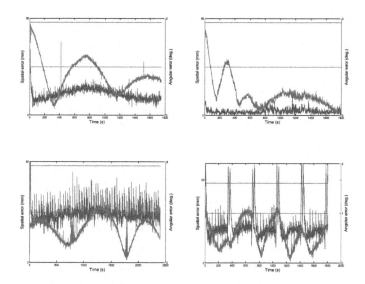

Fig. 12. Control errors in real time (in a left-right and up-down order): (1) Tracking; (2) Drift scan; (3) Basket weaving; (4) On the fly mapping (Color figure online)

During the test, some problems also exist in the measurement of total station, like sudden jump of data, possibly due to tracking failure of targets. It may cause serious incorrect operation of the robot. Currently a limit is set so that large jump can be neglected by robot controller and a previous value is replaced with it.

6 Summary

The FAST uses the largest cable robot in the world to drive its cabin for astronomical observation. Now the novel design is coming into being. This paper devotes to an overall introduction, including its mechanical composition, control architecture, system debugging and performance test. Actually each of these aspects involves in a lot of details that are not discussed here due to the limit paper length.

Performance test is always the key concerning focus in that it verifies the facts whether the cable robot is capable of astronomical tracking with desired positioning accuracy. The preliminary test results agree well with those of the end-to-end simulation. Some parameters like damping ratio are even better than previous estimation in simulations. The errors of the control test may not completely satisfy the required limit. Nevertheless the possible problems are visible, such as initial errors, unsmooth transition and data jumps of total station. They may be well cracked in the future work.

Acknowledgments. This work is supported by the National Natural Science Foundation (NNSF) under Grant No. 11573044.

References

1. Nan, R.D.: Five hundred meter aperture spherical radio telescope (FAST). Sci. China Ser. G **49**, 129–148 (2006)
2. Nan, R., Li, D., et al.: The five-hundred-meter aperture spherical radio telescope (FAST) project. Int. J. Mod. Phys. D **20**, 989–1024 (2011)
3. Kärcher, H.J.: FAST focus cabin suspension - simulation study. Internal Technical Research report of FAST (2007)
4. Li, H., Sun, J.H., Zhu, W.B., et al.: End-to-end simulation for the cabin suspension of FAST telescope. In: Proceedings of the 10th Asian-Pacific Regional IAU Meeting 2008, Kunming, China, pp. 361–363 (2008)
5. Li, H., Sun, J., Zhang, X.: Experimental study on the damping of FAST cabin suspension system. In: Proceedings of SPIE, Ground-Based and Airborne Telescopes IV, Amsterdam, Netherlands, pp. 1–12 (2012)
6. Kärcher, H.J., Li, H., Sun, J.H., et al.: Proposed design concepts for the focus cabin suspension of the 500 m FAST telescope. In: Proceedings of SPIE, Marsellie, France, pp. 1–9, June 2008
7. Li, H., Nan, R., Kärcher, et al.: Working space analysis and optimization of the main positioning system of FAST cabin suspension. In: Proceedings of SPIE, Ground-Based and Airborne Telescopes II, Marseille, 70120T-5, pp. 1–11 (2008)
8. Li, H.: A giant sagging-cable-driven parallel robot of FAST telescope: its tension-feasible workspace of orientation and orientation planning. In: Proceedings of the 14th World Congress in Mechanism and Machine Science, Taipei, pp. 1944–1952 (2015)
9. Li, H., Yao, R.: Optimal orientation planning and control deviation estimation on FAST cable-driven parallel robot. Adv. Mech. Eng. **2014**(1), 1–7 (2014)

10. Li, H., Zhang, X., Yao, R., et al.: Optimal force distribution based on slack rope model in the incompletely constrained cable-driven parallel mechanism of FAST telescope. In: Cable-Driven Parallel Robots, Mechanisms and Machine Science, vol. 12, pp. 87–102 (2013)
11. Ibrahim, S.R., Mikulcik, E.C.: A time domain vibration test technique. Shock Vib. Bull. **43**, 21–37 (1973)
12. Ibrahim, S.R.: Double least squares approach for use in structural modal identification. AIAA J. **24**, 499–503 (1986)

Author Index

A
Archambault, Philippe, 376
Ayala Cuevas, Jorge Ivan, 15

B
Badi, Abdelhak, 376
Baklouti, Sana, 37
Barnett, Eric, 319
Barrado, Mikel, 353
Berti, Alessandro, 207
Bonenfant, Michaël, 390
Bordalba, Ricard, 195
Boschetti, Giovanni, 292
Bouyer, Laurent, 390
Bruckmann, Tobias, 207, 307, 364
Bülthoff, Heinrich H., 254

C
Cabay, Edouard, 353
Campeau-Lecours, Alexandre, 167
Cardou, Philippe, 117, 128, 167, 390
Caro, Stéphane, 37, 117, 128, 268
Carricato, Marco, 207, 219
Caverly, Ryan James, 3
Choi, Eunpyo, 62
Courteille, Eric, 37, 280
Culla, David, 353

D
Dubor, Alexandre, 353

E
Eden, Jonathan, 50

F
Faure, Céline, 390
Forbes, James Richard, 3
Fortin-Côté, Alexis, 390

G
Garant, Xavier, 167
Gauthier, Guy, 376
Godbole, Harsh Atul, 3
Gosselin, Clément, 167, 219, 231, 319, 390
Gouttefarde, Marc, 128

H
Herder, Just L., 307
Hervé, Pierre-Elie, 353
Husty, Manfred, 97

I
Idá, Edoardo, 207
Izard, Jean-Baptiste, 353

J
Jadhao, Kashmira S., 307
Jiang, Xiaoling, 231
Jin, XueJun, 62
Jung, Jinwoo, 62

K
Khajepour, Amir, 243
Kim, Chang-Sei, 62
König, Markus, 364

L
Lambert, Patrice, 307
Laroche, Édouard, 15
Lau, Darwin, 50, 140
Laurendeau, Denis, 390
Lemmen, Patrik, 364
Lessanibahri, Saman, 128
Li, Hui, 402
Long, Philip, 268

M

Marquez-Gamez, David, 268
Martin, Antoine, 117
Masone, Carlo, 254
Mattern, Hannah, 364
McFadyen, Bradford J., 390
Meik, Michael, 364
Mercier, Catherine, 390
Merlet, Jean-Pierre, 180
Mottola, Giovanni, 219

N

Nelson, Carl A., 331
Newman, Matthew, 152
Notash, Leila, 26

O

Oetomo, Denny, 50

P

Pan, Gaofeng, 402
Park, Jong-Oh, 62
Passarini, Chiara, 292
Piao, Jinlong, 62
Piccin, Olivier, 15
Porta, Josep M., 195
Pott, Andreas, 73, 85, 106, 254

R

Rasheed, Tahir, 268
Reichert, Christopher, 364
Rodriguez, Mariola, 353
Rognant, Mathieu, 280
Ros, Lluís, 195
Rushton, Mitchell, 243

S

Saad, Maarouf, 376
Schadlbauer, Josef, 97
Schenk, Christian, 254
Schmidt, Valentin, 85
Shao, Zhufeng, 340
Shih, Albert J., 340
Song, Chen, 50
Spengler, Arnim, 364
Sun, Jinghai, 402

T

Tan, Ying, 50
Terry, Benjamin, 152
Trevisani, Alberto, 292

V

Vu, Dinh-Son, 319

W

Wang, Liping, 340
Wehking, Karl-Heinz, 73
Wehr, Martin, 73

Y

Yang, Qingge, 402

Z

Zaccarin, Anne-Marie, 319
Zanotto, Damiano, 292
Zhang, Zhaokun, 340
Zsombor-Murray, Paul, 97
Zygielbaum, Arthur, 152

Printed by Printforce, the Netherlands